A Centaur in London

INFORMATION CULTURES
Series Editors

Ann Blair
Carl H. Pforzheimer University Professor of History
Harvard University

Anthony Grafton
Henry Putnam University Professor of History
Princeton University

Earle Havens
Nancy H. Hall Curator of Rare Books & Manuscripts
Director, Virginia Fox Stern Center for the History of the Book
Johns Hopkins University

This book series examines how information has been produced, circulated, received, and preserved in the historical past. It concentrates principally, though not exclusively, on textual evidence and welcomes investigation of historical information both in its material forms and in its cultural contexts. Themes of special interest include the history of scholarly discourses and practices of learned communication, documentary management and information systems—including libraries, archives, and networks of exchange— as well as mechanisms of paperwork and record-making. This series engages with vital discussions among academics, librarians, and digital humanists, presenting a vivid historical dimension that uncovers the roles played by cultures of information in times and places distant from our own.

Also in the Series

Sailing School:
Navigating Science and Skill, 1550–1800 (2019)
Margaret Schotte

Leibniz Discovers Asia:
Social Networking in the Republic of Letters (2019)
Michael Carhart

Maker of Pedigrees:
Jakob Wilhelm Imhoff and the Meanings
of Genealogy in Early Modern Europe (2023)
Markus Friedrich

A Centaur in London

Reading and Observation in
Early Modern Science

FABIAN KRAEMER

Johns Hopkins University Press
Baltimore

For Alexandra and Nika

An earlier version of this work was published as *Ein Zentaur in London* (Didymos-Verlag, 2014). The English translation of that work was funded by Germany's Geisteswissenschaften International – Translation Funding for Work in the Humanities and Social Sciences, a joint initiative of the Fritz Thyssen Foundation, the German Federal Foreign Office, the collecting society VG WORT, and the Börsenverein des Deutschen Buchhandels (German Publishers & Booksellers Association).

Johns Hopkins University Press
2715 North Charles Street
Baltimore, Maryland 21218
www.press.jhu.edu

Library of Congress Cataloging-in-Publication Data
Names: Krämer, Fabian, author.
Title: A centaur in London : reading and observation in early modern science / Fabian Kraemer.
Other titles: Zentaur in London. English
Description: Baltimore : Johns Hopkins University Press, 2023. | Series: Information cultures | Adapted from Ein Zentaur in London: Lektüre und Beobachtung in der frühneuzeitlichen Naturforschung. | Includes bibliographical references and index. | In English, translated from the original German. |
Identifiers: LCCN 2022031461 | ISBN 9781421446318 (hardcover) | ISBN 9781421446325 (ebook)
Subjects: LCSH: Animals—Abnormalities—Research—History. | Monsters—Research—History. | Natural history—History. | Empiricism. | BISAC: SCIENCE / History | SCIENCE / Natural History
Classification: LCC QL991 .K66 2023 | DDC 599—dc23/eng/20230126
LC record available at https://lccn.loc.gov/2022031461

A catalog record for this book is available from the British Library.

Special discounts are available for bulk purchases of this book. For more information, please contact Special Sales at specialsales@jh.edu.

Contents

Figures

A Centaur in London

Introduction

Writings on natural history from the early modern period in Europe contain images of monsters, accounts of monstrous births, and statements of a general or theoretical nature about monsters that are strikingly similar. Once these had entered the discourse that concerned itself with rare and, as they were understood at the time, supernatural (*supra naturam*) or preternatural (*praeter naturam*) phenomena, they were used time and again by a multitude of authors. They multiplied, as it were. The phenomena in question, unbelievable though they may seem to the present-day reader, were thereby perpetuated, in writing as well as visually.

Some of the monsters that we encounter in these sources seem both familiar and plausible. Why should we not, for example, give credence to the report of the birth of a hermaphrodite baby in Zurich in the early sixteenth century? Other accounts sound less familiar and make a considerably less credible impression: the birth of a foal with the head and voice of a human or a race of centaurs living somewhere in the East. This second group of phenomena discussed under the generic term *monstrum* makes the circulation patterns of the textual building blocks in question an interesting problem for historians of science.

The "critical" exploration of earlier scholarly records that took place in the eighteenth century demonstrates a distinctly different approach to monsters. "Enlight-ened" naturalists removed individual phenomena from the category of *monstrum* and banished whole groups of phenomena that until then had traditionally been consigned under this generic term to the realm of "fables." This process requires explanation and has yet to be adequately examined by historians.

The self-presentation of the Enlightenment naturalists and much existing re-search literature is characterized by a dichotomy between early modern book-based scholarship on the one hand and enlightened empiricism on the other. This binary view does not do justice to the complexity of early modern studies in natu-ral history. In this book I offer an empirical examination of these changes to arrive at a more nuanced perspective, taking the following central questions as my focus: From where did naturalist writers draw their knowledge of monsters and similar phenomena? Which sources did they regard as particularly authoritative: book-based scholarship or empirical observation, if indeed these two modes of knowl-edge production can be juxtaposed so clearly? This examination of the practices used and of their conceptualization by contemporaries is the methodological per-spective of the study, which aims to contribute to a better understanding of the dramatic changes in the discourse on monsters.

The primary aim of the study is to achieve a nuanced understanding of the way early modern naturalists approached knowledge concerning oddities of nature. In addition, it aims to identify shifts in the more general *referencing structures* of writ-ings on natural history beyond the discourse on monsters. What I mean by the "referencing structure" of a text is the entirety of its references to other texts as well as to extratextual sources of knowledge such as observations of an object presum-ably conducted by authors themselves, for example.

The content of this study contributes to the history of monsters and of other phenomena that were understood from an early modern viewpoint to be super-natural or preternatural, which is to say, originating outside the habitual course of nature. The secondary literature published on this topic in recent years makes it possible to focus on the referencing structures of the sources, because many sub-stantive questions can be considered resolved. The field of research thus lends itself to a thematically narrow line of questioning.

The redundancy of the textual building blocks that we encounter in Renais-sance natural history writings on monsters—I refer to them, in Ann Blair's sense, as "factoids"[1]—offers a good starting point. The fact that author after author took up and reused not only the same examples of monsters but also images and statements of a general nature has hitherto been little discussed, much less satisfactorily explained. An examination of the circulation of such textual elements furthermore promises important insights into the scholarly practices that underlay them. The ways in which early modern naturalists read and wrote are of particular interest. Above all, however, this approach enables us to make well-founded statements about the relationship between direct observation and book-based scholarship in the concrete practice of these scholars.

My examination of the referencing structures of natural history writings on monsters is linked to analyses on two further levels: On the level of *Praktiken der Gelehrsamkeit*—"scholarly practices" as described by Helmut Zedelmaier and Martin Mulsow[2]—I consider these historically specific forms of scholarly reading and writing and determine how they related to the observational practices of the actors involved. How did early modern naturalists make excerpts from their reading? How did they order their excerpted material and edit it for use in their own writing practice? And how, in that context, did knowledge based on direct observation relate to that drawn from literature? Little attention has hitherto been given to the particular scholarly practices that underlie the images found in early modern publications. The present study examines these practices in a similar way to those relating to reading and to the ordering of material thus gathered. In doing so, it reveals interesting parallels that run counter to our intuitive association of image with observation and reading with supposedly nonempirical scholarship.

This book can be situated within a current line of research that is less concerned with discoveries, ideas, and theories, that rather looks to find fundamental changes in the sciences in places where, in many cases, they took root largely unnoticed: in the practices of the historical actors themselves. This approach to the history of science is characterized by expanding its concern beyond the scientific achievements of the few, the "geniuses." Its proponents study the spread of scientific practices and their impact on a broader scale.[3]

On the level of historical epistemology, I analyze the importance accorded by the actors to (their own direct) observation, *observatio* or *autopsia*, as compared to the *auctoritas* (authority) of the canonical ancient writers of medicine and natural history. Alongside these actors' categories, which are central to the analysis, I also investigate the rhetorical ideals of variety and plenitude, *varietas* and *copia*, which were influential in the Renaissance, and their impact on the scholarly discourse on monsters. The point of departure for this analysis can be illustrated with an example.

A Centaur in London

In the lectures that he gave on forensic medicine in Göttingen in the summer semester of 1751, the Swiss physician and polymath Albrecht von Haller (1708–1777)[4]
called into question several formerly canonical components of the early modern
discourse on monsters.[5] He took a critical view, for example, of the existence of
"perfect" hermaphrodites, whom he defined as hermaphrodites endowed with
functional reproductive organs of both sexes.[6] He also rejected monsters that were
said to possess both human and animal physical characteristics, which had formerly been attributed to intercourse between humans and animals.[7] Haller was not
alone in his stance: perfect hermaphrodites, human-animal hybrids, and many
other monsters that had been canonical in the early modern period disappeared
completely at this point from what Georges Canguilhem termed "the true."[8] They
were no longer of more than limited interest in intellectual debate, while this assessment itself was, in turn, put forth with unprecedented vehemence.

It is in this spirit that Haller refers to a rumor according to which a live centaur
had been sighted only recently: "In the forties of the century in which we live, the
rumor also circulated that there was a true, live centaur to be seen in London.
However, this legend was never confirmed."[9] More than almost any of his contemporaries, Haller epitomized the type of empirically oriented naturalist[10] called for
by the eighteenth century. In his view, physiological knowledge was to be based
primarily on anatomy and on physiological experiments.[11] It is unsurprising, therefore, that he emphatically rejected belief in centaurs. These views placed him in the
mainstream of naturalists at the time. But how was it possible for a rumor such as
this to come about in the first place?

Centaurs had made their way, if you will, along a trail of references, from text
to text, from ancient Greece into early modern Europe. If Albrecht von Haller considered the existence of the London centaur an impossibility, it was because references to writings by ancient authorities—which for large parts of the early modern
period had guaranteed the existence of such monsters in the eyes of many scholars—
by the middle of the eighteenth century, lost their weight. Haller rejected centaurs
and other hybrids because, first, he viewed their creation—that is to say, the successful mixing of two species as dissimilar as horses and humans—as impossible.
Second, though, he insisted on *reliable* accounts.

A naturalist such as Albrecht von Haller who perceived himself to be enlightened could not regard a rumor—understood as a statement of obscure origin—as
reliable. But neither were accounts by ancient authorities any longer authoritative

per se. According to Haller, "We do not have a single example of such a birth that has absolute historical certainty, for the fact that writers of antiquity claim that a live centaur existed in Egypt during the reign of Emperor Tiberius . . . far from settles the matter."[12] Whom did Haller have in mind? Among the "writers of antiquity" formerly regarded as trustworthy witnesses to the existence of centaurs was the Roman natural historian Pliny the Elder (ca. 23–79), who was much read in the sixteenth century in particular, and who, despite criticism of specific details, remained a highly valued source for naturalists into the seventeenth century.[13] Haller evidently had a passage from Pliny's *Natural History* at the back of his mind, though he did not necessarily have firsthand knowledge of it. Either way, he did not care to relate it to his students.

Let us jump, at this point, to the second half of the sixteenth century: Ulisse Aldrovandi (1522–1605) was a professor of natural history at the University of Bologna known far beyond the city itself, not least for his *museum*—his extensive natural history collection—which was said to include the preserved remains of a true dragon. In 1572, Aldrovandi began work on an enormous encyclopedia of natural history that was to be equal in every respect to that of his renowned Swiss colleague and contemporary, the polymath and naturalist Conrad Gessner (1516–1565). In his posthumously published *Monstrorum Historia* (1642), the volume of the encyclopedia concerned with monsters, Aldrovandi reproduces Pliny's statement in full and without contradiction: "This view [according to which centaurs derive from intercourse between humans and horses] seems to be favored by Pliny, who states the following in the seventh book of the *Historia Naturalis. Claudius Caesar writes that a hippocentaur was born in Thessaly and died the same day; and in his reign we actually saw one that was brought here for him from Egypt preserved in honey.*"[14] Claudius Caesar is the same Tiberius whom Haller was to mention to his students so briefly more than a century later.

Because Haller does not offer any information as to his ancient source in the case of the Egyptian centaur, we are required to take a detour via Aldrovandi to arrive at Haller's reference text. What Aldrovandi considered worthy of quotation—and does not comment on disparagingly—Haller later gave no more than a passing and dismissive mention. Whether Aldrovandi, for his part, believed in the existence of the Thessalian centaur and of other creatures that Haller placed firmly in the realm of the "fabulous" remains an open question, however.

By Albrecht von Haller's lifetime, Pliny's testimony was no longer enough to guarantee the centaur a place among the things of nature, not least because Haller considered earlier generations of scholars to have been just as "credulous" as many of his own contemporaries. "Credulity" is a key concept in Enlightenment criticism of both inherited and contemporary knowledge, not only concerning monsters. Both "unenlightened" contemporaries and the scholars of earlier, "unenlightened" eras seemed, because of this disposition, unsuited to reliable observation.

Ancient writings' loss of authority was not solely responsible, however, for the significant change in approach to monsters between Aldrovandi's and Haller's time. After all, it was not only ancient writers who vouched for the existence of many of the monsters called into question by the intellectuals of the eighteenth century. Numerous early modern European writers report sightings of very similar creatures in their own lifetimes. It would also be a misinterpretation of the discourse on natural history in the late sixteenth and seventeenth centuries to associate it with book knowledge and dispute its empiricism in so dichotomous a way. Indeed, it is perhaps more accurate to say that the two merged in the late Renaissance than that they were regarded as opposites. Reading was itself empirical. It formed an integral part of the various early modern expressions of empiricism and did so even in the eighteenth century. What was it, then, that separated Haller from Aldrovandi?

The "Scientific Revolution" and the "Naturalization of the Monstrous"

For reasons that are self-evident, we tend to answer such questions hastily in a way that is familiar to us. In his presentation of himself, Albrecht von Haller appears, seemingly in contrast to his Renaissance predecessors, as a paradigmatic embodiment of the Enlightenment naturalist: He carefully separates rumors from observations with a known, named origin and accepts only the latter. He appraises every observer-author and considers with care whether their account can be trusted. If in doubt, he relies on his own experience. Thus, he is well qualified to separate the wheat from the chaff in traditional knowledge about monsters. What remains is a verified body of knowledge about physical deformations attributable to natural causes in newborns that clearly belong to one species.

This self-presentation is persuasive to the present-day reader, too, because it corresponds in essential respects with two grand narratives in the historiography of science that remain highly influential despite the criticism to which they have been subjected in recent years and decades. The first, more general narrative is that the so-called Scientific Revolution of the sixteenth and seventeenth centuries put empiricism in the place of outdated book knowledge. This narrative was expressed influentially in the middle of the past century by Alexandre Koyré and Herbert Butterfield as a shift away from the knowledge of books toward the knowledge of things.[15] This interpretation has parallels in the rhetoric of many seventeenth-century naturalists, for example, in the much-used rhetorical figure of the "Book of Nature." Steven Shapin writes, "Here was one of the central rhetorical figures that the new philosophical practitioners used to distinguish themselves from the old. The proper object of natural philosophical examination was not the tradition-ally valued books of human authors but the Book of Nature."[16] *Nullius in verba*, the motto of the Royal Society of London, founded in 1660, seems to summarize the rejection of book knowledge well: distancing themselves from the authority of ancient masters of natural history such as Pliny and Aristotle, and abandoning the text-based knowledge practices of the older-style of scholar, the naturalists, it suggests, now dedicated themselves entirely to empiricism. Direct observation—*autopsy*—and experiment took the place of previous scholarly knowledge practices.

It is certainly no coincidence that we can begin to glimpse the modern natural scientist in this shift, and thus also the division between the sciences and the humanities that was proclaimed in the second half of the nineteenth century by Wilhelm Dilthey and further elaborated and lamented in the twentieth century in Charles Percy Snow's famous "two cultures." Albrecht von Haller fits entirely into this picture. Seen in the light of this narrative, he appears—when compared to Ulisse Aldrovandi—as a great observer and critic of the credulity that focused on the written (and, indeed, ancient) word.

The second narrative, one more specifically related to the discourse on monsters, is—to cite a term generally attributed to George Canguilhem[17] for an argument used by numerous authors—that of the "naturalization of the monstrous." This naturalization, it is argued, can be credited to early modern naturalists. They studied monsters with a view to identifying their natural origins and in doing so successfully challenged earlier supernatural explanations and, thus, theological interpretations of monsters as prodigies—divine portents or warnings. This narrative casts Haller as a great naturalizer.

Albrecht von Haller and his colleagues frustrate these categories. The present study represents an attempt to come to a better understanding of what separated Albrecht von Haller from Ulisse Aldrovandi—and what connected them. The period of time chosen for analysis corresponds largely with the period traditionally associated with the Scientific Revolution. The concept itself has been heavily criticized in recent decades, and from a present-day perspective, it no longer appears tenable as an event-like, self-contained historical phenomenon.[18] But its influence on our understanding of the processes of this period as they relate to the history of science nevertheless remains powerful.

In the present case, too, the interpretation offered by the Scientific Revolution continues to appeal. If we were to believe the historical actors, and with them older research in the history of science, the shift in the discourse on monsters between the careers of Ulisse Aldrovandi and Albrecht von Haller would be easily explained: earlier book knowledge, which was shot through with implausibilities rooted in credulity, had to be swept aside and replaced by carefully verified firsthand accounts.

Let us consider the naturalization thesis more closely: older research literature, of which influential publications by Jean Céard, Georges Canguilhem, and Lorraine Daston and Katharine Park[19] are particularly worthy of mention, argues that a "naturalization" of the monstrous took place in the early modern period. Despite some differences in detail, the various expressions of the naturalization thesis can be summarized as follows: they all assume that attitudes toward monsters—understood in every case as humans or animals with a physical deformity—developed linearly in the sixteenth and seventeenth centuries. Originally, they maintain, monsters were viewed as divine prodigies and were received with terror by the European population. This terror faded gradually over the course of the sixteenth and seventeenth centuries. Céard's and Canguilhem's expressions of the thesis, in particular, ascribe the sciences a central role in this process: naturalists abandoned theological explanations of monstrous births in favor of natural explanations, they claim, and this view prevailed.

Daston und Park's version of the naturalization thesis represents a more subtle expression of this line of argument with a three-phase model for the history of monsters in early modern Europe. Initially, they argue, monsters were regarded as prodigies in a sinister, religious sense. Over the course of the sixteenth century, they were increasingly viewed as natural wonders, associated with more positive emotions and indeed even becoming sources of delight and pleasure. Finally, they argue, in a process that reached its conclusion around 1700, monsters lost their connection with other phenomena considered prodigies, such as earthquakes and comets, entirely and became subjects of the medical fields of physiology and comparative anatomy.

Despite their differences, all three versions of the naturalization thesis have in common the assumption that an *older* theological paradigm regarding monsters was superseded by a *new* scientific one in a linear fashion. According to Daston and Park, "The principal line of development, from monsters as prodigies to monsters as examples of medical pathology, is clear."[20]

The analyses of the present study suggest a different view: What separated naturalists' approach to knowledge about monsters in the mid-eighteenth century from the sixteenth century was not, as one might imagine, a one-sided increase in the importance of empiricism at the expense of book knowledge or reduced credulity. Rather, first, the concept of experience itself altered during this period, and it did so in several respects: among the naturalists of the late sixteenth and early seventeenth centuries whose writings on monsters examined here, a concept of *experimentum* or *experientia* prevailed that referred to what was, in the Aristotelian sense, essentially a collective experience—not that of an individual, named author. What Haller, in the eighteenth century, considered to be a "rumor" and therefore not trustworthy could perfectly well be deemed an experience around 1600. The "authored" observation, which is to say one that could be traced back to an observer known by name, was only to become the central epistemic category that it was for Haller in the course of the sixteenth and, particularly, seventeenth century.

Second, the scholarly practices of the naturalists changed at the same time, though not in the way suggested by the grand narrative of the Scientific Revolution: Naturalists have *never*—and this remains true to the present day—gotten by without the observations of others, and in the early modern period observations existed for the most part in written form. Therefore, even for an eighteenth-century intellectual such as Haller, reading continued to be a central component of the study of nature. However, he read *differently* than did his predecessors in the Renaissance, because a "critical" form of scholarly reading that was less focused on comprehensive collecting had established itself in the intervening years. Nevertheless, for Albrecht von Haller, too, experience and reading went hand in hand. This is particularly evident when he himself repeated observations passed down in writing from earlier scholars: the observation itself was, in a sense, a part of the practice of reading.

When considered against the backdrop of the findings presented here, the naturalization thesis is similarly unconvincing. Long before the period proposed for naturalization, early modern naturalists already tended to focus on "natural" causes in their studies of rare phenomena. Their bias toward these "secondary" causes, however, did not imply a rejection of the "primary" ones, by which I mean the handiwork of God in the genesis of a monster. Rather, it was an expression of the division of labor between medicine and theology that was practiced in Europe for large parts of the early modern period: medicine was responsible for the secondary, which is to say natural, causes and left the primary causes to the theologians. It was only the experiences of the religious and civil wars of the sixteenth and seventeenth centuries that led, at quite different points in time in the states of Europe, to a vehement rejection by naturalists of belief in prodigies, for the most part in parallel with the secular and ecclesiastical elites.

As stated above, this study builds on substantial research on the early modern discourse on monsters. Let me briefly revisit some of its more pertinent insights and thus mark the starting point of the present study: As recent research has shown, early modern naturalists intensively studied monsters and other rare phenomena. The late sixteenth and early seventeenth centuries saw an enormous increase in natural history writing on monsters, both in relation to some of the more prominent individual phenomena discussed under this generic term—such as conjoined twins or hermaphrodites—and in relation to the subject as a whole. This was true for what is now Italy as well as for France, Switzerland, and the Holy Roman Empire.[21]

Naturalists were not alone in their interest in monsters. Their publications were part of a larger picture of the very widespread occupation with monsters and similar phenomena by learned and less learned writers and readers, particularly in the Protestant regions of Europe. In light of this, Jean Céard has described the mid-sixteenth century as a "golden age of prodigies" ("l'âge d'or des prodigies").[22] Many writers at the time were under the impression that divine omens such as the birth of babies categorized as monstrous were appearing more frequently than before. The relationship of vernacular, popular text forms to writing on monsters in the Latin, scholarly genres has yet to be adequately explained. Individual examples that have been examined in greater depth suggest, however, that transfers of textual building blocks in both directions were not uncommon.[23]

In relation to the naturalization thesis, a more nuanced perspective has increasingly emerged in recent literature. Significant impetus for this has come from Lorraine Daston and Katharine Park. Contrary to their earlier, linear three-phase model of scholarly engagement with monsters in the early modern period, they now postulate a more complex development: the interpretation of monsters as divine prodigies occurred into the seventeenth century in waves, they claim, which were often influenced by local political events and conflicts.[24] This line of argument has so far met with very little contradiction. Alan W. Bates likewise argues that many descriptions and visual representations of monsters from this period should be seen emblematically, rather than as prodigies, in the sixteenth and seventeenth centuries. They were furnished with a variety of meanings and, accordingly, elicited a variety of reactions.[25]

The engagement of early modern naturalists with phenomena that, in their view, now clearly fell into the category of preternatural was to culminate in the late seventeenth century: The journals of academies such as the Royal Society and the Academia Naturae Curiosorum (the present-day German National Academy of Sciences Leopoldina) favored discussions of preternatural phenomena. Often they occurred with reference to Francis Bacon's program for the reform of natural philosophy, or at least largely in accordance with it.[26]

As has furthermore been shown by recent research, intellectuals in the early eighteenth century increasingly shunned any occupation with preternatural phenomena; indeed, the category of the preternatural was itself called into question. This change was linked to the rejection of wonder as an emotion relevant to the study of nature and to criticism of the credulity of the presumably uneducated masses and earlier generations of scholars.[27]

Ulisse Aldrovandi and Albrecht von Haller define the limits of the present study in terms of time and, to some extent, geography. It moves between Italy in the south, England in the north, France in the west, and the Holy Roman Empire in the east. The *respublica literaria*, the Republic of Letters, was not restricted by the borders of individual territories. We therefore do well to be guided by the contours of the discourse under examination here itself: Which texts were connected by a referencing system, which is to say, who read or cited whom?

With a small number of exceptions, the sources examined date from the period between circa 1550 and 1750. This choice is suggested by the phenomenon under discussion itself. From the second half of the sixteenth century onward, naturalists from a variety of European countries began to concern themselves intensively with the subject of monsters. In terms of their content, the works that emerged from these studies in the decades that followed are quite consistently distinguishable by the circulation patterns described. Explicit criticism of an appreciable number of the factoids accumulated in this way—or of the objects under discussion themselves—can increasingly be found in the sources from the end of the seventeenth century onward. The process culminated in the removal of numerous phenomena—of which the centaur is just one particularly striking example—from the category of *monstrum*. Many disappeared completely from the "true" at the same time. This process had, to a large extent, concluded by the middle of the eighteenth century.

My analysis is composed of the following steps: The first chapter addresses the relationship between reading and observation in the natural history literature on monsters that was published in the late sixteenth and early seventeenth centuries.

I do this empirically using the factoids found in these texts and in critical interaction with the grand narrative of the Scientific Revolution. By means of three examples, the chapter demonstrates that case histories, theoretical statements, and visual representations of monsters were in wide circulation around 1600. The boundaries between natural history texts and other scholarly and less scholarly text types were remarkably permeable here, inasmuch as many of these textual building blocks crossed the boundaries between discourses and types of text.

After examining these three case studies, I consider the approach taken by the naturalists to the factoids that they drew from a variety of text genres and a variety of ancient and contemporary, learned and less learned authors. The range of writers consulted by naturalists grew significantly during the Renaissance and became more diverse. It was no longer primarily the accounts of ancient authorities of nature study that were reproduced and commentated. Strikingly often, naturalists combed through texts of a historiographical nature in search of factoids about monsters. In doing so, they favored accounts or images based on firsthand testimony, but not as a matter of principle. Rather, the value that they attached to observation varied according to object, their disciplinary background, and text genre. The Renaissance also saw the transfer of attributes of authorship from the *auctores*, the ancient authorities, to contemporary writers. However, naturalists of the late Renaissance did not yet—as their successors in the seventeenth century increasingly would—connect "authorship" with the demand that authors be the originator of their discourse in the sense that their texts record observations made directly by themselves.

In the second chapter we take a look into the engine room of the discourse that was described in chapter 1 from the perspective of its final, printed products and turn our attention to the scholarly practices that underlie it. My focus here is on the reading and writing practices of a prominent actor in this discourse, the Italian natural historian Ulisse Aldrovandi. The extensively preserved intellectual estate of Aldrovandi—and in particular an encyclopedic collection of reading notes hitherto neglected by historians—is examined with regard to Aldrovandi's reading practices, his approach to visual representations, and their relationship to his natural history writing on monsters. For Aldrovandi, there was no conflict between practices of reading and ordering that were also used by nonnaturalists and his own observational practice. Rather, they complemented one another. The reason for this was not least that the empiricism of early modern natural history was a collective one: a naturalist such as Aldrovandi was engaged in a (mostly written) exchange with learned contemporaries and, in a sense, also with members of earlier generations of scholars.

The inclusion in late Renaissance natural history encyclopedias, such as that of Aldrovandi, of monsters whose existence was considered questionable even at the time can be explained by this scholarly practice and not, for example, by the supposed "credulity" of their authors. Aldrovandi undoubtedly distinguished between "true" and "false" facts. But in preparing his volumes of natural history for publication, he made use of both observation and reading with the intention of surveying existing knowledge of the natural world as comprehensively as possible in the first instance—without excluding individual knowledge reservoirs prematurely. Even in the printed natural history itself, he included questionable factoids if in doubt and left his readers to decide for themselves what to make of them.

The third chapter addresses the intensive engagement with monsters by scientific academies in Europe. My focus here is on the publications of the members and correspondents of the Academia Naturae Curiosorum, the first academy north of the Alps devoted exclusively to the study of nature. The program of this long-neglected academy transformed fundamentally less than twenty years after its foundation in 1652. These reforms powerfully illustrate changes in the concept of authorship that were typical for seventeenth-century natural history as a whole: After the academy founded its journal, the *Miscellanea Curiosa*, in 1670, members of the Academia Naturae Curiosorum and external contributors who wished to place articles in the journal were required to master a new form of observing and writing. The *observatio*—the "epistemic genre" chosen for the articles[28]—required each author to isolate and document a *single* experience or observation. The natural history writer was no longer to be the *collector* but rather the *auctor*, which is to say originator of their discourse.

Rare natural phenomena were of particular interest to the journal. Nevertheless, there was soon opposition to the articles on monsters, which were considered too numerous. This reluctance cannot be attributed solely to oversaturation: it was also connected to the fact that monsters continued often to be regarded as divine omens embodying a message to be deciphered. The *curiosi*, as the members of the Academia Naturae Curiosorum called themselves, were in agreement with their sister academies in rejecting such interpretations. Intellectuals had become aware of the danger of belief in prodigies, not least through their frequent use in the propaganda of the conflicting parties in the devastating Thirty Years' War. Consequently, the *curiosi* attempted to enforce the *correct* observation of such phenomena.

The final chapter returns to the rumor of the centaur in London and clarifies the circumstances of its genesis. I then explore why numerous eighteenth-century intellectuals invested such efforts in criticizing the belief in centaurs and other beings and phenomena they viewed as "fabulous"—because they had, after all, been an integral component of the Renaissance discourse on natural history. An examination of the working methods of "enlightened" intellectuals such as Albrecht von Haller reveals that they did not read any less or observe any more than their predecessors. Their reading practice may have changed more than their observational practice. In a sense, they read even in their observation—when they repeated observations of others that were known to them from literature.

The allegation of "credulity" made by writers such as Haller of earlier generations of scholars was based only in part on changes in observational and reading practices. A further factor was a new temporality that accompanied the enlightened self-conception of these intellectuals: In their view it was not only their uneducated contemporaries who were in need of enlightenment—the scholars of earlier generations had not had the benefit of it either. From the perspective of Enlightenment naturalists, both their own contemporaries and the scholars of days gone by lived together, so to speak, in an unenlightened past. Criticism of their credulity served also to assert the Enlightenment agenda in the present.

CHAPTER ONE

Three Monstrous Factoids

Looking over two or three of the numerous scholarly treatises or summae on monsters written in the Holy Roman Empire, in Switzerland, or in Italy in the sixteenth and early seventeenth centuries will inevitably induce a repeated sense of déjà vu. Both the recounted cases of monstrous individuals and the general and theoretical statements made about them show striking similarities. The same is true of the accompanying woodcuts and etchings. Indeed, the representations often match exactly. Accounts were quoted, woodcuts copied. An examination of the connections between these texts confirms this impression. Furthermore, such an examination promises indirect insights into the practices employed by learned writers in the late Renaissance in preparing publications on subjects from the fields of medicine, natural history, or natural philosophy.

In this first chapter, I address literature on medicine, natural history, and philosophy written around 1600 in light of the observation outlined above. In critical interaction with the grand narrative of the Scientific Revolution—a concept now rarely to be encountered without quotation marks in literature on the history of science[1]—the aim first of all is to describe this superficial phenomenon empirically and then to inquire as to its causes. The examination of concrete case studies forms the basis for initial reflections on the approach of naturalists to their sources and on their understanding of "experience." What role did the authority of the ancient masters of nature study play in their discourse? Whose writings did naturalists consult in their search for accounts and images of monsters? And how did their own observation relate to that of other authors and, more generally, to reading?

As we will see, for the most part experience and book-based scholarship were *anything but* opposites for the writers involved. Accordingly, we cannot associate late Renaissance naturalists one-sidedly with book-based scholarship and dispute their empiricism, as the grand narrative of the Scientific Revolution might suggest. Before we can turn our attention to these questions, however, it is important first to summarize the naturalist discourse on monsters around 1600 in general terms.

A Veritable Explosion of Discourse

In the late sixteenth and early seventeenth centuries, a veritable explosion occurred in the medical, natural historical, and philosophical discourse on monsters. This applied both to monstrosities that were particularly intensively discussed, such as hermaphrodites,[2] and to the phenomenon as a whole. Whereas, following the invention of the printing press in the mid-fifteenth century, not a single text on monsters by a naturalist was published for more than a century, this changed significantly from the second half of the sixteenth century onward.

The beginning was marked by a public lecture on the origins of monsters given by the historian and humanist Benedetto Varchi (1502–1565) at the Accademia Fiorentina in 1548. It was printed in Florence in 1560 under the title *Lezzione sulla generazione dei mostri*.[3] In 1573, the first edition of the influential *Des monstres et prodiges* by the then-renowned French surgeon Ambroise Paré (ca. 1510–1590), *premier chirurgien* to Charles IX and Henry III, was published.[4] In 1595, Martin Weinrich (1548–1609) published his *De Ortu Monstrorum Commentarius*, a treatise about the causes of monstrous births.[5] The *Monstrorum Historia* by Johann Georg Schenck von Grafenberg (died ca. 1620) was published in 1609 and was translated into German a year later.[6] A monograph by Gaspard Bauhin (1560–1624) on hermaphrodites and other monstrous births, *De Hermaphroditorum Monstrosorumque Partuum Natura*, followed in 1614 and was—like the aforementioned texts—reprinted several times.[7]

In the decades that followed, the most influential treatise concerning monsters was perhaps Fortunio Liceti's (1577–1657) *De Monstrorum Natura, Caussis, et Differentiis Libri Duo*, first published in 1616.[8] The publication of *alte Dissertationen* on monsters that began around this same time—primarily but not exclusively at the

Protestant universities of the Holy Roman Empire—demonstrates that monsters became common during the same period as a subject for disputations *pro gradu* or *pro loco*—that is, for the academic debates required to attain an academic title or position.[9]

The numerous publications by naturalists on the subject of monsters cannot be explained in terms of a common agenda of "naturalization." Although it is true that the authors typically remain silent on the "primary," divine causes of monsters, this partial reticence has its roots chiefly in a long-standing division of labor between physicians and theologians. It does not, therefore, imply a rejection of "primary" causes. Even Fortunio Liceti made statements to this effect. His etymological derivation of the term *monstrum* might suggest he rejected the divine part in the generation of monsters on principle. He traces the term back to *monstrare* and thus to the idea that people showed monsters to one another—rather than to the idea that God uses monsters to show (*monstrare*) or warn (*monere*) humankind.[10] These texts came about, then, for different reasons. They point less to a shared naturalist agenda than to commonalities between learned writers from a variety of fields not limited to the study of nature.

This series of monographs written by naturalists was rooted in an intensive occupation with monsters that was widespread in early modern Europe—among scholars but also laypeople. In "l'âge d'or des prodiges,"[11] as Jean Céard has succinctly characterized the middle decades of the sixteenth century, there was a widespread impression that monstrous births and wonders were occurring more frequently than in previous eras.[12] This perception can be attributed chiefly to two factors, the first relating to the *spread* of information about monsters and the second to *interest* in this information.

The invention and ongoing development of the printing press had a great impact on the discourse on monsters. Books and particularly the less expensive broadsides on the subject could now be produced in large quantities and circulated comparatively quickly.[13] News of rare occurrences, such as the birth of a baby with organs of both sexes, for example, could now be spread relatively quickly over large distances. This technological development coincided with great interest in such phenomena on the part of readers. The widespread belief in divine omens, which were said to manifest themselves in rare natural phenomena, allowed the actors to interpret monsters in the light of political, religious, and military conflicts.

The insights into God's salvific plan that were gained in this way may well have offered orientation to early modern Europeans. In many cases, however, the conflicting parties took advantage of this need for orientation very deliberately for their own propaganda purposes: the famous Monk Calf and Papal Ass pamphlets of Martin Luther and Philipp Melanchthon are undoubtedly the best-known examples of this instrumentalization of prodigies—but they were by no means solitary cases.[14]

In addition, there was money to be made from readers' interest in monsters: prodigy books, which presented them primarily as terrifying divine portents and warnings, were sometimes great commercial successes. A good example of this is the *Prodigiorum ac Ostentorum Chronicon*, by the Protestant philologist and theologian Conrad Lycosthenes (ca. 1518–1561), which was first published in Basel in 1557, translated into German for a wider readership that same year, and later also published in English.[15] Conrad Wolffhart had adopted the Greek name Lycosthenes as an expression of his affiliation with humanism. The *Prodigiorum ac Ostentorum Chronicon* is, in the words of the contemporary German translation by Johannes Herold (1514–1567), a chronicle of prodigies and wonders "von anbegin der weldt, biß zu unserer diser zeit."[16] After its publication it quickly became a best seller in many European countries. Divine omens such as monstrous births, comets, earthquakes, and blood rain aroused the interest of large sections of society, particularly among Protestants. The numerous woodcuts in the two editions, both the Latin and the German, further increased the appeal of the work for potential buyers. With more than 1,500 woodcuts—often three to a page—it was among the most heavily illustrated books of its time.[17]

Naturalists, too, made use of the popularity of the topic very intentionally at times to reach a wider readership. This may have been true in the case of the aforementioned treatise *Des monstres et prodiges* (1573) by the French surgeon Ambroise Paré, for example. The publication, with its numerous woodcuts—and certainly not coincidentally in the vernacular—was of interest far beyond the limits of medicine.[18]

Natural historians and philosophers did not lag behind their contemporaries in their occupation with rare natural phenomena. The study of nature in the second half of the sixteenth and early seventeenth centuries was characterized by a growing interest in preternatural phenomena, which is to say phenomena that were believed to have originated outside the usual course of nature. The realm of the natural was bounded in the early modern period by four categories of the nonnatural: the supernatural, the preternatural, the artificial, and the unnatural. While the supernatural was attributed to God as its originator, the artificial to an artist or to the devil, and the unnatural to humans acting against God's will, the preternatural referred back first of all to nature itself. In its rare and extraordinary creations, it was thought, nature deviated for a variety of reasons from its familiar path.[19]

Alongside the increasing occupation of natural historians and philosophers with the preternatural, physicians began increasingly to take an interest in rare diseases and anatomical anomalies. The growing association of the emotion of wonder with the observation of nature contributed to this development. The prevailing understanding that nature had a comparatively large degree of freedom to deviate from its established paths in its productions also encouraged this interest on the part of physicians.[20]

The occupation of early modern naturalists with preternatural phenomena has attracted much attention from historians of science in the past four decades. The factors mentioned above are a significant result of research to date. Up to a certain point they can explain the interest of early modern naturalists in subjects such as monsters. The concrete form of the texts that emerged from this occupation, however—their specific weave, their *texture* in the sense of the Latin *textus* (their fabric or structure, also in a figurative sense)—still requires systematic analysis.

A good starting point for such an examination is offered specifically by the redundancy of their textual building blocks. Accounts and visual representations of individual cases, as well as general statements about monsters, were used time and again once they entered the naturalist discourse: they multiplied. There has hitherto been conspicuously little discussion of this fact, however. Neither can the factors to which historians attribute the growth of interest in monstrosities among naturalists adequately explain this phenomenon. It seems worthwhile to examine them in more detail, particularly because significant insights can be gained in this way into the practices of early modern naturalists—above all into those relating to reading and writing. In addition, empirically substantiated statements can thus be made about the respective epistemic values attached to direct experience (or observation) and book-based learning. And we can investigate whether it is indeed even possible to make a clear distinction between these two modes of knowledge production in the period in question.

The aim of this chapter is to describe the large-scale circulation of textual building blocks among the texts of naturalists (and other authors) throughout Europe and to inquire as to its causes. By way of an introduction to the phenomenon, three prominent textual elements—I refer to them as "factoids"—are followed on their journey from text to text. Since the factoid constitutes a central analytical category in my study, the term warrants clarification here at the outset.

Factoids

It is the repetitive character of the textual building blocks of early modern writings about nature that suggests the use of the term "factoid." In using it, I am following

Ann Blair, in whose work it refers to the "tidbits of knowledge" that the French scholar Jean Bodin (1529/30–1596) took from their original context and entered into his commonplace book—and later reused in his own writings.[21] I use the term in a similar way for the morsels of knowledge that circulated in the body of texts that I examine. However, I relate it to both textual and pictorial tidbits, the patterns of their circulation and the practices behind them justifying this parallelization.[22] It is well known that images found in early modern texts were often reproduced. But beyond isolated case studies, this phenomenon has not hitherto been systematically examined.

Additionally, the term "factoid" allows me to circumvent the epistemologically weightier concept of the fact.[23] A factoid is to be understood in this context as an autonomous knowledge unit in the sense that it appears repeatedly and in a seemingly endless stream of texts and text genres. It has an autonomy resembling that of the modern fact as discussed in recent research by historians of science.

The fact emerged as an epistemological category in the early modern period. It is not only autonomous in the sense that it can serve to support differing and even conflicting theories but was (and is) used ideally as a corrective to existing generalizations or as a basis for new ones.[24] In this sense every fact must be viewed as a factoid, but not every factoid is a fact.

My examination of the circulation of monstrous factoids takes as its starting point texts published by naturalists in the Holy Roman Empire and neighboring territories in the second half of the sixteenth and early seventeenth centuries.[25] It considers the following questions: From where did the naturalists' knowledge of monsters come? Which sources of knowledge did they favor—book-based scholarship or the empirical observation of (supposed) monsters? What conclusions can be drawn about the reading, observational, and writing practices of the authors from the circulation patterns of the monstrous factoids that they use?

To answer these questions, I focus on three monstrous factoids used frequently during the period in question. I have selected three representative types encountered by readers in almost every relevant early modern publication. Additionally, they represent different categories and geneses of monsters found in naturalist literature around 1600. And, finally, they include monsters that were—from the point of view of an enlightened, eighteenth-century intellectual—both "real" and purely "imaginary."

First, we consider the factoid of a headless baby, which represents an example of a *monstrum* that came into existence through lack of matter. Next, we address the case of a horse with the head of a human, an example of monsters that combine morphological characteristics from different species that were to be regarded as an impossibility by later, "enlightened" naturalists. Finally, we turn our attention to the hermaphrodite, which serves as an example of the form of monstrosity that was perhaps most frequently discussed during this period: sexual ambiguity.[26]

A Headless Baby

The first factoid, the case of a baby born without a head, is well suited to illuminating a general feature of the discourse on monsters during this period: the fact that individual monsters were often associated with corresponding "monstrous races"[27] or "exotics from the edge of the world." Since antiquity these peoples were believed to inhabit the fringes of the known world. Though the connection is not always explicit, it can be proved via parallels in descriptions and images.[28]

Individual cases of headless babies appear very regularly in the monographs on monsters published during the period in question. Johann Georg Schenck von Grafenberg, practicing physician in Haguenau in Alsace and author of several medical texts, brought together three such cases in his *Monstrorum Historia* under the heading "Acephali, or human monsters that strangely have no heads, clearly exposed."[29] He presents them briefly and matter-of-factly, without referring to the exotics from the edge of the world. All of the cases date from the sixteenth century and are located not in Africa or Asia, which were traditionally believed to be their abodes, but in Europe. They appear, then, to be empirically substantiated factoids, which seems consistent with the publication history of the *Monstrorum Historia*.

Johann Georg Schenck von Grafenberg based his *Monstrorum Historia* on the multivolume collection of *observationes* written by his father, Johannes Schenck von Grafenberg (1530–1598), who was himself the municipal physician of Freiburg im Breisgau. Large parts of the son's book read as a reissue of Johannes Schenck's *Observationes*, significantly shortened and focused on monsters. According to the foreword of the first volume, the father's collection pursued the aim of gathering new and wonderful things that had been observed by the most famous physicians (*clariß. medici*) by means not of teaching and learning (*doctrina*) but of experience (*experimentum*).[30]

Despite the epistemological parentage of Johann Georg Schenck's *Monstrorum Historia*, many of his factoids show a certain proximity to the topological tradition of the exotics from the edge of the world. This is demonstrated well by one of the three factoids about headless babies, which is reproduced here in full: "In the year of our Lord 1554, a baby without a head was born in Meissen that had the likeness [*effigie*] of eyes on its chest. Jobus Fincelius, de miraculis nostri temporis."[31] The factoid ends with a short reference to the source of the case report: a book about recent wonders, also frequently cited by other naturalists in their writings on monsters, whose author was the Protestant Hiob Fincel (also spelled Job Finzel or Jobus Fincelius).[32]

Hiob Fincel's date of birth is unknown. He received a *Magister* degree from the University of Wittenberg in 1549. In 1559, he became a professor of philosophy in Jena, and in 1562 the faculty of medicine also appointed him as a professor and assessor. In 1568, he became the municipal physician of Zwickau, where he spent the rest of his life.[33] He was, and still is, known chiefly for his work *Wunderzeichen: Warhafftige beschreibung und gründlich verzeichnus schrecklicher Wunderzeichen und Geschichten*, a three-volume chronicle of the wonders said to have taken place between 1517 and 1562.[34] Fincel interpreted these events as an indication that the end times were near; in this spirit, the biblical motto printed on the front page of the first volume prepares the reader for the contents of the work:

Apoc. 14. Fear God
and give honor to him
for the hour of his judgment is come
and worship
him
that made heaven and earth
and the sea
and fountains of water.[35]

The headless baby is only one of many occurrences from the year 1554 recounted by Fincel in his extensive chronicle. As was customary in his work, the author had the description of the event typeset as a separate paragraph: "IN the same year 1554. A baby without a head was born in Meissen / whose eyes were on its chest."[36] Fincel's account does not differ from that of Johann Georg Schenck von Grafenberg in terms of content, though it is not accompanied by a woodcut or etching, because Fincel's chronicle does without visual representations entirely.

The etching used by Johann Georg Schenck in the *Monstrorum Historia* to illustrate the case of the Meissen baby (fig. 1) constitutes a precise rendering of Fincel's textual description in pictorial form. It shows an upright baby with a completely "normal" body, without a head and with eyes set in its chest. The sex of the child is not discernible—which corresponds to the textual description, which provides no information in this regard. It is clear that the etching is based not on an inspection of the baby but rather on Fincel's description of it. At the same time, however,

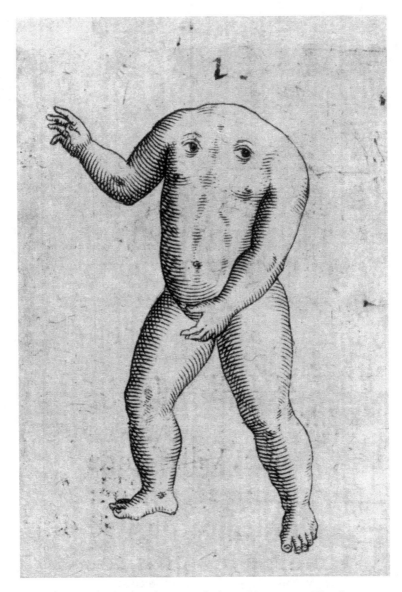

Fig. 1. Johann Georg Schenck von Grafenberg, *Monstrorum Historia Memorabilis* [. . .]. Frankfurt, 1609, 2; approx. 5.5 cm × 3.5 cm. Staatsbibliothek zu Berlin—Preußischer Kulturbesitz, Department of Early Printed Books, Shelfmark: Ke 7065 R.

it has great similarities to the iconographic tradition of those headless human races among the exotics from the edge of the world referred to as Blemmies (Blemmyae or Monopoli), as will be shown below.

The conversion of Fincel's description into pictorial form was not undertaken by an artist working for Johann Georg Schenck von Grafenberg. Rather, the etching can be traced back to an earlier version of the monster illustration in a third text, in just the same way as Schenck's account. For neither Johannes nor Johann Georg Schenck von Grafenberg reproduce Fincel's account of the headless baby *directly*—though both cite it as their source.[37] The connecting link between Fincel's version and that of the two Schencks was Conrad Lycosthenes's *Prodigiorum ac Ostentorum Chronicon.*

Lycosthenes's chronicle is arranged by years, which meant that the book presented a striking picture to readers at the time: not least because of the better access to information on recent prodigies resulting from the invention of the printing press, the cases compiled by Lycosthenes became increasingly numerous as they approached the present, creating the impression that the end times were imminent.[38] The edition of Julius Obsequens's fourth-century prodigy book published by Lycosthenes in 1552 was even distributed with eighteen blank pages at the back. In this way, readers were invited to continue the work and thus themselves to add to the effect described.[39] Lycosthenes took the prodigy book as a basis for his chronicle.

In the Latin edition of the chronicle, the section covering the year 1554 contains the same account of the headless baby found in the texts of Fincel and the two Schencks. With the exception of the reference at the end, it has been translated word for word into Latin: "IN Misnia infans natus est absque capite, oculorum effigie in pectore expressa, de quo idem Fincelius."[40] It is clear, then, that Johannes Schenck von Grafenberg made use of Lycosthenes's work without indicating the fact; instead, he referenced Fincel directly. Johann Georg, in turn, copied from his father.

The etching of the baby in Johann Georg Schenck's *Monstrorum Historia* can also be traced back to Lycosthenes. It is an almost exact reproduction of his woodcut (fig. 2). However, there is one significant difference between the two images: Schenck's illustration is a mirror-image reproduction of Lycosthenes's woodcut. Images reproduced in this way can be found quite frequently where woodcuts or etchings from earlier publications were copied. Avoiding this effect would have meant copying the template onto the printing block as a mirror image, which would have involved a little more work.[41] With the exception of this difference, their size, and slight variations in shading, the two images are identical—from the location of the

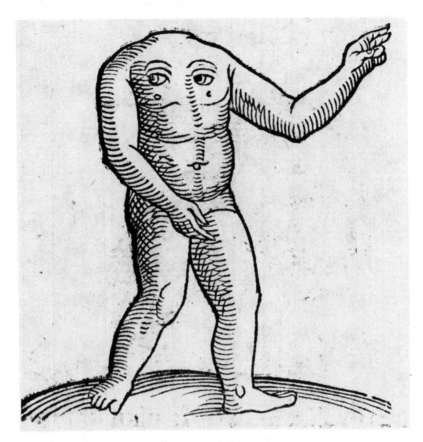

Fig. 2. Conrad Lycosthenes, *Prodigiorum ac Ostentorum Chronicon* [. . .].
Basel, 1557, 645; approx. 7 cm × 7 cm. Staatsbibliothek zu Berlin—Preußischer
Kulturbesitz, Department of Early Printed Books, Shelfmark: Bibl. Diez 2° 747.

eyes to the position of the baby's hands and fingers. The figure of the headless baby
is linked—not only iconographically but also by the context in which the woodcut
appears in Lycosthenes's text—to the depiction traditions of the exotics from the
edge of the world, for it resembles another woodcut employed not once but twice
in Lycosthenes's chronicle (fig. 3). It is typical of visual representations dating to the
Middle Ages of one of the monstrous races from the edge of the world: an upright
individual with no head or neck, whose eyes, nose, and mouth are located on its
chest. Such images can be found in the famous *Nuremberg Chronicle* written by the
humanist physician Hartmann Schedel (1440–1514) and other Nuremberg humanists, and in the *Cosmographia* by Sebastian Münster (1488–1552). In line with this

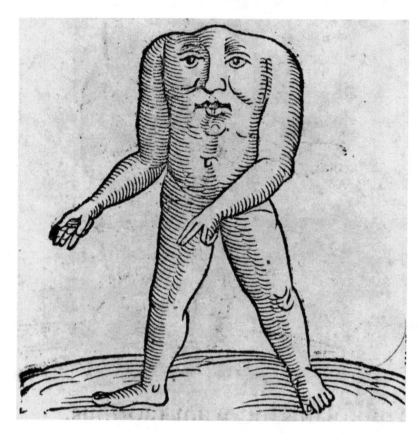

Fig. 3. Conrad Lycosthenes, *Prodigiorum ac Ostentorum Chronicon* [. . .]. Basel, 1557, 9; approx. 7 cm × 7 cm. Staatsbibliothek zu Berlin—Preußischer Kultur-besitz, Department of Early Printed Books, Shelfmark: Bibl. Diez 2° 747.

tradition of depiction, Lycosthenes uses the woodcut of the headless individual to illustrate an extensive list of monstrous races that God was thought to have created after the destruction of the Tower of Babel and the confusion of tongues: "According to the Blessed Aurelius Augustinus in book 16, chapter 8 of *The City of God*, after the destruction of the Babylonians' structure and the confusion of tongues, the Almighty God at once created monsters of a variety of human races. According to him they are located in Africa."[42] The list that follows this passage extends over ten pages and contains one short sentence describing the members of the headless race: their eyes, nose, and mouth, it claims, were on their chest and abdomen.[43] The same woodcut is used again in the appendix of the chronicle. Here, too, it illustrates a race, this time the so-called Monopoli. They are said to live in the forests of Asia, where they make a living by gathering and trading pepper.[44]

In the sixteenth century, the connection between individual monstrous newborns and the morphologically related monstrous human races of the topological tradition remained for the most part implicit, as in this present case. In Ulisse Aldrovandi's *Monstrorum Historia*, by contrast, it is immediately tangible. The Italian naturalist's encyclopedia of monsters, published posthumously in 1642, discusses the headless humans in two chapters. The first chapter—"Of Man"—serves, according to the introduction, to describe the perfect nature of humankind and thus to offer the reader a better understanding of the "errors" of nature that follow.[45] In one of its sections, Aldrovandi—or the posthumous publisher of the volume, Bartolomeo Ambrosini[46]—also deals in detail with the *differentiæ*, the differences, that could be observed between the human inhabitants of different parts of the world.[47] Here, the author engages critically with the exotics from the edge of the world, among whom he includes races whose members were said to have no heads.[48]

Citing a sermon (wrongly)[49] attributed to Augustine, Aldrovandi argues that there was indeed a people who had no necks, which created the impression that their heads were missing. But he is unwilling to believe in the existence of a truly headless race. It is seemingly only the authority of St. Augustine that prevents him from going further in his criticism.[50] Aldrovandi does not give reasons for rejecting an actually headless race of humans, but it seems likely that anatomical considerations had some part to play, alongside the reliability of the witnesses.

The second of Aldrovandi's chapters in which headless monsters play a part discusses *individually* occurring monsters. It is entitled "Of the Errors of Nature in the Formation of the Head" and follows immediately from chapter 2, an introduction that tells "of the monster in general." The concept of the *error naturae*, or error of nature, introduced in the chapter heading, implies that individual errors of nature, rather than monstrous races, are the focus here. Consequently, the chapter begins with a detailed list of individual cases of headless human monsters—including the baby born in Meissen in 1554 from Fincel's account. At the end of this list, before Aldrovandi discusses the causes of these cases, he returns to the subject of the monstrous race of headless humans:

> However, if Huldericus recorded in the account of his sea voyages that there was a race of headless humans living in the kingdom of Guinea . . . whose eyes & other facial features were found on their chest, we are not easily convinced to believe this, for it seems more fitting to fable than to the truth. However, Plinius

Secundus . . . wrote in his Historia naturalis as well that a race of humans with no heads lived to the west of the mountain of Milo in Asia. On this complex matter the reader may take recourse to the section on differentiæ in the first chapter of this Historia.[51]

It is unsurprising that Aldrovandi accords Pliny's *Natural History* an entirely different significance than a travel account from the Renaissance era.[52] What is striking, however, is that even within the discussion of *individual* headless monsters, he comes back to the subject of a corresponding monstrous race. After all, as we have seen, he took a critical view of the existence of a people whose members were entirely headless. In a similar way, in Johannes Schenck von Grafenberg's *observatio* on headless humans—*observatio* XXIV in the first volume of his collection of *observationes*—accounts of individuals sighted in Europe and statements about these kinds of monstrous human races stand harmoniously alongside one another. He ascribes the accounts to Pliny and Lycosthenes among others.[53]

Both writers thus bear witness to the strong connection that still existed in the late sixteenth century between individual monsters on the one hand and the exotics from the edge of the world on the other hand. This link was by no means restricted to headless humans but can be observed in the case of hermaphrodites, for example.[54] It is better understood against the backdrop of the scholarly understanding of authority and authorship, which is outlined below.

Authority and Authorship

The connection between the description and representation of individual monsters and the exotics from the edge of the world becomes clearer against the backdrop of the authority that Renaissance scholars ascribed to ancient writers such as Pliny and Augustine. But how, for these scholars, did the statements of ancient writers relate to the observations of contemporary naturalists?

Helmut Zedelmaier summarizes central features of scholarly knowledge in the early modern period as the "scholarly dispositive." Scholarly knowledge, he suggests, was book knowledge first of all, for books were the key medium by which knowledge was passed on. Furthermore, the scholarly dispositive points to the idea "that knowledge in the early modern period was linked in a particular way to 'inherited,' traditional knowledge, namely to texts from Greek and Roman antiquity, which set standards, and to the Bible, which as the 'Book of Books' . . . was read and interpreted as the epitome of knowledge."[55] The strikingly frequent references made by writers around 1600 to Pliny, Augustine, Aristotle, Hippocrates, or indeed

Galen were an expression of a central trait of scholarly knowledge in the early modern period.

Authority was a key aspect of the scholarly dispositive. In the Middle Ages and the Renaissance, the Latin term *auctoritas* denoted the words of an author—*auctor* in Latin. This did not include *every* writer, however, but rather, in Katharine Park's words, "one of the chain of especially revered and trusted writers or teachers, from the ancients to near contemporaries, whose texts established the framework within which questions in the learned disciplines were debated and explored."[56] Even into the eighteenth century, university education was devoted chiefly to the exegesis and application of a limited canon of books that for the most part were attributed to ancient authorities. Every academic field had its own Book of Books. Early modern fields of study were defined less by specific problems and methods than by their textbooks.[57]

Nevertheless, they were receptive to new knowledge: *loci*, or *topoi*, which were understood in the rhetorical and dialectical Topics tradition as both concepts and models, served to establish distinct fields into which knowledge was categorized. Newly acquired knowledge that did not come from ancient sources could thus be assigned a place within the topical order. This method assisted scholars in locating specific arguments when needed. It contributed indirectly to a certain constancy in the questions and subject matter of scholarly debate up until the eighteenth century.[58]

The reading and writing practices that were constitutive of scholarly knowledge were taught at schools and universities. Printed reading guides supplemented this intellectual training. "Paper technology," in particular the commonplace books kept by countless early modern scholars, embodied the learning techniques in question.[59] Ann Blair gives a good summary of the role played by the *loci* in Renaissance education: "Commonplacing was a practice of note-taking taught in Renaissance schools . . . [The commonplace book] was a personal notebook in which each schoolboy (and later, adult reader) was taught to enter and sort under subject headings interesting turns of phrase, opinions, or facts of all kinds encountered in reading, travel, and daily life, for later retrieval and use."[60] Notebooks organized according to the *loci* technique were intended to serve as a means of recording both the fruits of reading and direct empirical observations. In the long term, locating scholarly knowledge by means of commonplaces provided, in a sense, the model by which empirical knowledge would be managed. This, however, was a lengthy process, to which I will return later.[61]

The exegetical nature of scholarly knowledge and the significance of ancient authorities to that knowledge characterize the texts discussed here, too. It is true that in the long term, the so-called new science would—in the course of the seventeenth century—gradually distance itself from the other fields of study and increasingly establish itself *in opposition to* scholarly knowledge.[62] However, the rhetoric of the new that accompanied this process should be taken with a grain of salt.

Conversely, it was by no means the case that Renaissance naturalists followed their ancient predecessors in every respect, nor did they confine themselves solely to their works. Aldrovandi's oeuvre is a good example of the complex relationship of late Renaissance naturalists to the ancients. Aldrovandi, like many of his contemporaries, took upon himself the task of extending or even completing Aristotle's work on natural philosophy. He aspired to reproduce the entire Aristotelian corpus in his work.[63] And precisely because he saw his work as a continuation of Aristotle's, he certainly considered himself not a mere *compilator* but rather an *auctor*.

Even down to the level of individual questions, naturalists during this period continued to attach great significance to the statements of ancient writers. This is true of Aldrovandi, too, who invokes the authority of Pliny, Solinus, and particularly Augustine in his discussion of the exotics from the edge of the world. It is worth taking a closer look here, however: At the beginning of his discussion, Aldrovandi refers to Pliny and Solinus, who believe there to be headless humans—the so-called Blemmies—living in the forests of Asia. Aldrovandi, who in general takes quite a critical view of such reports, then goes on to cite Augustine: "*I had already become Bishop of Hippo and travelled onwards with several Christians to Ethiopia, to proclaim the Gospel of Christ to them. And there we saw many men and women who had no heads, but bore eyes on their chests. Their remaining parts, however, matched our own*: This same sentence by Saint Augustine on this matter was quoted by Fulgosius."[64]

In the canon of late Renaissance fields of study, it was rhetoric that was responsible for the evaluation of testimonies such as these, which could neither be proven nor disproven using the tools of Aristotelian logic. Based primarily on Cicero at this point, rhetoric taught the theory and practice of persuasive argument.[65] According to a widespread understanding in the early modern period that took its inspiration from a statement by Cicero, it was intended to serve the creation of *fides*—faith or trust.[66] A distinction was made between the trust that listeners placed in the speaker and the trust that they placed in what was said. Both were regarded as essential to the persuasive power of the speech.[67]

Aldrovandi's assessment of the above case can be better understood against the backdrop of this contemporary distinction: the content of the statement evidently seemed somewhat implausible to him. But because, as he writes, he classed Augustine as an exceptionally trustworthy author, he had no choice but at least to seek a compromise and to affirm confidently that this race of humans had no necks.[68] Statements by contemporary authors, however, were also taken up affirmatively in the books on monsters published around 1600 by physicians and other naturalists. A frequently used phrase, by means of which the authority of a writer was expressed or indeed established in discourse in the first place, was "clarissimus vir" or "clarissimus medicus." This phrase marked a scholar as exceptional in a positive sense and, as such, particularly trustworthy.

Johannes Schenck von Grafenberg, for example, uses the phrase "clarissimi medici" on the front page and in the introduction to the first volume of his *Observationes* to characterize the correspondents and authors to whom he owes the factoids he has compiled as particularly reliable sources.[69] Most of the authors he cites lived in the early modern period; ancient authors are clearly in the minority. Aldrovandi, too, applies the phrase "clarissimus vir" to contemporaries such as the Italian anatomist Realdo Colombo (ca. 1510–1559).[70]

There was, then, a transfer of authority to specific contemporary writers in progress that gradually endowed them with a significance comparable to that of the ancient scholars. Further indications of this process are found in the form of the *indices authorum*—lists of authors and texts consulted—contained in numerous texts. In many of the monographs on monsters published around 1600, both ancient and contemporary authors are cited in these *indices*. The bibliographical details of the contemporary texts and those of the texts of the *veteres*, the ancients, are frequently identical in form. Moreover, some *indices* dispense entirely with any typographical means of identifying the texts of the ancients and those of the contemporaries as two distinct groups.[71] These observations correspond with Roger Chartier's finding that a general transfer of the function of author[72]—and thus also of the authority of the author—took place in the early modern period: from the ancient *auctoritates* to contemporary writers, indeed even to those who wrote in their own vernacular.[73]

The concept of authorship that was broadened in the course of this process, however, was not identical to the one that, in the seventeenth century, was increasingly to be contrasted with the *collector*, the mere gatherer of factoids, and was to reject the tradition in which the authors of the late sixteenth and early seventeenth century discussed here continued to see themselves. For the time being, the authority of the ancients remained a defining characteristic of the "scholarly dispositive"— with direct implications for the discourse on monsters.

A Horse with a Human Head

It is no coincidence that the multiple variations of the second factoid that appear in the writings of different authors are reminiscent of the children's game telephone. The appeal of the game consists in a message whispered from one child to the next becoming increasingly distorted—or, framed positively, taking on a new form via creative mishearing—over the course of this chain of transmissions. A similar process of creative distortion seems to be responsible for the replication of many of the factoids in the texts discussed here.

Many of the scholarly texts on the topic that were published around 1600 contain a number of different accounts or visual representations of what was originally evidently one and the same *monstrum*. In part, this can be explained by the encyclopedic ambitions of authors like Aldrovandi, in light of which the collecting and documenting of factoids took on a value in their own right. From this perspective, each individual factoid seems, even if in doubt, to have been too valuable to ignore. The persistence of the different versions of the factoid under discussion here thus also allows us to draw conclusions about the way these writers approached contradictions among the sources they consulted.

In the culture of the late Renaissance, the boundary between humans and animals was more permeable and less clearly defined than in the modern era.[74] The character of Caliban in William Shakespeare's *The Tempest*, who exhibits both human and animal traits, is perhaps the best known of numerous personifications of this porous border regime in the drama of the time.[75] It is unsurprising, then, that texts about monsters from this period contain numerous beings whose monstrosity consists in a combination of human and animal traits or body parts.

The section of Lycosthenes's *Chronicon* that covers the year 1254, for example, tells among other things of the birth of a foal reminiscent of a centaur, with the head and voice of a human. In the light of the political landscape of the time, Lycosthenes interprets this *monstrum* as a divine omen:

> While the Florentines and the Pisans were engaged in a battle at Mount Altino, a mare in a field near Verona gave birth to a *monstrum*, a four-legged foal with a human head. The sound of its mutterings, squawking, and distressed roaring in a human voice brought an inhabitant of that rural area running and, thunderstruck at this unbelievable sight, he drew his mighty sword and killed the *monstrum*. When he was therefore brought to court and questioned as to the origin of the deformed creature and as to his reasons for killing it, he answered that he

had only done so out of fear and horror of the thing. After this had been heard, they relieved him of all suspicion.[76]

Like many other symbolic occurrences reported by Lycosthenes, this episode is illustrated by a woodcut (fig. 4). It follows immediately after the section of text and shows a horse with human facial features, standing on its hind legs with its front legs slightly raised.

Unlike many other woodcuts in this chronicle, the depiction is not used again to illustrate a second or third similar event elsewhere within the chronological system. The woodcut belongs unambiguously to this individual *monstrum*, then, and does

Fig. 4. Conrad Lycosthenes, *Prodigiorum ac Ostentorum Chronicon* [. . .]. Basel, 1557, 438; 7 cm × 7 cm. Staatsbibliothek zu Berlin—Preußischer Kulturbesitz, Department of Early Printed Books, Shelfmark: Bibl. Diez 2° 747.

not represent a class or species. In this respect, the depiction of this event corresponds with its status from the viewpoint of the author and many of his contemporaries: as a prodigy, the foal represented by definition an individual act of divine intervention. Both the precise form of the monster and the location and time of its appearance could hold decisive clues as to its significance, which meant that it had to be recorded carefully in all its individuality. Lycosthenes does not give any details of his source—in relation either to the textual or to the pictorial factoid. Other sections of his chronicle, by contrast, include short references to the source texts of the factoids they contain.

Quotations or paraphrased versions of this episode in Lycosthenes's text and reproductions of the accompanying woodcut can be found in many of the writings on monsters by academically trained physicians that were published around 1600. Johann Georg Schenck von Grafenberg, for example, the Haguenau municipal physician mentioned above, quotes the account in full in his *Monstrorum Historia* under the heading "A Horse with the Face of a Man." The quotation is followed, after a further heading ("A Mare Bore a Foal with a Human Face"), by a short account of the same incident written by Ambroise Paré, which differs slightly from the first in content. Here, the historical backdrop against which the symbolic event is said to have taken place is not the battle between the Florentines and the Pisans but rather a conflict that, according to Paré, had recently erupted between the Pisans and the Turks. The man who put an end to the *monstrum* in Lycosthenes's account is not mentioned.[77]

Both versions are found in the second part of the *Monstrorum Historia*. Although the larger first part of the book represents a compilation of the factoids about monsters in Johannes Schenck von Grafenberg's multivolume *Observationes*, the second part is an original contribution by the son, gathering factoids about monstrous animals.[78] In the first part, Johann Georg Schenck kept largely to the classification principle that underlies his father's collection of *Observationes*: it is ordered according to the *a capite ad calcem* ("from head to heel") schema that was often used during the same period in *practica* textbooks, works of practical medicine concerned with the diagnosis of particular diseases and their treatment.[79] For the second, however, it was necessary to choose a different system. He decided to begin with lions and horses, and then to proceed to successively smaller mammals. Monstrous birds followed, and then reptiles and amphibians.

Johann Georg Schenck's above factoids concerning the horse with a human face sighted in 1254 are not his only *historiæ* about a *monstrum* of this kind. They immediately follow a similar, undated factoid entitled "Monstrous Horse with the Head of a Man": "MONSTROUS HORSE, with a human head . . . I once saw a beast with a human head, born of a mare, which produced human noises but [was] otherwise a horse. Pictorius serm. convivial. lib. 4. p. 8. 5."[80] Schenck had clearly encountered this statement in the *Sermonum Convivalium Libri X.* of Georg Pictorius from Villingen an der Donau, who lived between ca. 1500 and the second half of the sixteenth century. Pictorius started out as a schoolteacher and later principal in Freiburg im Breisgau before studying medicine there. After completing a doctorate, he was offered a professorship, which he rejected in favor of a position as the municipal physician of Ensisheim in Alsace, beginning in 1550. His *Sermonum Convivalium Libri X.* were published in two editions, one in 1551 and the other in 1571.[81] Unlike the *monstrum* dated 1254, this beast is illustrated with an etching (fig. 5), though it is based solely on Pictorius's text passage.[82] The etching shows a rear-angle view of an upright horse whose childlike human head is turned toward the viewer.

Despite the conspicuous similarities between this first factoid and the two based on Lycosthenes's and Paré's writings that follow, it does not refer to the same event. Indeed, it originates from an entirely different text genre and, from a modern perspective, has a different epistemic status. For whereas the factoids in Lycosthenes and Paré's texts can be classified as two different accounts of the same *historical* occurrence, the statement quoted by Schenck has its origin in a non-historiographical text genre: Pictorius's work, which is divided into ten books, presents to the reader table talks held by scholars. Using intertextual references such as the appearance of Crito, for example—a pupil and friend of Socrates who figures in Plato's *Crito* as a dialogue partner of his teacher—Pictorius positions his text in the tradition of philosophical dialogue literature.

The sentence quoted by Schenck is spoken in book 3, in a conversation among Crito, Faust, and Hospes. It is Hospes who claims to have seen the *monstrum*.[83] The form of the citation, however, means that a reader with no knowledge of Schenck's source might easily get the erroneous impression that Pictorius himself had witnessed it. Schenck makes no mention of the character to whom Pictorius ascribes the statement in question. In a sense the etching reinforces this impression, suggesting as it does that the creature was seen by a (nonfictional) observer. In other words, the difference—which from a modern point of view is fundamental and irreducible—between the epistemic status of this and the other two passages is not reflected in the concrete presentation of the passage and is, as it were, leveled out.

Fig. 5. Johann Georg Schenck von Grafenberg, *Monstrorum Historia Memorabilis* [. . .]. Frankfurt, 1609, 99; approx. 7 cm × 6 cm. Staatsbibliothek zu Berlin—Preußischer Kulturbesitz, Department of Early Printed Books, Shelfmark: Ke 7065 R.

Two factoids from history and one from a "fictional" text written in dialogue form thus stand harmoniously alongside one another in a way that suggests that the author of these texts vouches for each of them.

In his immensely influential treatise *De Monstrorum Natura, Caussis, et Differ-*

entiis, first published in 1616, Fortunio Liceti, professor of theoretical medicine in Padua, also tells of two horses with human heads. The treatise was aimed at establishing a *scientia*, a science of monsters in the Aristotelian sense, which also implied a clarification of their causes.[84] To this end, Liceti assembled, among other things, a large number of individual cases and discusses them with regard to their classification and causes. In one chapter, which lists a large number of factoids with the intention of proving the existence of "multiform monsters" that have a part in more than one species within one kind (*genus*), Liceti also discusses the two horses.[85] The factoids cited in this chapter are taken in part from literature and in part from personal contacts. He presents them in loosely chronological order. Next to the first of the two accounts of horses with human heads, the year 1254 is printed as a marginal note. This first *monstrum*, then, is clearly the same one that appears in Schenck's account based on Lycosthenes's and Paré's writings. Unlike the two factoids in Schenck's text, however, the first of Liceti's factoids contains no details of the historical context of the monstrous birth. The reader can therefore no longer relate it to the conflict mentioned in this context by—for example—Lycosthenes. And the symbolic character of this monster is therefore no longer apparent in Liceti's text.

Liceti claims to have encountered this case in one of the main sources for his treatise—Ambroise Paré's *Des monstres et prodiges*:

> Another *monstrum* was, also according to Paré's observation [*observatione*], born in a field near Verona: but it was the four-legged offspring of a mare with a human head, at the sound of whose mutterings, squawking, and distressed roaring in a human voice an inhabitant of that region came running and, thunderstruck at this unusual sight, he drew his very large sword and killed the *monstrum*. When he was therefore brought to court and questioned as to the origin of the deformed creature and as to his reasons for killing it, he answered that he had only done so out of fear and horror of the thing. After this had been heard, the judges dropped all suspicion.[86]

What is striking about this passage is, first, that the report is presented as an *observatio*, an observation of Ambroise Paré's, although it occurred long before his birth. For Fortunio Liceti there was no contradiction in describing something that could only be known to that person from literature as a person's observation. This fusion of empiricism and reading in the late Renaissance, which is difficult to understand from a modern perspective, is something that this study encounters frequently.

Second, the account is not identical to the one that Schenck—also with reference to Paré—included in his *Monstrorum Historia*. In the version reproduced by Liceti, the factoid makes no reference to the war between Pisans and Turks that appears in Schenck's passage. How can this difference be explained?

The account quoted by Liceti is found in every edition of *Des monstres et prodiges*. It is cited by Paré as an example of a *monstrum* that came about as a result of the combining of human and animal seeds. Accordingly, the chapter in question is headed "Example of the Connecting and Mixing of Seeds." In the context of the line of argument that Paré follows here, then, the possible symbolism of this monster was simply not relevant. Thus he omitted the historical references to the divinatory interpretation that were superfluous to the discussion of the cause of such monsters' creation:

> In the year 1254, a mare near Verona gave birth to a foal that had a well-shapen human head and the remaining [body] of a horse. This monster had the voice of a human, at the sound of whose cry a local villager came running and, surprised to see such a terrible monster, killed it: therefore he was taken to court and questioned both about the birth of this monster and about his reason for killing it. He said that the horror and terror he had felt at the sight of it had caused him to do it. Thereupon he was acquitted.[87]

Beginning with the 1579 edition, however, this episode is also cited in another chapter in Paré's text—"Exemple de l'ire de Dieu" (Example of the wrath of God)— where it serves as an example of a prodigy created by divine intervention in the habitual course of nature. In this divinatory context, Paré mentions the war between the Turks and the Pisans, and it is this passage on which Schenck draws.[88]

Unlike Liceti's monograph,[89] Paré's text includes a visual representation of this monster. Paré's woodcut (fig. 6) resembles that of Lycosthenes, though the front legs of the horse in Paré's illustration are not raised as high.

Jean Céard's in-depth analysis of the sources used by Paré enables us to make a reliable statement about the origin of the woodcut and the account to which it belongs: Paré found this case among Claude de Tesserant's[90] additions to Pierre Boaistuau's (d. 1566) *Histoires prodigieuses*, a prodigy book first published in 1560 that was immensely successful, particularly in France.[91] It constitutes one of three main sources for Paré's monograph.[92] Tesserant, in turn—and so we come full circle—had encountered the above case in Lycosthenes's *Prodigiorum ac Ostentorum Chronicon*.[93] Lycosthenes's chronicle was also the starting point, then, for the circulation of this version of the account of the foal with a human head.

Liceti's second factoid of this kind is, despite some differences in detail, almost

Fig. 6. In the first and second editions (1573 and 1575) of Ambroise Paré's *Des monstres et prodiges*, this woodcut appears in the chapter that discusses the monsters attributed to the combining of animal and human seeds. From the 1579 edition onward, however, it appears in the chapter on monsters stemming from the wrath of God. It can thus be said to illustrate the two versions of the factoid in Paré's text equally. Ambroise Paré, *Des monstres et prodiges*, critical ed. with commentary by Jean Céard, Geneva, 1971 [1573], 7; approx. 6.7 cm × 11.7 cm.

too similar not to be based on the same event. As in Lycosthenes's version, the creature under discussion here was also born of a mare. In this case, too, the geographical point of reference is Verona. And once again it is reported that the monstrous foal produced terrible noises, drawing the attention of an inhabitant of that region who would eventually kill it. Again, this act had legal repercussions that ended well for the farmer. Nevertheless, the account has some distinctive features: The foal, it states, had not only the head but also the neck of a human. More significant again is the fact that this event is dated 1315 rather than 1254. Where, then, did this further version of the factoid of a horse with a human head originate?

Liceti cites as his source the work *De Gestis Italicorum post Henricum VII Caesarem*, written by the Italian historian and poet Albertino Mussato (1261–1330). It begins with the death of the Holy Roman emperor Henry VII and continues Mussato's *Historia Augusta Henrici VII*, a chronicle of his activities in Italy. Liceti had the opportunity to study the unpublished manuscript; Mussato's two-part work would not be printed until 1636.[94]

This second version of the factoid in the same work of natural philosophy is illuminating in two respects. First, its existence demonstrates that accounts and illustrations of monsters and other rare phenomena did not only multiply in the sense that they were quoted, paraphrased, or reproduced time and again once they entered the scholarly discourse. Sometimes they also multiplied in the sense that two or more divergent versions of a single factoid were present within one discourse or even one work, without this reduplication being addressed as a problem by the author. Fortunio Liceti does not seem to have noticed the similarities between the two accounts. Second, this example shows once again how closely the discourse of natural history and philosophy was bound up with that of historiography. I come back to this connection in greater detail later.

Finally, let us look at a third factoid in Liceti's text. Like Johann Georg Schenck, Liceti made use not only of the writings of other naturalists and of historical and chronicle literature, but also of the genre of table talks. He, too, found an account of a further horse-human hybrid in a text from this genre. In this case, however, the author was not Pictorius but the Greek biographer and philosopher Plutarch (ca. 45–125), who was highly regarded in the late Renaissance. The case is taken from his *Dinner of the Seven Wise Men*.

The being mentioned by Plutarch is even more reminiscent of a centaur than the versions of the factoid concerning the foal with a human head discussed above. Not only do its head and voice resemble those of a human; its hands are described as human as well. Liceti refers to this case at several points in his *De Monstrorum Natura, Caussis, et Differentiis*, including the chapter in which the above two cases are also mentioned: "Plutarch recalls a young shepherd who showed him the off-spring of a mare that had a human form above the neck and at the hands, and the form of a horse at the remaining parts of the body; it cried in the way that a new-born human baby does."[95]

The way in which Liceti reproduces the passage from Plutarch's text is reminiscent of Schenck's approach to the factoid he found in Pictorius's writing. It suggests that Plutarch himself was an eyewitness to the event recounted. Plutarch appears as both the subject of *memorare*, the person who mentions or recalls the event in question, and the indirect object of *monstrare*, that is, the person to whom the *monstrum* was shown. But just like Pictorius in the Schenck/Pictorius case, Plutarch, too, is merely the author of the table talks whose characters are discussing the being in question.[96]

Like Johann Georg Schenck some years earlier, Liceti clearly did not attach any importance to distinguishing between firsthand and secondhand observations. Liceti's indifference to this distinction also manifests itself in his reproduction, discussed above, of the Paré quotation concerning the foal allegedly born near Verona with the head and voice of a human. In this present case it is all the more significant from today's perspective because the eyewitness is not a historical person but rather one of Plutarch's characters, and Liceti makes it clear elsewhere in his monograph that he was well aware of this.[97]

Pictorius, too, leaves no doubt as to the usefulness of his text to its readers. This is emphasized even on the front page. Precisely how the reader might benefit from reading it is not explained there, but the front page does contain the information that, alongside matters of poetry, the table talks address historical and medical matters, too.[98] What is more, Pictorius did not simply invent the statements assembled in his work. Indeed, in many cases they are passages taken from the writings of ancient authors: in the first edition's dedication, Pictorius extols them as selected blossoms (*flores*) from the writings of the most distinguished men (*clarissimorum virorum*).[99] The catalog of authors cited, which likewise precedes the main text, reads accordingly like a Who's Who of ancient scholarship.[100]

Key elements in the information architecture of the text—the listing of topics discussed in the individual *libri* at the beginning of the volume and then again at the beginning of each book, for example—also demonstrate clearly that the work offered more than mere entertainment. It is a scholarly reference work, written in dialogue form, on a great variety of topics, from medical questions to questions of good living such as the *ars bibendi*, the art of proper drinking.

This work's embeddedness in the tradition of the table talk genre was also of some significance to its status. The author makes very deliberate use of the device: Pictorius devotes large sections of the foreword to the reader to positioning his text in the history of this genre, which from a humanist point of view was considered venerable.[101] From the perspective of early modern scholars, texts constructed as dialogues, such as both Plutarch's and Pictorius's table talks, were not simply fictional texts devoid of any authority. Dialogues gained in status in and through the humanist movement, not least because of the dialogue form's compatibility with humanist skepticism toward definitive truths.[102] Important authoritative reference texts and passages written in dialogue form included the works of Plato, Luther's table talks, and the Melian Dialogue in book 5 of Thucydides's *History of the Peloponnesian War*.[103] Furthermore, Plutarch's text belonged to a tradition of texts in which the so-called seven wise men appeared[104] and was held in correspondingly high regard among early modern scholars. The authority of the statement made in Plutarch's text and taken up by Liceti was, in addition, by no means diminished by the fact that the character to whom it is ascribed there—Thales—is inspired by the Greek natural philosopher of the same name.

In this context it is also relevant that a close intertextual relationship exists between the two passages in the texts of Plutarch and Pictorius: Plutarch's factoid reproduced itself here, if you will, inasmuch as it supplied the model for the corresponding factoid in Pictorius's text. The latter has Hospes—the character in his table talk who claims to have seen the *monstrum*—make a reference to Thales that reveals this relationship: "HOSPES. I once saw a beast with a human head, born of a mare, which produced human noises but [was] otherwise a horse. I wondered at it and muttered the words of Thales quietly to myself. For when Thales saw a similar beast in Periander's herd, he said to Periander: I exhort you not to employ any more grooms in the future: or, if you do employ them, to oblige them to marry."[105]

Some of Fortunio Liceti's contemporaries were more cautious in their approach to conspicuously similar factoids of monstrous occurrences. Ulisse Aldrovandi, for example, approaches the passage in Plutarch's text very similarly. He, too, includes the factoid in his monograph on monsters, in the chapter on monstrous animals with the heads of another species. He, too, presents it in a way that suggests that Plutarch himself had seen the *monstrum*.[106] And he treats this factoid in just the same way as one from a historiographical text type: it is placed in immediate proximity to the similar account by Conrad Lycosthenes discussed above. The latter, incidentally, is illustrated with a woodcut that is a copy of the one in Paré's work.[107] With regard to the two versions of Lycosthenes's factoid included by Johann Georg Schenck von Grafenberg in his *Monstrorum Historia*, however, Aldrovandi—or rather Ambrosini, the posthumous publisher of the volume[108]—takes a critical view. He observes that Schenck presents two different *historiae* of this monstrous horse, of which one corresponds with that of Paré and the other with that of Lycosthenes. He then goes on to express his view that the two *historiae* were aimed at the same goal.[109]

Like Liceti, Aldrovandi/Ambrosini includes in his *Monstrorum Historia* the case reported by the Paduan historian Mussato. And here, too, he proves to be the more critical reader: unlike Liceti, he suspects this case to be identical to the one that was generally dated 1254, even though Mussato places it in the fourteenth century.[110] It seems probable that these are the words of Ambrosini and that this passage in Aldrovandi's work is the expression of a sensitivity to such questions that developed *after* Liceti.[111]

Let us recapitulate what has been shown here: naturalist writings on monsters that were published around 1600 often contain differing versions of woodcuts and accounts that can ultimately be traced back to a single report or its visual representation in one source text. One factoid whose circulation began in the chronicle of Conrad Lycosthenes had a strikingly far-reaching impact, as did a similar textual building block that can be traced back to Plutarch's table talks. Both multiplied not only in the sense that they were reproduced time and again by a procession of writers in the early modern period but also in the sense that, in the course of the chains of transmission that developed in this way, different versions of these factoids came into being, reminiscent of the products of the children's game telephone. Such differing versions of the same phenomenon or event are found in both encyclopedic texts such as Johann Georg Schenck von Grafenberg's *Monstrorum Historia* and texts aimed at establishing a *scientia* of monsters. Neither Schenck nor Liceti problematized their similarities.

The source texts of the two factoids also represent two different text genres or discourses that were often consulted when early modern scholars with an academic medical background were planning publications on monsters: dialogic text genres such as table talks and chronicle and ephemera literature.[112] The first of the two text genres has been discussed sufficiently already. The second group of texts was of greater significance, however, and it is to it—and in particular to its relationship to medical and natural history literature concerning rare phenomena of nature such as monsters—that we now turn our attention.

Medicine and Historiography

The influential treatise *De Hermaphroditorum Monstrosorumque Partuum Natura* by the Basel municipal physician and university professor Gaspard Bauhin, which was published in at least two contemporary editions,[113] demonstrates again the close connection already observed in the works of other writers between medical and historiographical or chronicle literature. In chapter 35, "Examples of Hermaphrodites," in which Bauhin compiled *exempla* of hermaphrodites in chronological order, he primarily used the writings of four historians: "Now we will present the examples of different hermaphrodites from Livius, Eutropius, Obsequens, Lycosthenes, and others."[114]

It is interesting that he refers to the cases assembled here as examples—*exempla*. The term *exemplum* was often used in medical texts, particularly before the rise of the *historia*, to describe short narratives on individual cases.[115] The use of the term for narrative episodes was by no means confined to the medical field, however. Its importance in exemplary historiography as practiced by Theodor Zwinger is particularly noteworthy.[116] Medicine and historiography overlapped, then, in the term "example."[117]

In contrast to the Roman historians Livy and Eutropius, little is known about Julius Obsequens. We do know that, like Lycosthenes, he was the author of a prodigy book. Obsequens probably lived in the fourth century CE. His *Prodigiorum Liber* is a digest of the prodigies in Livy's work, though only those that are said to have occurred between 190 and 112 BCE are included. There was a close connection between Lycosthenes and Obsequens inasmuch as the former was responsible for the 1552 publication of a supplemented edition of the *Prodigiorum Liber* that was translated into Italian in 1554 and into French in 1555 and contributed significantly to the enormous popularity of this work in the sixteenth century.[118] The use of Theodor Zwinger's *Theatrum Vitae Humanae*, a historical work by the renowned Basel physician from which Bauhin took several *exempla* for his chapter 35, is further evidence of the close connection between medicine and historiography.[119]

As we have already seen, Bauhin's reading practice was not unusual: physicians or natural historians, who were academically trained as physicians for the most part, did not only consult the works of other naturalists in their search for material for their own publications on monsters. In addition to the ancient authorities of nature study, they favored chronicles of *mirabilia* or prodigies such as Hiob Fincel's *Wunderzeichen. Warhafftige beschreibung und gründlich verzeichnis schrecklicher Wunderzeichen und Geschichten* and particularly Conrad Lycosthenes's *Prodigiorum ac Ostentorum Chronicon*.[120] Both chronicles were written in the mid-sixteenth century by Protestant theologians—a point to which I will return.

The overlap between historical and medical or natural history literature that can be seen in the discourse on monsters is consistent with the close connection in the Renaissance between these two discourses and ways of writing. It was manifested most clearly in the "physician historians." As Nancy G. Siraisi has shown in her influential study, there is evidence of "extensive involvement of physicians in the reading, production, uses, and shaping of historical knowledge in the period ca. 1450–1650."[121]

The proximity of physicians to historiography was the result of factors both external and inherent to medicine: Elements of humanist culture such as its own particular sense of history, the new humanist historiography, and the growing interest in material evidence of the past all encouraged the study of history. As a part of this culture, medical discourse was so transformed in this environment that physicians possessed the right "intellectual formation" to succeed as authors of histories:

Such "historical" elements within medicine as narrative, empiricism, and attention to particulars and to material evidence also took on new prominence . . .
Thus, both the context of a polymathic and historically oriented humanist culture and features internal to medicine united to make it easy for some medical
men to turn without any sense of incongruity to the writing of civil history, pro-
fessional or other lives, or histories of sciences and to the study of antiquities and,
in short, of any aspect of the human past.[122]

In addition, academic training in all humanist fields offered—to those who had
the benefit of it—access to a world of learning that was as yet largely undivided into
distinct disciplinary discourses, and academically trained physicians could thus
claim an unrestricted place in the scholarly world. And, finally, the period's characteristic mobility of physicians between the epistemic spaces of the university and
the court may have contributed to their emergence as writers of chronicles and
other historical genres of a considerable number.[123] In the light of these research
findings, then, it is unsurprising that two of the authors of prodigy books mentioned above were scholars with a university background in medicine: Pierre Boaistuau and Hiob Fincel.

The boundary between natural and civil history was, furthermore, not particularly firm during this period. The natural philosopher and Lord Chancellor Sir
Francis Bacon (1561–1626) distinguished four kinds of history: natural, civil, ecclesiastical, and literary.[124] There is plentiful evidence that they were closely connected
in the scholarly practice of the late Renaissance, too. The broadsides on religious,
political, prodigious, and medical topics that were inserted into Hartmann Schedel's personal copy of his works demonstrate, for example, that he understood the
subject matter of history to include both *divine* and present and past *human* action,
but additionally medical and natural phenomena.[125] The "histories" of nature that
began to appear in Europe from 1540 onward, which from a contemporary perspective would fall within the rubric of natural history, were at first only very rarely
described as "natural histories" in their titles. One exception is Ferrante Imperato's
Dell'historia naturale from 1599.[126]

When "natural history" does appear as a category in the sixteenth and seventeenth centuries, it is usually not on front pages but in classifications of academic
fields such as Jean Bodin's threefold division of history into human, natural, and
divine, which also underlies Bacon's four categories.[127] Such classifications testify
to the fact that contemporary scholars distinguished different forms of history in
terms of their subject matter. But on many other levels, particularly with regard to

ethical and theological dimensions, for example, natural history had—in their view—commonalities with other forms of history.[128]

Broadsides as Sources for Scholarly Authors

Alongside scholarly histories, broadsides were often important suppliers of monstrous cases. Naturalists read (and wrote) broadsides themselves, or in some cases were familiar with their contents via the medium of chronicle literature. They played an important role for scholars gathering information about recent occurrences. Chroniclers and other compilers frequently drew on information from broadsides, and conversely, the authors of these ephemera made use of material from the writings of chroniclers and other scholars.[129]

For a long time, pamphlets and broadsides were wrongly regarded as part of the culture of the uneducated, or at least less educated, and thus as extrinsic to the culture of the learned elites.[130] As the example of Schedel has already shown, distinctions of this kind between elite and popular culture are problematic as far as these printed products are concerned. An illustration of this is the collection of pamphlets and other sources of information about wonders known as the *Wickiana*, which was created in the sixteenth century and is now kept in the Zurich Central Library. Its originator was the canon and second archdeacon of Zurich's Grossmünster, Johann Jacob Wick (1522–1588). Wick created his chronologically ordered collection of "many histories" and of "wonders"[131] between 1559 and 1588. His belief in wonders was, as Franz Mauelshagen has shown, sanctioned by the Reformation discourse supported by "an educated ruling elite."[132] Thus, the *Wickiana* collection was a product not of popular but rather precisely of elite culture.[133]

Unlike the scholarly literature consulted by writers such as Schenck or Aldrovandi, broadsides were often not named as sources. Nevertheless, both their textual components and their illustrations found their way, directly and indirectly, into scholarly literature—and vice versa.[134] Below, I investigate these transfer processes by addressing them on three levels of analysis: places of printing, publishers and printers, and authors.

The majority of the scholarly monographs on monsters published in central Europe—whether or not they were written by physicians—were conspicuously published in the Protestant (or, more specifically, Reformed) centers of printing. Basel and Zurich stand out particularly among them.[135] It is certainly no coincidence that most of the broadsides on wonders that are preserved in collections such as the *Wickiana* likewise come from these centers of Protestant (or Reformed) printing.[136] The *Wickiana* comprises twenty-five books bound together in twenty-four folio volumes of approximately six hundred pages each.[137] Aside from the 503 pamphlets and 431 illustrated and unillustrated broadsides that are included, it encompasses numerous handwritten records from correspondents, Wick's notes on accounts conveyed to him verbally, and drawings of his own. It represents "a historical, eschatologically oriented archive."[138]

As Wolfgang Harms and Michael Schilling have observed, the occupation with prodigies—both in pamphlet printing and in general—was more widespread in the Protestant than in the Catholic regions of the German-speaking parts of Europe.[139] And thus it can also be said of prodigy literature that it was primarily a Protestant phenomenon in German-speaking areas, though belief in wonders was widespread in the sixteenth century even across confessional boundaries.[140] Two factors contributed to this.

First, apocalyptic thinking was more widespread among Protestants than Catholics—although recent research indicates differences within Protestantism.[141] The Reformation, for example, was viewed primarily by the Reformers themselves as evidence that the end times were near.[142] This difference between Protestants and Catholics was due chiefly to Martin Luther's "alarming discovery" of 1520 that the pope was the Antichrist. Within a short time, his view became a "collective Protestant conviction."[143] Arno Seifert gives an apt summary of the effects of reconceptualizing his coming from the future to the past on the relationship of the two confessions to the end times: "The 'preteritization' of the Antichrist meant, from a Protestant point of view, that the present moved towards the very end of the schema outlined by the Biblical prophecies for the course of history after the life of Christ. The Protestants were living at a later point in the history of salvation than their Catholic contemporaries. In their view, the prophecies were as good as completely fulfilled and could therefore be used more extensively than before in the interpretation of historical events that had already occurred."[144] Prodigies in particular were often interpreted through the lens of eschatological expectations. Their increasingly frequent occurrence seemed to confirm the Protestants' location of themselves within the history of salvation.

Second, Catholics had a literary genre of their own that may have satisfied similar needs that prodigy literature did for Protestants: the *miracula* literature. It dealt with miracles—worked by God or by saints—whose time Luther and Calvin, by contrast, viewed as over. This literature reflected a merciful God, not a wrathful one.[145] However, as Rudolf Schenda has rightly pointed out, this constellation cannot readily be transferred to other parts of Europe. In France, for example, it was chiefly Catholics who wrote prodigy books.[146]

What is true of the authors of pamphlets and chronicles concerning prodigies in German-speaking areas is also true of those who wrote texts on monsters that fell more strictly within the category of nature study. Jacob Rueff is an example well suited to exploring the reasons for these parallels more closely. Rueff (ca. 1500–1558) was the municipal physician of Zurich—the home of Huldrych Zwingli. He published numerous books addressing both medical and other topics. His monograph on human conception and birth, published both in Latin and in German in 1554,[147] contains a chapter on monsters that was widely read at the time.[148] Many of the naturalists who published works on monsters in the second half of the sixteenth and early decades of the seventeenth centuries refer to this chapter.[149]

The fact that naturalists used chronicle literature and contemporary pamphlets as sources for factoids about monsters implies that the authors of these differing genres shared a common intellectual basis, with no intellectual gulf dividing them from one another. Though it runs counter to our intuition, which is shaped by modern natural science's criticism of prejudice, their views on the causes of monsters often proved to be a connecting element. Rueff, for example, shared the viewpoint of many authors of broadsides on monsters—and, we can assume, of many of their readers—that monsters partook in the supernatural (*supra naturam*).

In the eyes of many naturalists of the sixteenth and early seventeenth centuries, it was possible for both natural and supernatural causes to contribute to one and the same rare natural phenomenon. This was true for Jacob Rueff. Like Fincel, Lycosthenes, and many other contemporaries, Rueff viewed monsters as divine omens intended to warn or punish.[150] Indeed, he specifically explains the visual representation of many of the monsters he lists as an attempt to enable others to read the will of God in their appearance.[151] Nevertheless, he also names the "natural" causes of the monsters to which he refers.[152]

A discussion of "natural" causes, then, does not automatically indicate a rejection of the possibility of a divine ultimate cause. In this respect, Alan Bates's assertion that the birth of monsters was never regarded as supernatural but merely "outwith the normal course of nature"[153] is wrong in its exclusivity. Rather than categorizing them consistently as one or the other, many early modern naturalists considered monsters to be *both at the same time*—supernatural *and* preternatural, which is to say having their origins outside the usual course of nature.[154]

If we compare the situation in the Protestant (or Reformed) parts of central Europe with that in the Italian, predominantly Catholic territories in the sixteenth century, belief in prodigies and eschatological thinking were conspicuously widespread there too.[155] The printed results of scholarly engagement with monsters differ, however: Johann Georg Schenck, Jacob Rueff, and Gaspard Bauhin, for example, were practicing physicians who gave their attention first and foremost to individual cases of monstrosity. Varchi, Liceti, and Aldrovandi, by contrast, were not practicians. Their approach to the subject is characterized by greater emphasis on the systematization of the material. In Liceti's case in particular, this impetus was combined with a pronounced Aristotelianism, which as we know attached comparatively little importance to the observation of the object itself, though Paduan Aristotelianism[156] represented an exception in this respect.

Rueff also provides a good example of the role of printers and publishers in the transfers described here, which brings us to the second level of analysis. Both the German and the Latin versions of his *De Conceptu et Generatione Hominis, et Iis Quae circa Haec Potissimum Consyderantur Libri Sex* [. . .]. were published in 1554 by Christoph Froschauer the Younger (1532–1585, active as a printer 1552–1585). What was the publishing environment in which these two versions of his monograph appeared?

Froschauer did not only publish medical texts. The portfolio of his *Offizin* (printer's workshop) also included calendars and theological writings.[157] Christoph Froschauer the Younger had inherited the *Offizin* from his father, Eustachius Froschauer the Elder, who in turn had taken it over from his brother, Christoph Froschauer the Elder. We know that the Froschauers supported the Reformation in Zurich by publishing works by Luther and Zwingli. But Eustachius and his brother also published several broadsides that discussed prodigies. Among them is one written by Jacob Rueff, in which he informs his readers about a set of conjoined twins and also a hermaphrodite born in Zurich in 1519 and interpreted as a divine omen, which will be the focus of the next section.

Rueff was familiar with this pamphlet, included the case in his monograph, and thus, along with other authors, ensured that it found its way into the scholarly literature.[158] It does not seem improbable, then, that Rueff may have become aware of the pamphlet via his publisher. But even independent of this particular question, it is clear from publishing programs such as this that such workshops had the potential to be connection points at which vernacular and learned, Latin culture came into contact. Broadsides and scholarly medical texts could have their origins in the same house.

Jacob Rueff is of interest in relation to the third level of analysis, too. He not only published works on medical topics but also wrote plays, some of which addressed religious topics. In addition, he was responsible for at least two broadsides. As already mentioned, a broadside written by him and published in 1543 dealt with conjoined twins born in that year in the Swiss town of Schaffhausen. In it, Rueff attributes births of this kind to "natural" causes. He mentions several factors that might cause lasting damage to the delicate seed in utero: extremes of emotion on the part of the mother as well as unspecified vicissitudes (*casibus*) and the varying effects of medicines. Even in the habitual course of nature, they would thus inevitably, he argues, result in a "monster, or monstrous birth." But he also asserts that "prodigies, portents & marvels of every kind" such as comets, for example, did not occur for no reason: they were meant to be interpreted because they represented an admonition to repentance from God.[159] According to Ulla-Britta Kuechen, the fact that this pamphlet was composed in Latin indicates that Rueff's intended readership was other physicians and the senate of the city of Zurich.[160]

By contrast, Rueff's second broadside, from the following year, was published in German. In it, Rueff interprets a ring-shaped light in the sky in the context of other unusual celestial phenomena and concludes that it represents a warning.[161] Given that Jacob Rueff was the author of broadsides on divine omens, it is unsurprising that he used the two described above in his *De Conceptu et Generatione Hominis, et Iis Quae circa Haec Potissimum Consyderantur Libri Sex* [. . .]., too.[162] In the authorial person of Rueff himself these differing text genres were united.

The network of correspondents who contributed to the collection of reports on wonders created by Wick in Zurich is similarly illuminating. Although he was not a physician, Wick did something very similar to Johannes Schenck von Grafenberg, who used his acquaintanceships to collect *observationes* of rare and monstrous medical phenomena for his publication project. In his foreword, Schenck stresses that his *Observationes* collection was possible only because of an extensive network of correspondents, who sent him both information on rare published *observationes* and unpublished *observationes* of their own.[163] Wick's collection, too, benefited from letters and from other contributions passed on to him by learned contemporaries.[164] The list of these contributors includes, among many others, the Basel prodigy book author Conrad Lycosthenes, who was in close contact with several Zurich scholars, and the natural historian and bibliographer Konrad Gessner.[165]

In summary, the world of broadsides and chronicles—particularly those concerning divine omens—and the world of scholarly nature study were closely connected. This is especially true of vernacular and Latin literature on monsters and other phenomena regarded as divine signs. Indeed, several of the texts discussed in this chapter were published simultaneously or almost simultaneously in Latin and in German, and the two versions were typically produced by the same printer.[166]

However, the writers of vernacular pamphlets in particular were not regarded by scholars as authors in the sense that these texts were identified as sources. The transfer of authority from the ancient *auctoritates* to contemporary writers by and large sidestepped them. Two factors appear to have played a part in this. First, many members of the *respublica literaria*, the early modern scholarly world, held bookselling in disdain.[167] This contempt was rooted in part in a peculiarity of their authorial function, which was "built on a value system largely unconcerned with the monetary rewards promised by the author's propriety."[168] The anti-authorial practices that in the view of many scholars were widespread among printers and booksellers further contributed to their negative perception of these trades. As Adrian Johns has shown, these practices encompassed far more than the unauthorized reprinting of entire volumes: the unauthorized publication of shortened versions or translations of already published academic texts, for example, was also viewed extremely critically by scholarly authors.[169] It seems likely, therefore, that a commercial printed product such as the broadside—ephemeral because of the centrality of its news value—was met with certain reservations, all the more so because it was often published anonymously.

Second, the broadside authors who are known to us by name were frequently not members of the *respublica literaria* and often belonged to a different social class.[170] Thanks in particular to Steven Shapin's work, we know of the importance that the social rank of witnesses played in the credibility of their testimony in the early modern study of nature.[171] That rank may also, then, have been of some significance with regard to the recognition of authorship. Furthermore, broadsides were not usually written in any of the three languages of scholarship—Latin, Greek, or Hebrew—which also excluded them from this discourse. It is revealing in this respect that Konrad Gessner did not include any vernacular broadsides in his *Bibliotheca Universalis* but at the same time referred repeatedly to pamphlets in his works.[172] And, finally, most broadsides did not fit the short bibliographical reference format that was customary among scholars, because they did not usually bear an author's name or a title. Citing them caused formal problems, as well.

A "Perfect" Hermaphrodite

The third example of a monstrous factoid in scholarly literature around 1600 shows us exactly what happened when a factoid was copied from one text to be reused in another. Such a transfer does not only imply a change in the context in which the factoid appears. Its epistemic status, too, may alter radically in the process.

Our third factoid is a report—reproduced frequently from the late sixteenth century on—of a so-called "true" or "perfect" hermaphrodite. The concept of perfection referred to the relative proportions of the male and female sex that qualified an individual as a hermaphrodite. Naturalists linked to this the question of whether complete sexual duplication might be possible in an individual or animal. For early modern physicians, this implied the ability to reproduce both as a man and as a woman.

Even those physicians who considered the occurrence of perfect human hermaphrodites to be possible in principle regarded them as extremely rare. Many other authors already denied in the sixteenth century that fully hermaphrodite humans could exist at all. The boundary between the two groups ran, roughly speaking, between the followers of the Galenic-Hippocratic *generatio* theory on the one hand and the followers of Aristotelian teachings, who took a more critical view of hermaphrodites, on the other.[173] It was the presumed rarity or disputability of perfect hermaphrodites that led to reports about such humans attracting an especially large amount of attention.

A particularly prominent case of a perfect hermaphrodite was sighted in France. It can be found in many medical, natural history, and natural philosophy texts of the time and was—at least in the late sixteenth and early seventeenth centuries—a commonplace. Almost every physician who published anything on hermaphrodites during this period mentions it.[174] The French hermaphrodite entered scholarly medical discourse in central Europe primarily via Johannes Schenck von Grafenberg's multivolume collection of *observationes*.

Schenck's declared aim was to gather new and wonderful things that had been observed by preeminent physicians and were thus based on experience (*experimentum*) rather than teaching and learning (*doctrina*). From a modern perspective it may have a certain irony that to achieve this aim he had to rely primarily on the tools and practices of book-based scholarship: he combed through a vast number of treatises pertaining to the study of nature in search of passages that were of interest for his collection. However, as we have already seen in the cases of Fortunio Liceti and Johann Georg Schenck and in contrast to the modern understanding of empiricism, scholars of the late Renaissance did not necessarily see any contradiction between *experimentum* and reading.

For Johannes Schenck von Grafenberg, *experimentum* did not mean firsthand experience. Rather, his work reveals an older use of the term, as found in the medieval *florilegia*. In practical medicine, the *florilegium* format was employed primarily in collecting formulas for remedies. As a rule, such collections do not name an author or text to verify individual experiences or concrete formulas. Here, *experimentum* or *experientia* stood for a collective experience that did not need to be attributed to a named author.[175] Schenck, too, was interested in this kind of collective experience, acquired at different times and in different places, though he does specify the sources of his factoids. His use of the term *experimentum* also corresponds entirely with the medieval usage in terms of the subject matter he addresses: it generally referred to knowledge of "singular, specific, or contingent phenomena that could not be grasped by deductive reasoning, as well as the process by which this knowledge was obtained."[176]

Schenck's collection is impressive for its sheer size: the seven octavo volumes each contain between 229 and 418 *observationes* on 450 to 803 pages. The collection is a striking example of the commonplacing method and its importance in the reading and writing practices of early modern naturalists, as examined by Ann Blair on the basis of Jean Bodin's work.[177] It makes a large number of factoids—drawn by Schenck from numerous and widely varied sources—easily accessible to the reader. Schenck was by no means alone in this endeavor. Collections of *observationes* of rare medical phenomena were much in demand in the late Renaissance.[178]

Both the work as a whole and the individual volumes were organized largely according to the *a capite ad calcem* schema mentioned above. Accordingly, the first volume, published in Freiburg in 1584, contains the *observationes* relating to the head. The title of the volume promises the reader *exempla* of diseases of the head, their causes, and so on, taken from the writings of distinguished (*clarissimi*) physicians both ancient and contemporary.[179]

One of the paratexts, the *catalogus authorum*—a list of these distinguished physicians—helps to clarify the identity of the authors from whom Schenck drew his *observationes*, and the proportion of ancient and more recent scholars among them. It shows us that Schenck paid far greater attention to contemporary physicians. Only eleven of the authors mentioned fall into the categories of *graeci* and *latini*, which were used in the *catalogus* for ancient Greek and Roman authors. By contrast, the category that covers the early modern physicians, *recentiores docti & elegantes*, contains eighty-four.[180] The ratio of ancient to early modern writers in the six subsequent volumes is similar. The impression that Schenck was concerned first and foremost with contemporary *observationes* is reinforced by the fact that he also includes observations of his *own* in the work.[181] At least one *observatio* in the first volume is identified as his.[182]

Schenk does not use the term *autopsia* in connection with these direct observations of his own, either, because he was not concerned with distinguishing between autoptic experience and the experience of others. It is consistent with Schenck's concept of *experimentum* that he identifies both his own observations and the experiences of other authors taken from literature by means of a short reference. So it seems only logical that Schenck should appear—as "Ioan. Schenckius"—in his own *catalogus authorum*.[183] It was immaterial to Schenck's ideas of *observatio* and *experimentum*, then, whether the experience occurred in the distant past or in the present, and, furthermore, whether he had observed the phenomenon himself or knew it only indirectly through literature. This is typical of the period around 1600. Over the course of the seventeenth century, however, the use of these terms was to shift clearly toward *direct* experience or observation, as will become clear in chapter 3.

How might such a text have been read by its author's contemporaries? The body of the text consists of a numbered series of *observationes* that begins with the observations concerning the head (the hair first of all). An index at the end of the volume summarizes the contents by listing the title and number of each *observatio* in its original order. An unknown early modern reader of one of the surviving copies of the volume[184] was dissatisfied with this overly simplistic search engine: to make it easier to find *observationes* on specific topics, he wrote the Latin names of the subjects addressed in the margin of the index, immediately adjacent to the first *observatio* of each subject, which in one case might be a disease and in another a part of the head or a phenomenon of a completely different nature. Thus, for example, he wrote *epilepsia* at the start of a large group of *observationes* concerning the disease. Apparently, this reader was not alone in his desire for greater clarity. Beginning with the second volume, this shortcoming in the debut volume was systematically rectified: Schenck's *observationes* were now organized according to generic terms that closely resemble those used by the anonymous reader. *Observationes* concerning the same disease, the same body part, or other common subjects were marked as a group by a corresponding secondary heading.

This improvement gives indirect insight into how the work was meant to be read: Schenck's printed *florilegium* was not intended to be worked through from cover to cover. Rather, it was meant to enable the readers—predominantly learned and practicing physicians—to find the specific *observationes* that were relevant to them at any given moment as easily as possible. Thus, thanks to this information architecture, the reader searching, for example, for *observationes* on hermaphrodites quickly came upon the account of the French perfect hermaphrodite.

The fourth volume contains the observations gathered by Johannes Schenck von Grafenberg on the human reproductive organs. In the part on the position of these organs, the following *observatio* can be found: "I myself saw a hermaphrodite who was believed to belong to the lesser or female sex, and who had married a

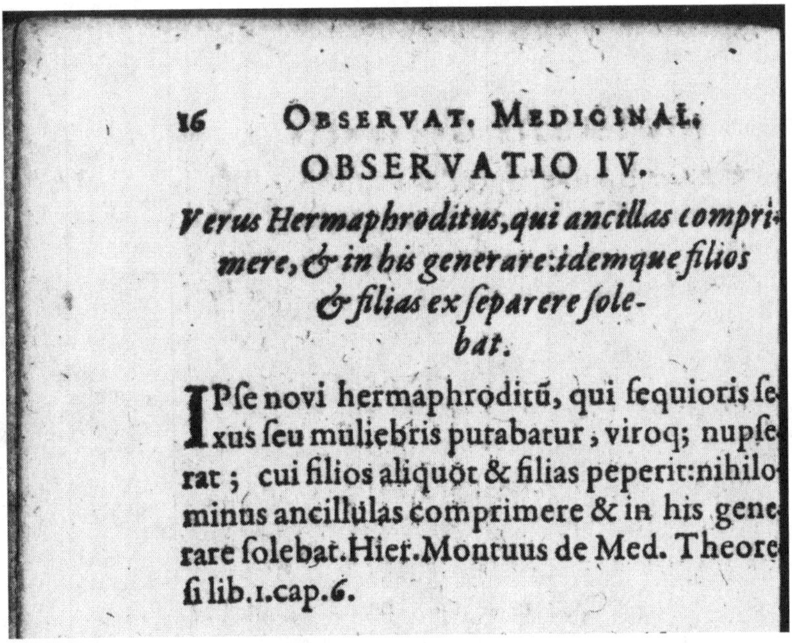

16   OBSERVAT. MEDICINAL.

OBSERVATIO IV.

Verus Hermaphroditus, qui ancillas compri-
mere, & in his generare: idemque filios
& filias ex separere sole-
bat.

IPse novi hermaphroditū, qui sequioris sexus seu muliebris putabatur, viroq; nupse-
rat ; cui filios aliquot & filias peperit: nihilominus ancillulas comprimere & in his generare solebat. Hier. Montuus de Med. Theoresi lib. 1. cap. 6.

Fig. 7. Johannes Schenck von Grafenberg, *Observationes Medicæ de Capite Humano* [. . .]. Basel, 1584, 16. Staatsbibliothek zu Berlin—Preußischer Kulturbesitz, Department of Early Printed Books, Shelfmark: Jc 3410 R.

man; he[185] bore him several sons and daughters; nevertheless he was wont to defile and impregnate young maidens." This observation occupies a section of its own and is headed "Observatio IV" and "True hermaphrodite, who was wont to defile and impregnate maidens; and the same hermaphrodite bore many sons and daughters." At the end of the short section there is a short reference to Schenck's source for this factoid (fig. 7). This is followed by a new section that contains the fifth *observatio* of the volume—on a different topic.

The text layout of the *observatio* concerning the perfect hermaphrodite—it was set as a separate section, with a number and heading of its own and a reference to its source text—assigns it the epistemic status of a discrete, autonomous factoid. Schenck arranged all individual observations in separate sections. But many head-ings cover several sections or morsels of knowledge that do not each have a number of their own. Since the terms *historia* and *exemplum* often appear in the headings in the plural, whereas Schenck always uses the term *observatio* there in the singular, it can be reasoned that, in his epistemology, *exemplum* and *historia* constitute smaller knowledge units than the *observatio*. An *observatio* might be composed of more than one *historia* or *exemplum*.

Schenck's use of the three terms was typical of his time. The use of *observatio* to describe several individual observations (*exempla* or *historiae* in his terminology) is reminiscent among other things of Francis Bacon's concept of *observationes majores*. Unlike Bacon, though, Schenck does not form syntheses from the individual morsels of knowledge collected in an *observatio*. Bacon's *observationes majores* are the result of the comparison of *several* individual observations and, as such, are of greater generalizability and significance.[186]

The factoid of the French perfect hermaphrodite, meanwhile, is presented by Schenck as an observation sui generis. Removed from its original textual setting, it awaits discovery by the reader and use in a new context. Its special status becomes apparent when we compare Johannes Schenck von Grafenberg's version of this factoid with the original.

Aside from minimal variations in punctuation and the abbreviations used, Schenck's factoid is an exact quotation of the corresponding passage on the two pages below (fig. 8). It begins on the fourth-to-last line of page 35 and ends on the first line of the following page. These two pages are found in Jérôme Monteux's medical textbook *De Medica Theoresi*.[187] Monteux was court physician to Henry II of France.[188] As a text in its own right, the book was published in a single edition only, also in 1556.[189]

The comparison reveals immediately that the passage quoted by Schenck von Grafenberg had a different status originally. In *De Medica Theoresi*, the report of the perfect hermaphrodite is not set as a separate section, let alone given a number or heading. Rather, it is incorporated into a larger chapter. Its immediate context is a discussion of the teachings of the ancient authorities on the differences between the sexes in humans and animals. In the course of the discussion, Monteux touches on the subject of hermaphrodites. According to him, both sexes and their respective forces (*vis*) are present in hermaphrodites (or androgynes, as he also calls them). He also mentions that he himself had seen a perfect hermaphrodite. The statement serves, then, as an *exemplum* in the traditional rhetorical sense: as an example intended to illustrate the *doctrina*, the teachings and learnings, of ancient authorities. The observation itself does not stand alone; it is not a factoid in its own right independently of the context of the argument.

Monteux's statement was reproduced by numerous scholarly writers of the early modern period, who present it as a firsthand account. This can arguably be inter-

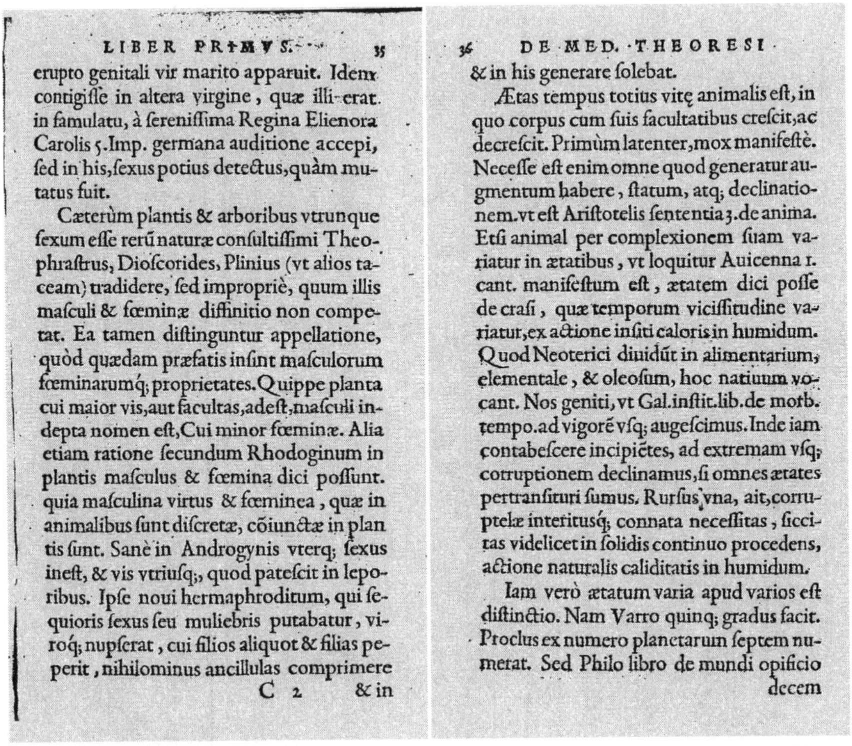

LIBER PRIMVS.   35

erupto genitali vir marito apparuit. Idem
contigiſſe in altera virgine, quæ illi erat.
in famulatu, à fereniſſima Regina Elienora
Carolis 5.Imp. germana auditione accepi,
fed in his,fexus potius detectus,quàm mu-
tatus fuit.

Cæterùm plantis & arboribus vtrunque
fexum eſſe rerū naturæ conſultiſſimi Theo-
phraſtrus, Dioſcorides, Plinius (vt alios ta-
ceam) tradidere, fed improprie, quum illis
maſculi & fœminæ diffinitio non compe-
tat. Ea tamen diſtinguntur appellatione,
quòd quædam præfatis inſint maſculorum
fœminarumq̃; proprietates.Quippe planta
cui maior vis,aut facultas,adeſt,maſculi in-
depta nomen eſt,Cui minor fœminæ. Alia
etiam ratione fecundum Rhodoginum in
plantis maſculus & fœmina dici poſſunt.
quia maſculina virtus & fœminea , quæ in
animalibus ſunt diſcretæ, cōiunctæ in plan
tis ſunt. Sanè in Androgynis vterq̃; ſexus
ineſt, & vis vtriuſq̃;, quod pateſcit in lepo-
ribus. Ipſe noui hermaphroditum, qui fe-
quioris ſexus ſeu muliebris putabatur, vi-
roq̃; nupſerat , cui filios aliquot & filias peperit, nihilominus ancillulas comprimere

C 2   & in

36   DE MED. THEORESI

& in his generare ſolebat.

Ætas tempus totius vitę animalis eſt, in
quo corpus cum ſuis facultatibus creſcit,ac
decreſcit. Primùm latenter,mox manifeſte.
Neceſſe eſt enim omne quod generatur au-
gmentum habere, ſtatum, atq̃; declinatio-
nem.vt eſt Ariſtotelis ſententia 3.de anima.
Etſi animal per complexionem ſuam va-
riatur in ætatibus, vt loquitur Auicenna 1.
cant. manifeſtum eſt, ætatem dici poſſe
de craſi, quæ temporum viciſſitudine va-
riatur,ex actione inſiti caloris in humidum.
Quod Neoterici diuidūt in alimentarium,
elementale, & oleoſum, hoc natiuum vo-
cant. Nos geniti,vt Gal.inſtit.lib.de morb.
tempo.ad vigorē vſq̃; augeſcimus.Inde iam
contabeſcere incipiētes, ad extremam vſq̃;
corruptionem declinamus,ſi omnes ætates
pertranſituri ſumus. Rurſus vna, ait,corru-
ptelæ interitusq̃; connata neceſſitas , ſicci-
tas videlicet in ſolidis continuo procedens,
actione naturalis caliditatis in humidum.

Iam verò ætatum varia apud varios eſt
diſtinctio. Nam Varro quinq̃; gradus facit.
Proclus ex numero planetarum ſeptem nu-
merat. Sed Philo libro de mundi opificio
decem

Fig. 8. Hieronymus Montuus, *De Medica Theoresi Liber Primus* [. . .]. Lyon, 1556, 35–36.
Göttingen State and University Library: 8 Med Miscell 122/39 (5).

preted as an expression of the high regard in which empirical observations were
held. The rise during the same period of the *observatio*[190]—and more generally of
the *historia*, both as a text genre and as a form of knowledge[191]—in medicine and
similar fields of early modern knowledge can be interpreted as further evidence of
this. As the example of Johannes Schenck has shown, however, observations that
had been made centuries or millennia earlier were described as *observatio* or *historia*, too. *Historia* also encompassed knowledge that was based on the sensory perception of *others*.[192] Thus, there was not necessarily any contradiction from Schenck's
point of view in creating his collection of experiential knowledge primarily by means
of extensive reading and tireless note taking.

Once removed from its original context and made widely accessible via Schenck's
Observationes, the factoid of the French perfect hermaphrodite was frequently taken
up and used by other naturalists as a textual building block in new publications.[193]

It thus became part of the text-based practices of naturalist knowledge generation, which Ann Blair describes with some justification as circular: "To make natural knowledge was to transmit, sort, explain, and modify the definitions, facts, and arguments accumulated previously, producing texts that following generations of scholars would process in much the same way."[194] Printed *loci communes* collections like that of Schenck facilitated this pattern of writing.[195] Such works made factoids about monsters more readily available to contemporary and later writers and thus contributed to their reproduction.

Not all factoids about monsters were presented, as the one above was, with the claim that they were rooted in observation. For example, Gaspard Bauhin's monograph on hermaphrodites and other monstrous births and Ulisse Aldrovandi's *Monstrorum Historia* contain whole chapters—on questions concerning the (predominantly Greek and Latin) etymology of the various names given to monstrosities, for example—that forego observation-based knowledge.[196] Furthermore, they frequently refer to opinions attributed to ancient authorities such as Aristotle, Galen, Hippocrates, and Pliny—and indeed less well-known authors from that period—that represent general or theoretical statements rather than observation-based knowledge. There were, however, differences from author to author, from text genre to text genre, and from field to field with regard to the epistemic value assigned to observation. The subject itself played a part, too. These aspects will be considered in greater depth below.

The *Observatio* and Its Subjects

Some monsters were examined in greater depth than others. The existence of true or perfect hermaphrodites, for example, was a highly controversial question around 1600. One of the most discussed cases of hermaphroditism in early modern Europe was that of the chambermaid Marie le Marcis from Rouen in France. She was accused in 1601 of committing sodomy (in the sense of having a sexual relationship with a woman) with her female lover. In her defense, Marie le Marcis stated that she was not a woman but a man with a hidden member. However, two medical commissions came—on the basis of purely external physical examinations—to the conclusion that this was not the case. When le Marcis appealed to the parlement de Rouen, a third medical commission was appointed to examine her.

This third commission, too, was convinced at first that le Marcis was a woman, but then the tide turned. Direct observation played a large part in this. One member of the commission, a physician from Rouen named Jacques Duval (ca. 1555–ca.

1615), "inserted his finger into Marie's vagina, found a hidden member, and filed a dissenting opinion declaring her a predominantly male hermaphrodite."[197] Duval's verdict saved Marie from the death penalty, and she—or rather he—lived as Marin le Marcis from that time on.

Autopsy, "seeing for oneself,"[198] was also to play a significant role in the academic debate—conducted in French and with a far-reaching impact—that followed the publication of the *Traité des hermaphrodites, parties genitals, accouchements des femmes*[199] by Duval. At the heart of the controversy was the case of Marie/Marin le Marcis. Duval's opponent was the Parisian professor of anatomy and botany Jean Riolan (1580–1657). Unlike Duval, Riolan was a proponent of the Aristotelian viewpoint that all supposed hermaphrodites were ultimately women or men who at most were endowed with a surplus, nonfunctioning reproductive organ. Riolan responded to Duval's text with his *Discours sur les hermaphrodites*, published only two years later. He judged le Marcis to be a woman and therefore guilty of sodomy. Duval, in turn, reacted with a polemic that was published under the title *Responce au discours fait par le sieur Riolan*.[200]

The two opponents could scarcely have been more different. It was not only their views on hermaphroditism that were in stark contrast. Duval was a relative outsider, both in geographical and in academic terms, while Riolan was a member of the Paris medical establishment. Duval published his texts in French, whereas Riolan generally wrote in Latin. In his dispute with Duval, admittedly, he deviated from his usual choice of language. And yet they both agreed that knowledge of the inner structure of the genitalia was pivotal in this case—though Riolan never had the opportunity to examine le Marcis himself.[201]

The conflict concerning Marie/Marin le Marcis demonstrates clearly why first-hand inspection was of particular importance in the case of hermaphrodites. Because access to various rights in early modern Europe differed significantly according to biological sex, it was necessary for each individual to be assigned one. The precise form of the sex organs of a sexually ambiguous individual was central to assigning that person's gender. Another example of a type of monster typically examined in detail were conjoined twins. Physicians were very eager to carry out dissections on them. The reason for this seems not least to have been that the number of organs present promised insight into whether the case in question could be deemed one individual or two.[202] The significance that authors ascribed to observation was also influenced by the field of study to which they were contributing, however, as I aim to demonstrate below.

Realdo Colombo's Emphatic "I": The *Observatio* in Anatomy

Of the scholars involved in the discourse on monsters, it was the anatomists who took particular pride in basing their knowledge for the most part on having seen or touched their subject matter themselves. When they used the first-person singular pronoun in their writings, it often came with a pathos peculiar to the profession. In some cases, the "I" was directed explicitly against inherited knowledge. From the mid-sixteenth century at the latest, there was an enormous increase—at least on a rhetorical level—in the value placed on firsthand inspection.[203] This can be seen clearly in Realdo Colombo's *De Re Anatomica Libri XV* (1559), for example, which I consider in detail below.

Colombo witnessed the renewal of anatomy firsthand. His epistemology was shaped by the fact that he had studied in Padua under Andreas Vesalius and had often served him as a *sector*, that is, the person cutting the body, in public dissections. For a time in January 1543, he held anatomy lectures on his teacher's behalf while Vesalius was in Basel arranging the publication of his *Fabrica*. In October 1544, Colombo was eventually appointed to be Vesalius's successor as professor of surgery and anatomy. Between 1546 and 1548 he taught at the University of Pisa, before moving to Rome, where he worked as personal physician to Paul III and taught at the university.[204] In his posthumously published treatise *De Re Anatomica Libri XV*, Colombo's declared aim—entirely in the spirit of the new anatomy—was to correct errors made by both Galen and Vesalius in the description of the human body.[205]

It was integral to the "scientific persona"[206] of the anatomist during the anatomical renaissance of the late fifteenth and sixteenth centuries in Italy—and increasingly north of the Alps, too—to have personally conducted a large number of dissections. The anatomists of the sixteenth century saw their endeavor as characterized by a "disciplined seeing" schooled at the dissecting table.[207] This had concrete material repercussions: the rapid spread of the new epistemic virtues of the anatomists in sixteenth-century northern Italy resulted in an enormous increase in the number of bodies required for dissection.[208]

Though the anatomists of the sixteenth century did not venture any great distance from the teachings of the ancient authorities, they did acknowledge in principle, at least, that observation must be accorded more weight than authority.[209] This increase in the importance of the *observatio* came at the expense of knowledge acquisition through teaching and learning (*doctrina*) based on formerly authoritative texts and of knowledge generation through rational thought (*theoria*). And conflicts with traditional anatomical scholarship were inevitable whenever observation based on the new epistemic virtues produced results that contradicted established doctrine.

A notable testimony of this constellation is provided by lecture notes written by the German student Baldasar Heseler. He documented the first public dissection conducted by Andreas Vesalius (1514–1564) in Bologna, which took place in 1540, and accompanied an anatomy lecture held by Matthaeus Curtius (1474/5–1544). Heseler's notes interlace Curtius's anatomy lecture and Vesalius's anatomical demonstrations. They show that Vesalius was far from subordinating himself unquestioningly to Curtius or to the teachings of which he was a proponent. While Curtius insisted on the validity of the authoritative literature, Vesalius was adamant in his reliance solely on his own observations. In the course of the lecture, therefore, a whole series of conflicts arose.[210]

In disputes such as those between Vesalius and Curtius, we see a sharp growth in the importance of observation (or of *historia* acquired through observation) as compared to *doctrina* and *theoria* that was characteristic of much of the ambitious anatomical research of the sixteenth century. This cannot readily be said of other fields. Andrew Wear's study of the method and epistemology of William Harvey (1578–1657), who discovered the circulation of the blood, has revealed the emergence of an independent "way of the anatomists." Harvey "followed what was by then established anatomical practice by regarding his work on the circulation of the blood as fundamentally a matter of *looking* in the right way ('autoptic' experience)."[211] Over the course of the sixteenth and early seventeenth centuries, anatomists increasingly began to regard observation as a more reliable source of knowledge creation than rational argument. In the context of this development, observation-based knowledge—which in Aristotelian and Galenic anatomy, too, was traditionally merely a step on the way to *scientia*—increasingly became an end in itself.[212]

The final chapter of Realdo Colombo's anatomical treatise *De Re Anatomica Libri XV*, published in Venice in 1559, devoted to rare anatomical phenomena, exemplifies the central importance of observation-based knowledge in anatomy at that time. For one thing, Colombo stresses in a way that is typical of post-Vesalian anatomy that his expositions are based on his own studies. And for another, the corresponding sections of his monograph were much cited by naturalists around 1600, including precisely those discussed here.

Let me first clarify the intellectual context in which this chapter, entitled "On the Things That Are Rarely Found during a Dissection," was written. The anatomical discussion of rare variations in human anatomy took place against the backdrop of early modern anatomy's emergence as a distinct field of study. This process made it necessary, not least for university teaching, to establish the subject matter of anatomy, the "normal" body. Around the middle of the sixteenth century, however, an understanding of nature became prevalent—both in anatomy and elsewhere— that gave generous scope to jokes of nature, *lusus naturae*,[213] and tended to emphasize the *varietas*, the diversity or variety, of its productions, rather than their uniformity. Many anatomists found this to be problematic for the reason mentioned above, and because of this the question of anatomical variations became a very sensitive one.[214]

In this discussion, Realdo Colombo took a position in favor of the uniformity of the human body and the rarity of anatomical variations. This positioning is indicated by the composition of his *De Re Anatomica Libri XV*, which devotes itself largely to the "normal" structure of the body. It was merely as a corrective that Colombo devoted a separate chapter at the end of his book to the rare anatomical anomalies that he had encountered in "countless" dissections.[215] Toward the end of this chapter, he recounts his observations in relation to hermaphrodites. Let us take a closer look at this passage, which was cited particularly frequently by his contemporaries.

At the very outset, Colombo emphasizes the rarity of this phenomenon in humans. Using three verbs in the first-person singular, he underlines at the same time that he has a well-trained eye in such matters: "Among all the astonishing and rare things that I have observed [*observavi*] at different times in the structure of the human body, I judge [*censeo*] nothing to be more astonishing, nothing to be rarer than what I have carefully investigated [*investigavi*] concerning beings of neutral sex."[216] Colombo stresses his well-trained sight several more times in the course of the lengthy passage.

First, he gives a detailed account of the dissection of a hermaphrodite that he conducted himself. Few early modern physicians could boast of even having examined a living hermaphrodite. So it is understandable that Colombo reports his valuable observations in detail. The entire dissection report is written in the first-person singular, and the frequent use of sight-related verbs is also striking. It seems appropriate, then, that Colombo finishes by stressing his firsthand experience one more time: "And let that be enough about the hermaphrodite that I dissected [*secui*] as a dead body."[217]

Colombo follows this with his account of two examinations of living hermaphrodites, of which he classes one as predominantly female and the other as predominantly male. This account, too, is written in the first-person singular. Again, Colombo stresses at the outset that he himself saw what he reports: "I have, moreover, thoroughly examined [*consideravi*] two living hermaphrodites."[218] And here, too, the passage does not end without Colombo stressing again: "And these are the hermaphrodites that I have seen [*vidi*]."[219] The pride that the post-Vesalian anatomists took, then, in the fact that their knowledge was based largely on examinations that they carried out personally pervades the entire passage.

The Summary and Collective Nature of the *Observatio*

Colombo's terminology has a further notable aspect. The way he uses the verb *observare* differs from that of the other verbs of sight and examination in his work. The passage's introductory sentence, quoted above, is the only one in which *observare* appears. Interestingly, the verb refers here summarily to *all* of the individual observations of rare anatomical phenomena that the writer has made at different times—presumably over the entire course of his career as an anatomist.

Observatio, then, was the sum total of his experiences with rare and astonishing phenomena such as hermaphrodites. All other verbs of sight or examination that are used describe *individual* acts of seeing or examining: *investigare* (examine), *videre* (see, look at), *pervestigare* (examine completely; here: dissect), *secare* (here: dissect), *considerare* (examine thoroughly, observe), and *inspicere* (inspect, examine). *Observare*, by contrast, encompassed all the individual acts of seeing focused on a single object. Correspondingly, *observatio* was the sum of all individual observations relating to one phenomenon and thus stood for the entirety of the knowledge that an anatomist acquired though his daily contact with living and dead bodies. In the work of late Renaissance writers, the summary nature of the *observatio* often existed alongside its *collective* character, as will be shown below.

This *summary* understanding of *observatio* can also be found in the work of other contemporaries. It manifests similarly clearly—to give just one further example—in the work of the Freiburg municipal physician Johannes Schenck von Grafenberg. As has been shown above, a single *observatio* in his printed *loci communes* collection often includes several individual observations, which are described, for example, as an *exemplum* or *historia*. In Schenk's diction, too, then, the term *observatio* encompasses all of the observation-based knowledge that existed on a specific rare medical phenomenon. Unlike Colombo, however, he was explicitly concerned not only with his *own* experiential knowledge.[220] His concept of *observatio* was at once summary and *collective*.

There is evidence of the term being used in this summary sense as early as the Middle Ages. In engagement with classical terminology, medieval writers distinguished between experience (*experimentum* and *experientia*) on the one hand and observation (*observatio*) on the other. The two activities focused on differing phenomena and implied different forms of nature study.[221] Katharine Park summarizes the core meaning and typical use of the twin terms *experimentum* and *experientia*: "At the heart of the medieval concept of experience lay the notion of test or trial . . . More broadly, the term corresponded to knowledge of singular, specific, or contingent phenomena that could not be grasped by deductive reasoning, as well as the process by which such knowledge was obtained."[222] It related, then, to an *individual* experience of particular phenomena, whereas *observatio* typically described a cumulative activity conducted over a longer period of time, which focused on nonparticular phenomena.

Although medieval authors from Augustine onward redirected the meaning of *observatio* to religious purposes,[223] they retained the conceptual division of experience and observation from classical authors such as Cicero and Pliny. They associated *observatio* primarily with the observation of the celestial bodies and the seasons and the interpretation of symbolic natural phenomena, and thus identified it chiefly with medicine, navigation, and agriculture. Accordingly, they viewed observation as a "collective and largely anonymous process, associated with the early, originary phases of natural knowledge, long ago and far away."[224]

To some extent, the ways Colombo and Schenck used the term were a continuation of this tradition, inasmuch as they also regarded *observatio* as a cumulative process or as the sum of individual observations. In their works, however, the observers were no longer anonymous. In this respect, they were part of the development that began in the late fifteenth century by which *observatio* became an epistemic category—that of an authorized observation, which is to say one linked to a named scholar—firmly rooted in natural philosophy. This version of the concept was modeled neither on the anonymous Aristotelian experience, *empeiría*, nor on the also anonymous Plinian (and Ciceronian) *observationes*.[225]

During the course of the early modern period, the character of *observatio* also changed in a second respect: collectively made observations became coordinated to an unprecedented degree. Efforts in this regard by scientific academies are just one of many examples; individuals, governments, and trading companies should be mentioned here, too.[226] Salomon's House in Francis Bacon's posthumously published utopian text *New Atlantis* (1627) contains what is probably the best-known and—for the academy movement—most momentous vision of this kind of coordinated collective observation, in the form of a "centralized, state-financed, hierarchically organized corps of observers subordinated to the 'Interpreters of Nature.'"[227]

The Place of Experience-Based Factoids in Literature on Monsters (ca. 1600)

The rise of observation shaped much of the scholarly literature on monsters but to differing degrees. Compared to anatomy, other fields of nature study in which works on monsters were published around 1600 tended to ascribe a lower value to observation. In his treatise *De Hermaphroditorum Monstrosorumque Partuum Natura* (1614), Gaspard Bauhin, for instance, attaches great importance to observation-based knowledge. He distinguishes clearly between dissection findings and other kinds of case reports concerning monsters. The latter are referred to as *historiae* or *exempla*. This was no coincidence: Bauhin was not only the municipal physician of Basel and a professor of botany and practical medicine but also a professor of anatomy.

It is interesting to consider his work in terms of this distinction. The full title—*Two Books on the Nature of Hermaphrodite and Monstrous Births, according to the Theologians and Lawyers, Physicians, Philosophers and Rabbis*—gives a first indication of the kind of text with which the reader is dealing here: the work presents a compilation of factoids taken from literature.[228] The size of the collection—the book contains almost six hundred pages in octavo format[229]—further indicates its encyclopedic aspirations.

Like the above-discussed chapter 35 ("Examples of Hermaphrodites"), chapter 34 ("Some Cases That Explain the Nature of Hermaphrodites") presents a compilation of individual cases taken from literature. Chapter 32 ("On the Inner Structure of Hermaphrodites, Revealed by Dissections"), by contrast, is devoted almost exclusively to dissection reports on the internal anatomy of hermaphrodites. Here, Bauhin, like many of his contemporaries, quotes the reports of other naturalists at length. Among them is Realdo Colombo's account of his autopsy of a hermaphrodite, reproduced along with his statements about the living hermaphrodites he had examined. Bauhin makes a point of mentioning that the report reflects a firsthand observation.[230] At the same time, he feels obliged to justify drawing on the observations of others: "Now that the causes of the creation of hermaphrodites have been set out, the order demands that we set out their inner anatomy, knowledge of which was obtained by means of dissections: however, as we have not had the opportunity [*contigerit*] to dissect one, we will substitute [*subjiciemus*] what Colombo recorded and observed [*observavit*]."[231]

Like Andreas Vesalius, Bauhin, professor (not only) of anatomy, attached great importance to direct observation and rejected the opinions inherited from Galen on some questions. Unlike William Harvey, however, he did not go as far as to deny the authority of the ancient canonical authors Hippocrates, Galen, and Aristotle completely or to attempt to replace book knowledge entirely with knowledge based on direct sensory perceptions. The two together—the Book of Nature and the teachings of both older and more recent authors—he refers to as the foundation of his *Institutiones Anatomicae*: "non solum ex lectione Veterum et Recentiorum, sed ex ipso naturae libro ex Dissectionibus plurimis."[232] In cases where these two sources of knowledge produced differing results—where direct observation contradicted the teachings of the ancients, for example—Bauhin, by his own account, favored observation-based knowledge, provided it was consistent with what others had seen. For the truth was declaredly more important to him than the authority of these authors: "For Hippocrates is a friend, Aristoteles is a friend, Galen is a friend, but a still bigger friend is TRUTH."[233]

Other naturalists who published work on the topic of monsters distinguished far less scrupulously between direct observation and knowledge from other sources. One illustration of this are the *Observationes* of the municipal physician Johannes Schenck von Grafenberg, which also represent an encyclopedic collection of morsels of knowledge taken predominantly from literature. He, too, quotes Colombo's account of his examinations of one dead and two living hermaphrodites in full, but combines it without differentiation with a mixture of empirical descriptions and theoretical statements by other writers under the general heading "Observatio III: Hermaphrodites, Which Once [Were Called] Androgynes."

The sources of the factoids that Schenck gathers here could scarcely be more different. The collection begins with a statement that seems only partially consistent with his announcement in the foreword of his intention to make experience-based knowledge available to the reader. The statement in question is Aristotle's rejection of the concept of hermaphroditism in his *De generatione animalium*, which was cited with great frequency by early modern authors. In it, Aristotle argues that the duplication of the genitals that occurs in some animals is deceptive inasmuch as one of the two sex organs is always nonfunctioning (*irritum*) and therefore akin to a swelling (*abscessus*).[234] Schenck follows this morsel of knowledge with the definition from Pliny's *Natural History* (VII, III), which in contrast to Aristotle affirms the dual sex of hermaphrodites.[235] There follows a general statement on a monstrous race of hermaphrodites in Africa, also a quotation from the *Natural History* of Pliny (VII, II), who in turn attributes it to Calliphanes or Aristotle.[236] Only after these factoids, which can only to a limited extent be traced back to *experimentum*, let alone to an individual observation, does Schenck cite a concrete case of hermaphroditism (in the animal kingdom). But by no means does this factoid represent the start of a list of case reports based on experiential knowledge. Rather, two further statements of a general nature follow the report; namely, a four-part classification of human hermaphrodites ascribed to Caelius Rhodiginus and a short note on terminology.

It is only after this note, and without any typographical indication of a new section, that a longer list of factoids addressing individual cases of hermaphroditism in humans begins. Among them is the long Colombo quotation. Schenck also identifies the observers of several of the other factoids. And in addition to Colombo's account, he reproduces a number of other case descriptions in the first-person singular. For the most part, however, these factoids are not testified to by specific observers.

Schenck's text contains numerous factoids in other places that cannot readily be accepted as observation-based knowledge. In contrast to what was observable among anatomists of the time, he was clearly not concerned with distinguishing observation-based knowledge from knowledge that was owed to deduction or other sources, including the teachings of the ancients. In the case of his *observationes* on hermaphroditism, he clearly considered other criteria a more obvious basis on which to organize them: The only factoid on the subject that he does not subsume under *observatio* III but gives a heading of its own is Jerôme Monteux's above-discussed account of his observation of a perfect hermaphrodite. What distinguishes this factoid from the others is the alleged perfection, the completely dual sex of the individual in question, and not the epistemic status of the account.[237]

The texture of Schenck's *observatio* III on hermaphroditism can be viewed as an impactful manifestation of an essential feature of early modern empirical research that Gianna Pomata and Nancy G. Siraisi have defined in the concept of "learned empiricism":

> There is no doubt that the empiricism of the early modern *historia*, in all of its varieties, must be qualified as erudite or textual in nature. Direct observation was preceded and accompanied by laborious compilation, based on the culling of information from earlier texts . . . We are dealing here with a highly scholarly or learned variant of empiricism—an *empirisme érudit*, we may call it . . . Scholars have found it hard to reconcile the emphasis on direct observation, in Renaissance anatomy for instance, with the enormous baggage of philological skill and antiquarian learning that Vesalius and his peers brought to the dissecting table. This philological and antiquarian apparatus has been seen mostly as a handicap, an

oppressively constraining theoretical filter that limited and distorted observation—and in some cases it undoubtedly did. But there is also evidence to the contrary, evidence, namely, that the linguistic sophistication and tremendous familiarity with ancient texts that were the hallmark of humanist training could be harnessed to the cognitive goal of direct observation so as to complement or even enhance them.[238]

Against the backdrop of this early modern expression of empiricism, it is easier to understand how the practicing physician Johannes Schenck von Grafenberg was able to include factoids in his work that were evidently not derived from sensory experience at all.

However, it does not explain why Schenck did not separate observation-based factoids from those that did not meet that description. As we have seen, anatomists such as Realdo Colombo and Gaspard Bauhin certainly *did* attach importance to this distinction. For the academically trained municipal physician Johannes Schenck von Grafenberg, by contrast, the abundance of his collection was evidently more important than strict adherence to his own requirement of collecting experience-based factoids: *copia*, plenitude, took precedence over observation. He was not alone in this. In chapters 2 and 3, I address the influence of the ideal of plenitude on the discourse on monsters in greater depth. In doing so, I show that plenitude was a central category not only for the rhetoric of the Renaissance but also for the study of nature. As is shown not only in Schenck's work, the collecting of factoids on rare phenomena constituted a value in its own right around 1600, to the extent that even factoids of doubtful veracity were included in such collections. I will pick up this thought in chapter 2 and expand on it on the basis of Ulisse Aldrovandi's *Monstrorum Historia*.

But let us first continue our examination of the influence that the field to which authors circa 1600 meant to contribute had on their approach to experience- or observation-based knowledge. It is striking in this context that, although Johann Georg Schenck von Grafenberg (Johannes Schenck's son and, like him, a municipal physician) changed some of the details in his father's original text when writing his *Monstrorum Historia*, in relation to the two *observationes* on hermaphrodites the alteration consists solely in the fact that Johann Georg subsumed almost all of the factoids that his father presents under the heading "Observatio III," including the long Colombo quotation, under the misleading caption "A Person Missing the Genitals of Both Sexes." In terms of the relationship between observation-based knowledge and knowledge from other sources, he follows his father's example entirely, presenting factoids based on experience and those that come from other sources of knowledge without distinction and in immediate proximity to one another.[239]

Like the two Schencks, Fortunio Liceti and Ulisse Aldrovandi were not anatomists. And like the work of Johannes Schenck von Grafenberg and, to a more limited degree, that of his son Johann Georg, their monographs on monsters are comprehensive collections that do not restrict themselves to individual observations. While Liceti taught philosophy in Padua and elsewhere, Aldrovandi was a professor of natural history in Bologna. The same standardizing approach to factoids on monsters that we have seen in the work of the two Schencks is evident in their work, too. Aldrovandi's history of animals subordinates individual morsels of knowledge to an overarching interpretative framework. In this respect it resembles the *Historia Animalium* of his rival Konrad Gessner.[240] Aldrovandi's approach to Colombo's hermaphrodite accounts is interesting in this context.

Aldrovandi's *Monstrorum Historia* mentions Colombo twice in connection with hermaphrodites. The first mention is found at the end of a section on *differentiæ* in the chapter "Of Man," in which, alongside other exotics from the edge of the world, the story of an alleged hermaphrodite race is discussed. In this context, Aldrovandi paraphrases both Jérôme Monteux's account of the perfect hermaphrodite and Colombo's dissection report. He uses the two observations to convey general information about hermaphrodites to the reader. Their setting within his argument is revealing: Aldrovandi assembles statements by ancient authorities and those of more recent naturalists, completely independently of whether they are rooted in the concrete experience of an author.[241]

The same is true of the section "The Variety of Hermaphrodites" in chapter 5, in which Aldrovandi returns to the subject of Colombo's hermaphrodites. The chapter is devoted to the "misshapen" or "deformed" structure of the abdomen and the reproductive organs. Alongside a discussion of general questions such as, particularly, the classification of hermaphrodites, the section includes a list of all the cases that Aldrovandi was able to find in literature—including the three hermaphrodites examined by Colombo.[242] Most of these factoids are not labeled as firsthand accounts; like the *observationes* in the works of Schenck von Grafenberg senior and junior, they are for the most part no more than short notes that follow the formula "a hermaphrodite was born in place X on date Y, says Z." One might argue that this list of distinct case reports amounts—in keeping with the summary nature of the *observatio*—to *one* observation. But in this section, too, the clearly case-related factoids mix without distinction with morsels of knowledge that cannot be traced back to acts of observation at all.

Fortunio Liceti's Aristotelian approach to monsters assigns individual observations an even lower epistemic status; they are clearly subordinated to the work's Aristotelian framework. The causes of monsters according to Aristotelian teaching and a classification of them that is oriented to it constitute the organizing principles of his treatise. He mentions Colombo's three hermaphrodites in chapter 53 of the second book, devoted to the "multiform monsters of the same species" that he outlines briefly at the end of the first book. What is meant are monsters that combine various forms but that can be assigned to the same species.[243] Among them, Liceti includes women with men's heads, also described as bearded women, and hermaphrodites. The chapter presents a whole series of case reports but also quite a number of theoretical passages in which Liceti defines and discusses different species of monster that he assigns to this class.

Such passages contain numerous references to the ancient authorities of the study of nature. Liceti thus invokes both Aristotle and Pliny to lend greater weight to his view that perfect hermaphrodites were not monsters.[244] The factoids linked to individual cases function in Liceti's discourse merely as *exempla* that serve in particular to validate the category of multiform monsters of the same species that he introduced. Only to a limited extent are they presented as an empirical basis from which the category was developed.

Whether these individual factoids are firsthand accounts is therefore of no significance to Liceti. He usually presents them synoptically instead of quoting them in full. It remains unclear whether a report stems from an eyewitness. Indeed, the first case of a bearded woman presented by Liceti does entirely without any reference to a source,[245] and it is no exception in this respect: though most of the case-related morsels of knowledge are given a source reference, factoids with no reference are found repeatedly in Liceti's casuistic lists, too.

Liceti's illustrative use of such accounts is demonstrated by the case of Monteux's perfect hermaphrodite: Liceti defines perfect hermaphrodites as those with complete and properly positioned sex organs of both sexes, and therefore he does not regard them as monstrous. The individual witnessed by Monteux serves Liceti merely as an illustration of this view.[246] By contrast, he cites one of the three hermaphrodites mentioned by Colombo as an example of a monstrous hermaphrodite. The other two, Liceti states, are of no interest in this context because they do not conform to his definition of "monstrous."[247] Thus, the statements attributed to the ancient authorities on the one hand, and case-related factoids on the other, have differing, complementary functions within the chapter "Multiform Monsters of the Same Species." Liceti does not—in relation to the treatise as a whole—consider firsthand observation to be the privileged foundation of the *scientia* of monsters that he develops.

The epistemic value of *observatio*, and more generally of forms of knowledge based on sensory perception, experienced a general increase, then, in the discourse on monsters in the second half of the sixteenth and early seventeenth centuries. This discourse thus reflects a wider development in nature study. However, the valuing of experience-based knowledge varies depending on the subject matter and the field in which authors were trained and to which they wanted to contribute. There are also differences in the epistemologies of Gaspard Bauhin, Johannes Schenck, Johann Georg Schenck, Ulisse Aldrovandi, and Fortunio Liceti that cannot simply be attributed to their proximity to or distance from the study of anatomy. Although, for example, Aldrovandi and Johannes Schenck—who had not studied in Italy—had encyclopedic aspirations for their work and accordingly attached great importance to amassing as great an abundance of relevant factoids as possible, the Aristotelian Liceti proceeded more selectively with regard, for instance, to the cases documented in literature. The individual text genres also differ, not least with regard to the relationship between individual observations and the overarching interpretative framework.

∽

In this chapter I have analyzed the circulation patterns of monstrous factoids in the naturalist literature of the late sixteenth and seventeenth centuries, and traced them back to central elements of the epistemology on which they are based. As has been shown, a strong connection existed around 1600 between the depiction and description of individual monstrous births and the exotics from the edge of the world that were vouched for by ancient authorities. This had an impact on the concrete form of the factoids in question. In addition, it has been demonstrated that naturalists drew their factoids about monsters from a wide variety of sources. In their search for text passages on and images of monsters and other preternatural occurrences, they did not by any means confine themselves to texts on natural history and philosophy. The circulation of factoids was not, therefore, restricted by boundaries of text genre or field of study, which were not yet academic disciplinary boundaries in the modern sense. The close connection between medical and historiographical discourse is particularly conspicuous.

How do the results of the above analysis relate to the grand narrative of the Scientific Revolution? From the perspective of an eighteenth-century intellectual, the texts discussed here may have appeared to be distinguished by an excess of book-based scholarship and a lack of empiricism. However, experience and observation played a central role in the discourse described. Both categories, though, were for the most part defined in a way that encompassed both one's *own* experience and that of *others*. Against the backdrop of such an epistemology, experiential knowledge and book knowledge appeared *not* to be opposites at all but to be intrinsically connected. As we have seen, however, the relationship of one's *own* observation to that of others (and to *theoria* and *doctrina*) varied, and it did so particularly according to the field of study to which the respective author was contributing but also according to subject matter and text genre.

Chapter 2 traces the superficial phenomenon described here, and the epistemology correlating to it, back to fundamental characteristics of scholarly practice of the time. Taking as a basis the works and intellectual estate of Ulisse Aldrovandi, it examines more closely the *Praktiken der Gelehrsamkeit* (scholarly practices) that were employed in the writing of the texts discussed in this chapter. We turn now to the practices associated with the collecting and ordering of factoids.

CHAPTER TWO

Ulisse Aldrovandi's Twofold "Pandechion"

Collecting Knowledge about Monsters

Why were reports and visual representations of preternatural phenomena so persistent—particularly those that would be classed as supremely implausible just a few generations of scholars later? Were the Renaissance naturalists really "credulous," as their successors in the eighteenth century would have us believe? An analysis of the predominant "scholarly practices"[1] of the early modern period, particularly those of reading and writing, may help to shed light on these questions. Ann Blair has rightly pointed out that note taking should be regarded as a "central but often hidden phase in the transmission of knowledge."[2] So let us bring it into view.

In addition to the ordering of material gathered by reading, I examine the approach taken to images. I accord the same attention in the following to both forms of embodying knowledge—namely, text and image—with the work and working methods of the Italian natural historian Ulisse Aldrovandi (1522–1605) serving as a case study. We find interesting parallels in Aldrovandi's approach to images and to texts: despite repeatedly emphasizing the centrality of direct observations to his working methods, Aldrovandi mostly gathered both textual and pictorial factoids about monsters and other rare natural phenomena from literature, arranging them afresh in publications of his own. In the course of this appropriation, they underwent transformations—not only in relation to their specific form but also with regard to their meaning.

Much like his later English colleague Francis Bacon, Aldrovandi endeavored first of all to document every case of an anomaly in nature that he encountered. He did this using paper technology such as the "Pandechion Epistemonicon"—a comprehensive handwritten encyclopedia—and his natural history collection. Not all the factoids thus gathered made their way into his printed natural history, which, unlike the "Pandechion," did distinguish between "true" and "false" facts. Nevertheless, when in doubt, the Italian naturalist appears to have opted *in favor* of a factoid, to avoid ending a discussion prematurely. As a result, modern readers may get the impression that Aldrovandi was overly credulous. As we shall see, however, this is a misconception encouraged by Aldrovandi's working methods.

Ulisse Aldrovandi's natural history operates with a broad concept of *monstrum*, encompassing not only deformed animals and humans but also rare natural phenomena in the plant and mineral kingdoms and in the sky. In the following analysis, I will concentrate on two exemplary factoids from Aldrovandi's natural history that fall within the fields of anatomy and zoology, since these fields were already central to the concept of *monstrum*. The first example involves the description of a hermaphrodite, and the second the image of a centaur.

As the analysis will demonstrate, Aldrovandi cannot easily be ranked among the heroes of the Scientific Revolution. As we shall see, he is at odds with this interpretive model as, in his practice, empiricism and book-based scholarship did not constitute the contradiction that, in my view, it presupposes.

Who Was Ulisse Aldrovandi?

A portrait by the Florentine artist Lorenzo Benini that now hangs in the Biblioteca Universitaria di Bologna (fig. 9) depicts the Italian naturalist as a dignified elderly scholar in academic robes.[3] Ulisse Aldrovandi was born in Bologna in 1522. He was the son of a noble Italian family with close ties to the future pope Gregory XIII (1572–1585), who was likewise from Bologna and supported Ulisse's scholarly career. As a young man, Ulisse studied mathematics, Latin, law, philosophy, and medicine in Bologna and Padua—two universities that were highly regarded, not least by physicians, even beyond the Italian peninsula.

A trip of his to Rome featured a famous episode in the history of science: he met the French physician and naturalist Guillaume Rondelet (1507–1566), who at the time was preparing his *De Piscibus Marinis, Libri XVIII.* (1566). Aldrovandi is said to have accompanied him to fish markets to study the various species of fish, and

Fig. 9. Lorenzo Benini, *Portrait of Ulisse Aldrovandi*, 1586–1588, BUB. Alessandro Tosi, *Portraits of Men and Ideas: Images of Science in Italy from the Renaissance to the Nineteenth Century*, Pisa, Edizioni Plus—University Press, 2007, 61.

in so doing to have developed an interest in these things, subsequently deciding to study natural history himself. At this time he also began a natural history collection of his own.

He was awarded a doctorate in medicine in Bologna in 1553. His admission to Bologna's *collegio dei dottori* authorized him to practice and also licensed him to teach at the university. However, he never practiced as a physician. Instead, he was soon appointed to teach logic at the University of Bologna, too. Of greater importance to his main interest was his appointment in 1561 as the first professor of natural history in Bologna or, to be more precise, as professor of the knowledge of the simples (*semplici*).[4]

Historians of science know Aldrovandi both for his *museum*, which was already famous during his lifetime,[5] and for the gigantic encyclopedia of zoology on which he began work in 1572. When Aldrovandi died in 1605, just four of the planned thirteen volumes had been published. A fifth volume completed by Aldrovandi himself appeared in 1606. All the remaining volumes were successively published posthumously.[6] Aldrovandi's impressive collection of drawings is also increasingly receiving the attention it deserves.[7] It encompassed approximately eight thousand images bound in several volumes, some three thousand of which have been preserved. As Giuseppe Olmi has observed, these drawings were more than mere illustrations: "The paintings, in watercolour or tempera, were essential tools for Aldrovandi."[8] In many instances they formed the basis for the five thousand woodcuts he commissioned for his natural history encyclopedia.[9] In recent years, his library has also become a focus of research.[10]

Trattata Copiosamente: Aldrovandi and the Monsters

An engagement with the preternatural, and thus with monsters, was central to Aldrovandi's natural history. A conspicuously large number of his drawings depict monsters. Monstrous births to humans and animals of all kinds figure here, as do the monstrous races of humans and other beings viewed as belonging to the realm of fables by intellectuals from the eighteenth century at the latest. These images might easily give the viewer the impression that nature predominantly produces exceptions to the rule.[11]

Aldrovandi was considered an expert in this field, as an episode from 1572, compellingly recounted by Paula Findlen, demonstrates: On May 13, 1572, "the very day that Ugo Buoncampagni had chosen to return to his hometown [Bologna] to be invested as Gregory XIII (1572–1585), a fearsome dragon appeared in the countryside near Bologna, an omen of terrible times to come."[12] The dragon, a large serpent with two legs, was brought to the city for further examination, where the senator Orazio Fontana was responsible for the further proceedings. He handed it over to his brother-in-law Aldrovandi—not only because Aldrovandi was a cousin of the recently elected and likewise Bolognese pope but also because he had at this point already made a name for himself as "a collector of strange and wonderful things and an expert in draconology."[13]

Aldrovandi incorporated it into his natural history collection, where it attracted a large number of visitors.[14] As an explanation for the appearance of the serpent on the very day of the investiture of the new pope Gregory XIII, he suggested an elegant and, for the Vatican, convenient interpretation: "By explaining away the serpent as an example of nature's fecundity rather than a diabolical catastrophe, he diffused its saturnine implications, scientifically securing the foundation of the new papacy for his patrons in Rome."[15] It is worth taking a look here at the precise wording of Aldrovandi's interpretation as he describes it in his *Discorso naturale*. The *Discorso* is a kind of intellectual autobiography he wrote at the age of fifty for Giacomo Boncompagni, the illegitimate son of Gregory XIII. While describing his collection, he touches on the subject of the dragon: "Let us now leave the water and proceed to the crawling animals of the land, namely to the serpents, which by nature have no legs, until the precise time and day on which Gregory XIII, the excellent Pope, was elected to his office, when in our countryside surrounding Bologna two two-legged serpents and dragons were seen, one of which came into my possession. I have presented their story at length [*copiosamente*] and expressed my opinion."[16]

The concept of *copia* as it appears here is a subject to which we shall return. Here, it refers to the plenitude of his discourse. Aldrovandi hastened to write an extensive Latin treatise about the dragon and other serpents. He informed influential people of this undertaking and instructed his artists to produce a drawing of the dragon. Ulisse Aldrovandi thus did his best to turn the *monstrum* that had fallen into his lap to his advantage. Almost immediately, he was veritably bombarded from all sides and all corners of Italy with questions about the monstrosity that only its custodian appeared capable of answering.[17] If, then, the fact that Aldrovandi was entrusted with interpreting the Bologna dragon shows that he was already considered an expert in this field at the time, he was subsequently able to make further symbolic capital of the circumstance.

The *Draconologia*, Aldrovandi's treatise on serpents, was published posthumously in 1639. This was partly because his request for patronage of the pope's family had not met with success.[18] By the time of its publication, the dust had settled. It was too late for the work to attract the kind of attention that it might have immediately after the sighting of the dragon.[19] But monsters were central not only to the *Draconologia* and to Aldrovandi's likewise posthumously published encyclopedia of monsters, the above-mentioned *Monstrorum Historia*, published in 1642. Earlier volumes of his natural history also devoted significant space to monsters: for example, the volumes on solid- and cloven-hoofed quadrupeds, *De Quadrupedibus Solidipedibus Volumen Integrum* and *Quadrupedum Omnium Bisulcorum Historia*.[20] These, too, were published posthumously, though somewhat earlier, in 1616 and 1621.

The Place of Monsters in Aldrovandi's Natural History

The volumes on solid- and cloven-hoofed quadrupeds bring together the knowledge available to Aldrovandi about the animals of the respective species, including the known monstrous specimens. The first chapter of *Quadrupedum Omnium Bisulcorum Historia*, for example, which deals with cattle, contains a section with the heading "Omens. Prodigies" ("Auspicia. Prodigia"). It encompasses the factoids available to Aldrovandi, both textual and pictorial, about monstrous cattle.[21]

William W. Ashworth coined the term "emblematic natural history" for the natural history embodied by Aldrovandi and Konrad Gessner. By this he means that zoological texts written in the Renaissance included not only information fundamental to modern zoology, such as the an animal's morphological features, habitat, and diet, as knowledge about the animal, but also information like emblems, sayings, and fables in which the animal featured.[22] Wolfgang Harms expresses a similar conclusion more clearly, observing that in the standard zoological works by Konrad Gessner and particularly Ulisse Aldrovandi, the meaning of a thing was part of the thing itself. Even more definitely than Gessner, who was firmly anchored in the scholarly tradition of the *artes liberales*, Aldrovandi takes it as a given "that the factual description is embedded in systems of meaning, though he pays close attention in all the descriptive passages to more recent works that aspire to a strictly empirical zoology."[23] In addition, Aldrovandi is more definite than Konrad Gessner in naming the individual *artes* whose proponents might benefit in various ways from reading his natural history. Possible uses included recognizing *exempla* in certain animals for improved human conduct in ethics, economics, or politics. In all of this Aldrovandi made a point of including not only pragmatic uses but also uses "that have been obtained from biblical exegesis and moral example in more recent interpretive genres such as hieroglyphics, emblems, and fables."[24]

The system by which Aldrovandi ordered knowledge about each animal reflects this encompassing aim. It varies to a certain extent from one animal to another, depending on the type of knowledge available to him about a species. In light of his broader aim, it seems logical that his system is more flexible than Gessner's eight-fold grid.[25] And it is precisely the philological eighth part of each of Gessner's animal descriptions (*philologica*)—the one rooted most clearly in the *artes liberales*—that is larger in Aldrovandi's work and is split across various headings.[26] Thus, in contrast to Gessner's work, his reflections on the meaning and symbolic nature of the animal are no longer clearly separated from the other parts of his description, leading Wolfgang Harms rightly to conclude, "For Aldrovandi, the meaning is not merely an extra, added dimension, nor is it associated with the animal only with regard to the broader factual context of the entire *artes*, but rather the dimension of meaning is understood here as a part of the individual *res*, as something insep-arable from the individual animal."[27]

Angela Fischel's blunt distinction between the metaphysical and empirical as-pects of the system used by Aldrovandi to order his history of animals—Fischel calls them "conceptions and observations" ("Anschauungs- und Beobachtungswerte")—is therefore unconvincing.[28] Michel Foucault rightly observed, "Aldrovandi was meticulously contemplating a nature which was, from top to bottom, written."[29] In a way, Aldrovandi was still *reading* as he observed—also in the sense that his own direct observations were by no means clearly differentiated from factoids taken from literature.

Cases of monstrosity in animals are consequently included in Aldrovandi's nat-ural history because of their significance for humans. They are typically found under the heading "Prodigia."[30] In addition to "Synonyms," "Age," and "Emblems," "Prodigies" is one of the categories we regularly come across in the chapters on in-dividual animals in Ulisse Aldrovandi's work. The centrality of meaning and sym-bolism can be seen particularly clearly in the exceptionally large chapter on horses[31] that opens the volume *De Quadrupedibus Solidipedibus Volumen Integrum*. It has fifty-seven sections on subjects such as "love, gratitude, and loyalty of the horse to its master," its "use in the fruitless sacrificial rites of the heathens," and "proverbs" in which horses play a role. The sections correspond to the categories of the grid.

Aldrovandi's all-embracing approach to natural history should not, however, be misinterpreted to mean that he made no attempt to distinguish between "true" and "false" factoids. Headings such as "Fables" ("Fabulosa") in the chapter on horses indicate that this distinction undoubtedly did matter to him. But if he came to the conclusion that a being, a supposed fact, was "fabulous," he did not exclude it from his work: in his conception of natural history, the fables surrounding a particular animal represented legitimate and valuable knowledge.

The central importance of monsters to Aldrovandi's natural history is particularly apparent on the level of the illustrations. Woodcuts are employed disproportionately frequently in the respective sections dealing with *prodigia*. Thus, in the otherwise sparsely illustrated chapter about cattle in *Quadrupedum Omnium Bisulcorum Historia*, the reader suddenly encounters a large number of them in the section "Omens. Prodigies." Sixteen woodcuts, each stranger than the last, jostle for attention within a small space.[32] It is striking that Aldrovandi uses sixteen *different* woodcuts: Unlike the woodcuts that Lycosthenes, for example, used to illustrate the prodigies in his *Prodigiorum ac Ostentorum Chronicon*[33]—or that Hartmann Schedel used for the towns mentioned in his *Nuremberg Chronicle*[34]—these illustrations do not recur in Aldrovandi's work. Each woodcut thus pertains to a *specific* bovine monster. In this respect, Aldrovandi's natural history already shows the first signs of a kind of comparative seeing.

This striving of monstrous animals for attention was successful, too, in relation to Aldrovandi's oeuvre as a whole. In the regular volumes of his natural history, such as those mentioned above, he frequently devoted entire sections to them. Many of the monsters that had already been mentioned or depicted visually were to appear for a second time in the *Monstrorum Historia*, a separate volume devoted specifically to monsters.[35] The separate publication thus grew out of the endeavor as a whole. Numerous as they already were in the regular volumes of the encyclopedia, in the *Monstrorum Historia* Aldrovandi gave space and scope to the monsters in a book of their own, thereby visibly elevating them to an object of natural history sui generis.

Monsters as an Object Sui Generis: Aldrovandi's *Monstrorum Historia*

The *Monstrorum Historia* was published in 1642, almost forty years after the death of its author. Bartolomeo Ambrosini (1588–1657) was the man responsible for its publication. Ambrosini was not a student of Aldrovandi's, but he was one of the administrators of the Studio Aldrovandi, where the intellectual estate that the naturalist had bequeathed to the senate of his hometown was stored. It was also here that the publication of the works that Aldrovandi had not published himself during his lifetime was organized.[36] In the case of the *Monstrorum Historia*, Ambrosini did not adhere religiously to Aldrovandi's plans, in several instances introducing his own scholarship into the text.[37] Despite the changes he made—often in the light of more recent literature—the treatise was conceptually outdated almost as soon as it was published and received just one further edition (1658).[38]

Nevertheless, the work was to prove highly influential over the course of the latter part of the seventeenth century, particularly among physicians in the Holy Roman Empire. Numerous references to it in the *Miscellanea Curiosa*, the journal published by the Academia Naturae Curiosum from 1670 onward, are eloquent testimony to this. They also shed light on *how* the volume could still be profitably read even in the late seventeenth century: authors of articles published in this journal generally referred to the *Monstrorum Historia* when they wanted to cite earlier observations that were similar to a recent observation of a rare phenomenon of nature.[39] Neither the overall approach nor the "emblematic" parts of the work were an obstacle to this kind of selective reading.

Despite the contributions that can verifiably be traced to Ambrosini, it is not disputed that Aldrovandi himself compiled most of the material on the subject. The dedication to Ferdinando II de' Medici (1610–1670), the Grand Duke of Tuscany, accordingly attributes the volume to him: "The entirety of the monsters that a fault of the elements & the blind error of the causes has [*sic*] produced, has been brought together with tireless diligence in a felicitous form within this one volume by our Aldrovandi, the everlasting pride of Bolognese Athens."[40] Moreover, almost all of the printing blocks used for the more than one hundred woodcuts that make the *Monstrorum Historia* an extraordinarily impressive volume visually were commissioned during his lifetime.[41]

The preface introduces the subject of the volume as topics omitted from previous volumes of Aldrovandi's history of animals:

> As the customs [*moribus*] & nature [*naturis*] of all animals have been made known, the wondrous things of nature [*physica miracula*] that are to be presented, and which have generally been referred to by the name of monsters, are now coming to the public's attention; it is known as a verified truth that they sometimes come to pass, & predominantly when, contrary to the usual rule of nature, either the entire body of the genitor or one of its parts (whether it be man, or it be animals, or it be plants that are spoken of) is imprinted with some strange form [*figura*] or an unusual character [*character*].[42]

As this introductory sentence itself shows, Aldrovandi[43] based the volume on a relatively broad definition of *monstrum*. Unlike Johann Georg Schenck von Grafenberg in his *Monstrorum Historia*, Aldrovandi deals not only with monsters in humans and animals, but also with monstrous plants. But even this does not capture the full scope of the book, because chapter 13 discusses the "monsters of the skies": celestial omens such as animals or battling armies sighted in the clouds, as well as comets and meteors.

The second sentence of the introduction specifically underscores the fact that the author regards monsters in principle as real phenomena: "Indeed, these are not cloudings of consciousness or dreams on the part of the authors, who wish to prattle on about monsters with feigned conviction and according to their own whims: for daily observation [*autopsia*] makes us more certain of them."[44] This statement demonstrates two points. First, at the time Aldrovandi was preparing his *Monstrorum Historia*, monsters were not yet an established subject of natural philosophy. This explains the explicit emphasis on their existence. Second, it reveals a basic feature of the epistemology of the *Monstrorum Historia*. It does not pit *autopsia* against earlier records; rather, the two complement each other. In the following we look more closely at the working methods corresponding to this epistemology by means of two examples.

A Hermaphrodite Birth

Our first example, the report of a monstrous birth, is found in the *Monstrorum Historia*: "According to Lycosthenes [*iuxta mentem Licosthenis*], a hermaphrodite was born in Zurich in Switzerland on January 1, 1519 that had a fleshy swelling on its navel and, a little below it, the genitals of a woman; whereas the male member was located in its proper place [*suo loco*]."[45] How did this case come to be included in Aldrovandi's natural history? First, let us briefly recall the discursive conditions that made its appearance in Aldrovandi's work possible set out in detail in the first chapter: The report on the hermaphrodite birth appears in the context of the scholarly discourse on monsters. Aldrovandi's *Monstrorum Historia*, a folio volume of almost eight hundred pages full of reports and images of individual monsters, is itself an impressive manifestation of the tremendous increase in attention afforded the subject in the sixteenth and seventeenth centuries—in Italy and elsewhere in Europe. The boom in the discourse on monsters was thus a central precondition for this report's appearance in Aldrovandi's book.

How did Aldrovandi deal with this factoid? The case is presented in the section "The variety of hermaphrodites" in chapter 5, which discusses the "misshapen" or "deformed" structure of the abdomen and reproductive organs. One central concern of the section on the variety of hermaphrodites was to classify them. After a short general introduction to the subject, Aldrovandi devotes himself to this project, developing his classification—very much in the spirit of his time—on the basis of the position of the genital organs. After naming the classes thus developed, however, he is compelled to admit that hermaphrodites are often *difficult* to classify. Their variety (*varietas*) is very great (*multiplex*), such that they cannot be categorized with certainty. This is attributed to two factors: their great number (*ob copiosum numerum*) and the lack of information as to the position of their external sex organs provided by authors who had something to say about them.[46]

Only after this assertion—as though it were now a question of empirically substantiating the impossibility of classification in the face of this *copia rerum*—does the author list the cases of hermaphroditism he has found in literature, including a baby born in Zurich. Case histories and other more general statements both by ancient authorities and by later authors are set out here for the reader's perusal as an overview of the available knowledge concerning this phenomenon.[47] Following this and a passage on hermaphrodites in the animal kingdom, the section ends with a lengthy discussion of the causes of their births.

How should we interpret the structure of the argumentation in this section? Aldrovandi wanted to come to grips with the anomaly represented by the hermaphrodite by first comprehensively surveying the available knowledge about such phenomena. Only after making his way through all obtainable documented cases did he then proceed to a classification, which in view of their great variety and the shortcomings of the reports proved virtually impossible. This approach is typical of nature study at the time. It is perhaps no coincidence that Aldrovandi's hermaphrodite passage calls to mind the lists of rare *observationes* related by the natural philosopher and Francis Bacon in the second book of his *Novum Organum*, which assumed such a central position in the program for the reform of natural history and philosophy that he developed between 1605 and 1620. In his *Novum Organum*, Bacon explains the necessity of ordering the observations in the form of lists as follows: "But *Natural* and *Experimental History* is so various (*varia*) and scattered that it may bewilder and distract the intellect unless it be set down and presented in suitable order. So we must fashion *Tables*, and *Structured Sets of Instances*, marshalled in such a way that the intellect can get to work on them."[48] Bacon's point of departure, then, is an understanding of nature that corresponded—in terms of the diversity of natural phenomena—with that of Aldrovandi. If we compare Bacon's proposed method with that already being practiced by his Italian colleagues, it becomes apparent that certain aspects were not as radically new as the rhetoric of the Englishman repeatedly suggests.

The collating of documented cases had already been instrumental for Aldrovandi in helping to clarify the phenomenon. Yet both in the section on hermaphrodites and elsewhere in the *Monstrorum Historia*, "the Erasmian ideal of eloquence through *copia rerum* or abundance of material"[49] seems at the same time to have hampered, if not undermined, the attempt to order the material. The sheer number of carefully compiled individual cases, such as that of the hermaphrodite born in Zurich, and several general statements by ancient authorities that Aldrovandi quotes in this section, disrupted the order he sought to impose on his subject. *Copia* sometimes took precedence over classification.

This observation supports Brian Ogilvie's critique of the long-prevailing view according to which early modern natural history was concerned chiefly with the taxonomization of its subject matter. Ogilvie regards the description of the object to be the central focus, not its classification.[50] Fundamentally, the plenitude of descriptions (and illustrations) that Aldrovandi collected served to survey, describe, and order nature's productions.[51] But he was careful—and here, too, he shows a similarity to Bacon—not to jump to conclusions. In addition, pure pleasure in the *varietas*, the great variety of nature, may also have played a role, as we shall see further below.

To a naturalist like Aldrovandi, the factoids about monsters that were available to him were so precious that, apart from a few exceptions, he included them in his printed works. Their value derived not least from the rarity of these phenomena. The example of hermaphrodites serves to illustrate this. Not one of the textual elements of which his passage on hermaphrodites is composed is based on Ulisse Aldrovandi's own experience. It is true that observations of his own come up repeatedly in the *Monstrorum Historia*, typically among the factoids taken from literature—and in this regard we turn later to Aldrovandi's studies of monstrous fowl—but hermaphrodites were extremely rare, and Aldrovandi had not had the good fortune to see one himself. This did not prevent him, however, from addressing the subject at length in his natural history. This was the case in part because of just how well the subject, how well the perceived diversity and abundance of nature's productions, fit in with the intellectual culture of the Renaissance in general.

Copia and *Varietas* as Epistemic Categories

Terms from the semantic fields of *copia* and *varietas* appear time and again in the *Monstrorum Historia*. The introduction, for example, ends with the remark that wonders (*admiranda*) sometimes occur in inanimate things, too, with no indication as to their cause. By way of example, Aldrovandi mentions "various [*varia*] likenesses of eagles or other animals, or of knights in the air."[52] In his opinion, then, not only hermaphrodites but monsters in general occurred in great plentitude and variety.

The headings also express this idea. This is the case not once but twice in chapter 13, which discusses the monstrous phenomena of the skies: One heading announces a section on the monstrous *varietas* of fiery inanimate things, particularly comets,[53] and the second a section on the astonishing variety of meteors.[54] According to the headings, then, the varied nature of these phenomena is at the heart of the sections devoted to them.

As we have seen, *copia* is closely linked to *varietas* in Aldrovandi's work. The former concept is a familiar one to historians of the early modern period, primarily from the field of rhetoric. *Copia* was a key concept in rhetorical instruction during the Renaissance and related to the material available to the speaker. As Terence Cave has demonstrated with reference to French literature, Erasmus of Rotterdam's influential *De Duplici Copia Verborum ac Rerum* (1534) was central to the Renaissance understanding of *copia*. This textbook was designed to teach students how to express a thought in a variety of ways and to develop and apply argumentative strategies. The focus was less on an abundance of words or speech per se than on the ability to express a thought or idea—termed *res* by Erasmus—aptly by selecting the appropriate expression from an abundance of options. Erasmus had originally written his book for St Paul's School in London, but it was to have a great impact on the writing style of European authors in the sixteenth century because of its widespread use in classroom instruction on speaking and writing techniques.[55]

As a stylistic concept, *variatio* was also a central component of rhetorical instruction in the Renaissance and closely linked to *copia*. Textbooks such as the one by Erasmus of Rotterdam again played an important role: "In the wake of the emphatic return to the ancient ideals of rhetoric and poetry, exercises in style that trained *variatio* in written and oral usage occupied an important place in early modern school curricula. Recommendations for the v[ariation] of a subject are found in the sixteenth and seventeenth centuries in the textbooks titled 'Copia verborum et rerum,' which provide methods for the *expolitio* and *amplificatio* of a speech."[56] *Variatio* was also present as a stylistic principle in sixteenth-century poetics, for example, in Julius Caesar Scaliger's *Poetices Libri Septem* of 1561. Here, Scaliger replaces the ancient *virtutes elocutionis* with four *virtutes poetae*, with *varietas* taking the place of *ornatus*, or rhetorical ornamentation.[57] Both *copia* and *varietas* were thus known to humanist authors from the field of rhetoric as skills of a good orator and as literary stylistic virtues—but the influence of these concepts was not limited to rhetoric.

The two concepts also played a major role in the epistemology of many sixteenth- and seventeenth-century naturalists. This is little researched as yet but unsurprising given the central importance of the two concepts in aesthetics and epistemology, and thus the humanist culture of the Renaissance.[58] Marie-Dominique Couzinet even postulates an omnipresence of *varietas* in the natural philosophy of the Renaissance, which she explicates with reference to the works of Gerolamo Cardano (1501–1576), who was held in very high regard by Aldrovandi[59] and Jean Bodin. She writes that the concept of *varietas* corresponded to a view of nature as a place of diversity that would attain perfection in variety.[60]

This positive concept of *varietas* is one of the reasons that the particular in nature was held in such regard at this time. Thus, the success of the medical genre of *curationes* or *observationes*, for example, may be explained by the "huge appetite for *varietas*, which was a marked trait of Renaissance intellectual taste."[61] There is a great overlap between the subjects of early modern collections of *curationes* and *observationes* and the contents of treatises and encyclopedias on monsters in the same period. Physicians were avid collectors of the rare and monstrous cases in particular, even into the eighteenth century.[62]

The more the publications of early modern naturalists teemed with rare and singular phenomena, the stronger the impression grew that nature was characterized by *varietas* and *copia*, and that the book of nature was, therefore, written according to the stylistic rules of contemporary rhetoric. So although the motive for collecting a plenitude and variety of the rare phenomena may have been to discern an order in nature, at the same time such evidence confirmed and reinforced an understanding of nature that resisted these attempts at classification. I investigate this view of nature in greater depth in the following chapter, on the Academia Naturae Curiosorum.

A natural world distinguished by plenitude and diversity, and an epistemology that focused on precisely these characteristics of nature, necessitated technologies that enabled the naturalist to master this abundance, this variety, and make productive use of them. This takes us back to the practical question of how the case of the hermaphrodite born in Zurich in 1519 found its way into Aldrovandi's work, and how Aldrovandi ordered notes on this and other cases.

Christen Ursely: A Chain of Transmission

How did Aldrovandi learn of the hermaphrodite born in Zurich in 1519? Through reading. The wording "iuxta mentem Licosthenis" suggests that Aldrovandi's knowledge of the case stems from Conrad Lycosthenes's *Prodigiorum ac Ostentorum Chronicon*. The vast majority of the examples that Aldrovandi compiled in his *Monstrorum Historia* originate from earlier publications. Given that Aldrovandi's concern was to undertake a comprehensive survey of known cases, this was inevitable. But let us take a closer look at whose texts Aldrovandi was reading, and how.

Among Aldrovandi's various reference texts, Lycosthenes's *Chronicle* played a particularly prominent role. For most scholars writing about monsters, his chronicle was the key source for reports and images of historical preternatural events.[63] It is worth mentioning in this context that the first edition of Theodor Zwinger's *Theatrum Vitae Humanae*—an important example of early modern exemplary historiography, that is, historiography organized by example, again written by a physician—was based on an extensive collection of *exempla* that fell to Zwinger after the death of his stepfather, Conrad Lycosthenes.[64] And despite their differing information architecture, Lycosthenes's *Chronicon* and Zwinger's *Theatrum* are, in parts, quite similar in content.[65] Aldrovandi made a careful reading of his copy of Zwinger's 1586 collection of examples, as he had of his copy of Lycosthenes.[66] This is evident from the handwritten note "perlegi" (read in full) on an empty page of the *Theatrum*, as well as a two-volume handwritten index he created of the work.[67]

Like Aldrovandi's *Monstrorum Historia* itself, Lycosthenes's *Chronicon* was primarily a comprehensive compilation of information from the writings of earlier authors. In the case of the report of the hermaphrodite in Zurich, then, there is a whole chain of transmission at play. Lycosthenes's source was a broadside printed in Zurich in 1519, written by one Balthasar Spross, the son of a wealthy Zurich family (fig. 10).[68] It tells of the monstrous birth and interprets it as a divine warning. Alongside the baby's hermaphroditism, the broadsheet mentions a red growth ("a lump of red flesh") near the navel, which on the page in the *Wickiana* was hand-colored red. Apropos these unusual characteristics, the author emphasizes the vices of gluttony and unchastity in particular as scourges afflicting his age.[69]

Lycosthenes presents his brief passage on this hermaphrodite, accompanied by a woodcut, in the section covering the year 1519. As is typical in his book, a printed marginal note provides information on the year in which the event took place. Lycosthenes disregards the broadside's theological expositions and briefly summarizes the sparse information it contains on the baby's date of birth and external features: "Tiguri Helvetiorum ad Calendas Ianuarij hermaphroditus natus est, propè umbilicum magnum crude carnis tumorem, & paulò inferius muliebre, suo autem solito loco virile membrum habens."[70] There is no need to translate this sentence: we already know its contents, because they are identical to those of the above-cited passage in Aldrovandi's work. Lycosthenes's woodcut also closely resembles that of the illustrated broadsheet and appears to be based on it.[71]

By the time Lycosthenes's *Chronicon* brought the case of the Zurich hermaph-

Die kraſſt nieman für kein tod ſund mer
Es ſyg ritter knecht oder herr
Das hand vnſſer vordren nie gethan
Sunder vff frumkeit vnd eer ſil ghan
Aber yetz strept niemans nach manheit
Allein vff füllen vnd vnlurerkeit
Ein bider man stellen vff ſin wiß vnd kind
Sölichs yetz manlich taten ſind
Das zeigt diß figur hie klar vnd gantz
Der frowen ſcham ſtat ob dem ſchwantz
Dazü das die groß vnkünſcheit
Hatt nider truckt alle manheit
Vnd all ſig hafft lüt über wunden
Söllich Jn macrobio wirt funden (macro:
Hañibal ward ſie durch vfürt (li:z.ſartz
Vnd all ſin volck mit Jm zerſtört
Jn campania gantz nider gleit (val.max.
Alls das vallerius von im ſeyt (li:9.
Durch wibſche vnkünſcheit das bſchach
By Olyfernes man das ouch ſach
Verlor den ſtrit võ vnkünſcheit wegen
Das mag man von vns ouch wol ſegen
Die vnſer manheit iſt yetz gleit
Vff braſſen ſchlemen vnd vnkünſcheit
Da durch wir haʼß wider worden ſind
Als man ettlich volck Jn africa find
Die ir weſen vermiſcher tribent
Alls plinius von inen tür ſchribe (pli:li:7
Androgyne iſt der ſelben nam
Jch förcht es werd nütz gütz drus kan
Gott der natur das hat anzeigt (Aug9.de.
Jm wid wertig ſy die vnkünſcheit (vo.chri.
San Sodomot vnd Gomorra die zwo ſtett
Das helſch für vom himel verbrent hert
Von ir wüſten vnlurerkeit wegen
Darum wirt vns gott das nit vertregen
Das mag vns diß figur bedüten
Als einiſt by der Römeren zitenn
Gſchach durch ein kind ſechs manott alt
Dz vkünt die zerſtörig des römſche gwalt
Als vns titus liuius ſchript (li:li:23.bell:
Da kein miſſerat vngeſtrafet blipt (pu:li:zo
Noch eines mißo Jch Jechen
Man hat für am himel geſehen
Jn dem Burigen nüwen Jar
Das ſond jr nit verachten gar
San lucio Junio kam zü vnſtatt (val.li:j.
Dz er die wunder zeichen verachet hat
San bald Jm mer ſin güt verſanck
Vnd er perſonlich ſelberrtranck
Darum wend jr ſin vor kumer vnd leid
So fliechend die ſund der vnkünſcheid
Vnd vff tugend vnd manheit tringen
Das wirt veh glück vnd heil bringen
Hie vnd dört ewencklich
Sölichs vns allen gott verlich
Amen   ξ Getruckt zü Zürich ξ

Billich verwundert ſich Jung vnd alt
Ab diſem kind vnd ſeltzamer gſtalt
Liplich geborn do man zellt für war
Tuſend fünffhundert vnd nüntzeche Jar
Jm Jenner uff dem erſten tag
So yeder man der fröiden pflag
Ein nüwes Jar frölich zempfah
Rücht diſs kind an die welt zü kan
Sölichs zü Zürich iſt beſchechen
Piderb lüt hand es geſechen
Redent für war on allen zorn
Ob ſiner weiche ſy es geborn
Subtil mit glidmas hüpſch vnd gantz
Nithalb hab es ein manen ſchwantz
Einer frowen ſcham ſtünd nach da by
Ein knollen rotfleiſch ouch da ſy
Sölich figur iſt geborn zwar
An dem Burigen nüwen jar
Nach kriechiſcher ſprach iſt der nam ſin
Androginos vnd ouch Jn latin
Wirt es genant Hermaphrodit
Dar von Van Ouidius ſchript (Oui:4 met
Das ſölich veneris vnd mercurij kind
Durch die waſſer götter erzogen ſind
Durch das man eigentlich mag verſtan
Gros vnkünſcheit ſyg vnder vns kan
Mit zü trincken füllen vnd üppikeit
San einer almal zum andern ſeit
Von Braſſen Bülen vnd zü triben
Es ſyg von röchtren alo e wibren
Kupplen ars welben vnd dero ley
Das iſt allenthalb yetz das gſchrey
Wir füré nun ein ſeltzen orden
es ſind allein vier houpt ſünd wordē
Dañ vnkünſcheit vnd füllery
Vnd ouch groſſe hoffart dar by

Fig. 10. Young and old, all wonder, and rightly so (Incipit). Zentralbibliothek Zürich, Department of Prints and Drawings / Photo Archive, PAS II 12/15.

rodite to a wider audience, the factoid had already outlived its reporter. We learn this from the very person in whose collection of hundreds of contemporary pamphlets the broadsheet described above was preserved; namely, the canon and second archdeacon of the Grossmünster, Johann Jacob Wick. The broadside is found in volume 24 of the *Wickiana*. In volume 13, Wick included a copy of the page along with a pen-and-ink drawing based on the woodcut. He also notes the hermaphrodite's name, Ursula Christen (Christen Ursely). Wick adds that the individual in question had been "subsequently drowned in the 44th year on account of a transgression."[72] Later in the same volume he inserted a woodcut depicting the execution.[73]

Wick was not alone in collecting material. Later scholars, particularly academically trained physicians like the Schweinfurt municipal physician Johann Laurentius Bausch (1605–1665), also considered broadsides important historical sources. Bausch, the founder of the Academia Naturae Curiosorum, to whom I return in the following chapter, kept a handwritten chronicle for a number of decades in which he also included broadsides and other material.[74]

The longevity of factoids such as the one about Ursula Christen is typical of the discourse on monsters in this period. And it was not unusual that this chain of transmission began with a broadside. As we have seen, learned contemporaries like Wick, who had studied theology,[75] saw this genre as a valuable information source. The same was true of natural historians and physicians like Bausch. However, they tended not to name them as sources in the author catalogs or indexes of their publications.[76] As they were acquainted,[77] Lycosthenes may have learned of the birth of the Zurich hermaphrodite—or the broadside in question—from Wick, or vice versa.

The "Pandechion Epistemonicon"

How did Aldrovandi read and use the passage in Lycosthenes's book? Lycosthenes's account was copied from the volume by one of the three amanuenses working for Aldrovandi,[78] by his second wife, or by a correspondent, and then cut out and pasted into an extensive handwritten encyclopedia (fig. 11).[79] An analysis of this long-neglected collection of reading notes, the "Pandechion Epistemonicon,"[80] as Aldrovandi himself called it, affords us some insight into the role played by the scholarly practices of reading and ordering in the compilation of the *Monstrorum Historia* and the other volumes of Aldrovandi's natural history. It can help us understand how Aldrovandi organized—or had assistants organize—morsels of knowledge gleaned from countless texts.

Our approach to the history of natural history has been changed by the ongoing

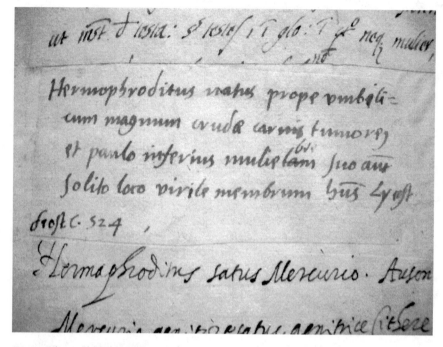

Fig. 11. Ulisse Aldrovandi, "Pandechion Epistemonicon," vol. H–HIRUN, fol. 504r. Biblioteca Universitaria di Bologna: Ms. Aldrovandi 105.

discussion about the effects of humanist practices of reading and ordering on the form of the texts written with their help, notably the use of commonplace books. The same is true of other early modern fields of study that engaged in these practices.[81] Anke te Heesen has contributed the fruitful concept of "paper technology" to this debate, and she underscores with it the materiality of the tools used in the early modern and modern eras to organize excerpts gathered by reading. Along with accompanying tools such as ink, pencils, and pens, paper was suited to tasks that could not be accomplished with other materials. This led, she argues, to the close connection between paper and note taking in both senses of the phrase— both as "taking note *of*" and as "taking notes *on*."[82]

Early modern commonplace books and other contemporary techniques for making reading notes predefined the form and arrangement of the factoids recorded in them, thereby shaping the way a scholar approached a given subject. At the same time, the specific reading and observational practice that was, to a certain extent, always particular to the scholar in question also inscribed itself into their approach.[83]

So what specific form did Ulisse Aldrovandi give to his "Pandechion," how did it contribute to its function, and what epistemic effects did it have?

Aldrovandi himself describes his gargantuan handwritten encyclopedia in a brief overview of his works that he sent to Ferdinando I de' Medici, the Grand Duke of Tuscany (1549–1609), in 1588: "This is a sum of 64 volumes, which I have thus named, namely universal forest of knowledge, by means of which one will find there whatever the poets, theologians, lawyers, philosophers, and historians have written on any natural or artificial thing that one might wish to find or write about ... and one will find other [things] mentioned by the authors [and] that have come to my attention through many documents, in a variety [*varietà*] of places, and through a plenitude [*copia*] of authority of authors."[84] The "Pandechion" was thus intended as an all-encompassing resource for *inventio*, a finding aid.[85]

Not all of the entries in the "Pandechion" deal directly and obviously with aspects of nature; many relate to religious topics or proverbs. Others are of a philological nature, such as a brief note in one of the volumes covering the letter C, stating simply that, according to Plautus, *collum*, the Latin term for "neck," can also end in -*us*: "Collum, et Collus Dici possit <u>Plautus</u> ... 231."[86] The wide range of subjects covered led Sandra Tugnoli Pàttaro, one of just a handful of researchers to have studied the "Pandechion," to class it as "a kind of dictionary ... wherein the most disparate subjects are examined in alphabetical order with a wide range of references and information."[87] Given the way it was actually used, however, this characterization seems unsatisfactory.

With regard to the genre of the text, the metaphor used by Aldrovandi of the *selva* (Latin: *silva*, *sylva*), the "forest," is illuminating. It points to the tentative nature of the knowledge gathered together here. It is a forest that, just like unfelled timber, has yet to be put to its specific use. In classical Latin, *silva* already had the meaning not only of "forest" but also of abundant "material," of "provisions" awaiting their use. Cicero repeatedly uses the term for the "speech material" or "building material" of the orator. And *silva* makes an appearance in the titles of classical Latin writings, too, where it was used to indicate the great variety of the material addressed.[88]

The fact that Aldrovandi uses the metaphor of the forest without further comment is consistent with Wolfgang Adam's observation that learned readers beginning in the middle of the sixteenth century could be assumed to be familiar with the term *silva* and its generic implications.[89] Like Aldrovandi, other early modern authors adopted this classical use of the term. One need only think of Francis Bacon's

Sylva Sylvarum, published posthumously in 1627—his collection of natural history observations and experiments, numbered and divided into centuries, which was based both on reading and on direct observations of his own.[90] As the editor of this work from Bacon's intellectual estate explains in his preface, they were intended— as distinct from earlier natural history collections—as "materials for the Building" of the new natural philosophy that was to be erected:

> For those Natural Histories which are extant, being gathered for delight and use, are full of pleasant Descriptions and Pictures; and affect and seek after Admiration, Rarities and Secrets. But contrariwise, the scope, which his Lordship intendeth, is to write such a Natural History, as may be fundamental to the erecting and building of a true Philosophy: For the illumination of the Understanding; the extracting of Axioms, and the production of many noble Works and Effects. For he hopeth by this means, to acquit himself of that, for which he taketh himself in a sort bound; and that is, the advancement of Learning and Sciences. For having, in this present Work, collected the materials for the Building; and in his Novum Organum . . . set down the Instruments and Directions for the Work; Men shall now be wanting to themselves, if they raise not knowledge to that perfection, whereof the Nature of Mortal Men is capable.[91]

Aldrovandi's "Pandechion Epistemonicon" was also a repository for building material, though the material stored in it was even more basic than that of Bacon's *Sylva Sylvarum*. While Bacon's consecutively numbered observations and experiments constituted a natural history collection in themselves, and were as such the precursor to a new natural philosophy, the "Pandechion" was a handwritten precursor to writings of all kinds and in this respect more open in terms of subsequent uses. It was a tool, and not itself intended for publication. As such, *everything* belonged in there, whereas when it came to his history of animals, which *was* intended for printing, Aldrovandi was altogether more selective.

Many of the notes in Aldrovandi's "Pandechion Epistemonicon" are factoids about monsters in general or about specific types of monsters. The "Pandechion" contains hundreds of entries solely on the term *monstrum*.[92] The entries on specific kinds of monsters, including the "monstrous races" or exotics from the edge of the world, are fewer, though still impressive, in number. It is not uncommon for several dozen notes to deal with a single kind of monster. Well over one hundred notes are devoted solely to centaurs and closely related species: There are 129 entries on *centauri* in volume CAR–CER and seventeen in CAB–CIB. In addition, five notes on the term *hippocentauri* (horse centaurs) are found in H–HIRUN and two more on *onocentauri* (ass centaurs) in OLIB–ORO. On this level of his work, in his "Pandechion," the importance of this subject to Aldrovandi's study of nature is thus already evident. We turn our attention later to the place of the centaur in his printed natural history.

In choosing to name his handwritten encyclopedia "Pandechion," Aldrovandi was alluding to Pliny's list of Greek titles. Pliny discusses what he calls *pandektai*, "hold-alls," in the introduction to his *Natural History*.[93] Both the Greek *pandectai* and the Latin *pandectes* were used for encyclopedic collections, for example, the Pandects of the *Corpus Iuris Civilis* compiled by order of the Eastern Roman emperor Justinian I in the first half of the sixth century. The Pandects, the third part of the *Corpus*, were an ordered compendium of binding law that was, or claimed to be, complete; they documented, commentated, and discussed legal disputes. The use of the term by Aldrovandi's Swiss colleague Konrad Gessner is closer to that of Aldrovandi himself. Gessner named the second, systematically structured part of his *Bibliotheca Universalis* "Pandectae" after the Pandects.[94] The text is more than a bibliography ordered by topic: as Helmut Zedelmaier has shown, this part of the *Bibliotheca Universalis* proves "on closer inspection to be a draft for a grid for the storing of reading matter."[95] Closer still to Aldrovandi is the use of the term for a printed and, with the exception of the heads, empty commonplace book by John Foxe, which bore the name *Pandectae* in its title from the significantly altered second edition onward: *Pandectae Locorum Communium* (1572).[96]

Ulisse Aldrovandi's use of the term was not limited to the realm of reading. Like his reading notes, he called his *museum* a *pandechion*—a "pandechion of nature." This, too, was an encyclopedic collection—though one of things, not words.[97] It encompassed not only the objects themselves but also printing blocks for woodcuts and Aldrovandi's collection of drawings of *cose di natura* (things of nature). We will come back to these parallels later.

Given the origin and meaning of the term *pandechion*, it is no surprise that the work thus described assumed enormous proportions: in its present-day form it comprises eighty-three heavy tomes containing thousands of handwritten pieces of paper pasted onto their pages in alphabetical order. Each slip of paper contains a note beginning with its keyword.[98] The "Pandechion" is thus a collection of secondary reading notes: it was only later that notes originally made by Aldrovandi himself or by his amanuenses in the *studio* were sorted into the books. The opus was finally completed in 1589.[99]

According to Sandra Tugnoli Pàttaro, Aldrovandi continuously made reading notes on loose pieces of paper, writing on one side only. They were initially kept in no particular order, before being spread out on burlap sacks, with one sack for each letter of the alphabet. Finally, he arranged them into strict alphabetical order and glued them into the volumes of his "Pandechion."[100] Not until 1568, he writes, did he perfect this method following many experiments, the first of which he conducted while still a law student.[101]

Figure 12 shows the first page of entries on the term *monstrum* in the volume of the "Pandechion" covering terms beginning with the letter M. The first note on each subject was glued in at the top left, with space left between it and the next subject. The space thus demarcated was then to be filled gradually with further notes on the subject. It seems that once the left half of the page was full, notes would be pasted onto the right-hand side. Aldrovandi estimated the space required with varying degrees of accuracy: Many of the pages are full to the point of overflowing, while others—like the one reproduced above—have notes only on the left half of the page. And many are completely empty, bearing witness to an overestimation of the space required for a particular subject or series of letters.

All commonplace books and other early modern collections of material gleaned from reading that were bound in book form[102] presented their users with the same problem: estimating the amount of space one would require for a particular term or particular subject and allocating the available space accordingly. A miscalculation inevitably resulted in a problem. Aldrovandi's cut-and-paste technique[103] could not resolve this issue completely, but it did at least give the option of removing and rearranging the notes. This is evident in a number of places in the "Pandechion" where traces of glue show that notes pasted onto the page were removed; not infrequently, new notes replaced them.[104]

A statement by the English natural philosopher Robert Hooke (1635–1703) sheds light on this aspect of cut-and-paste. Hooke advises writing the precise observations as to how a specific experiment was conducted on a small piece of very high-quality paper. To store these notes, he recommends gluing them into books. The type of glue was crucial. It was, he writes, "very convenient to have a large Book bound after the manner of those that are very usual for keeping Prints, Pictures, Drawings, etc. to preserve them smooth and in order: On the sides of which, in the same manner as those Pictures are kept, it would be convenient to stick on with Moth Glew . . . But they may at any time, upon occasion, be presently remov'ed or alter'd in their Position or Order, that which was plac'd first may be plac'd middle most, or last."[105] In an age in which material was organized in bound book form, this technique, in combination with a glue that made it possible to remove notes

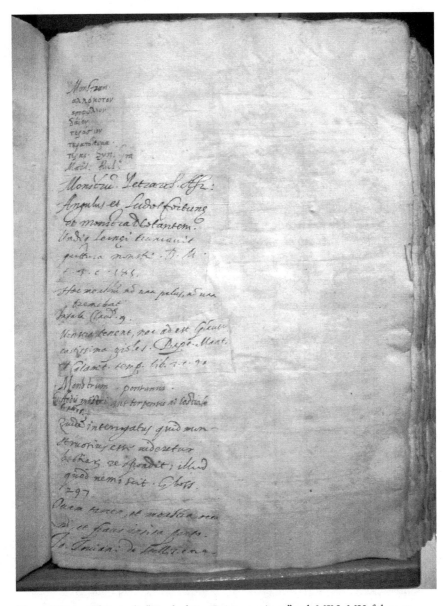

Fig. 12. Ulisse Aldrovandi, "Pandechion Epistemonicon," vol. MIN–MU, fol. 234r. Biblioteca Universitaria di Bologna: Ms. Aldrovandi 105.

after they had been pasted onto the page, was ideal for ordering an unknown and potentially unlimited number of entries. This was undoubtedly a significant reason for its widespread use. In addition, the technique made it easier to carry out the process from writing a reading note to the complete "Pandechion" in a number of steps, allowing the labor to be divided.

The technique that Aldrovandi used was widespread among scholars during his lifetime. Konrad Gessner and the Italian polymath Gerolamo Cardano are among the better-known examples.[106] Aldrovandi corresponded with Gessner, and Cardano was one of the authors whom he held in highest regard.[107] It was not only in preparing the "Pandechion Epistemonicon" that Aldrovandi employed this technique. Ann Blair describes another example: "In another hint of some practice of this kind, Ulisse Aldrovandi . . . thanked his wife for putting together his five-volume *Lexicon of Inanimate Things*. Most likely this meant arranging and fastening in the correct order for the printer a vast number of notes on slips of paper, such as those she also contributed to taking for her husband's *Pandechion Episto-monicon* . . . Both cases involved primarily or exclusively passages cut and pasted from personal manuscript notes."[108] And the visitors' book of his natural history collection also followed this principle.[109]

Aldrovandi's book-based practice for ordering material gathered by reading is consistent with a general feature of knowledge systems in this period; namely, the status of the book, alongside the library, as the central knowledge space. But how does this practice relate to its topographical organization? As Helmut Zedelmaier stresses, knowledge in the early modern period was conceived in topical categories:

> Knowledge appears in the image of its topography; the epitomes of knowledge are the book and the library. Knowledge begins with its inclusion in the book and the systematically ordered library, and new knowledge inscribes itself into this system. Past knowledge is not conceived as a chronological process of development in which earlier stages are *overturned* by later ones, the former now being of merely historical interest. Rather, inherited knowledge appears as a uniform, topically ordered constellation established by objectively defined relevance criteria that are not relative to the respective historical context. Texts are not considered *witnesses* to their times but are, rather, *loci* of knowledge.[110]

This form of organizing knowledge and material—in topical form and chiefly book bound—was embedded in the scholarly dispositive outlined in the previous chapter. The book was the foremost medium for communicating scholarly knowledge. And as such it also indirectly determined the practices employed in ordering material and knowledge.[111]

Reading guides, which were published in large numbers during the early modern period, are a particularly instructive type of source in relation to these questions. They repeatedly address the threat posed by non-bound forms of reading to the memorization of knowledge: "The excerpted material is to be entered into bound excerpt books according to fixed categories and heads, as memorization requires firmly established structures. A freely arrangeable filing system with loose notes, on the other hand, is discussed as a threat to recollection, because, as a mere instrument for storing and locating knowledge, it cannot be mastered by the natural memory."[112] Accordingly, evidence of flexible, non-bound knowledge management technologies being used for personal reading or for library catalogs in the seventeenth century is scarce.[113] Such practices did not come into greater evidence until the eighteenth century. And it is certainly no coincidence that the topical organization of knowledge was in dissolution during the same period—as a result of the concept of the critique of prejudice and the historicization of knowledge.[114]

As was usual in the early modern period, Aldrovandi organized his factoids in a book-based form. However, he decided to order them purely alphabetically rather than topically (which is to say using a system arranged according to *topoi* or *loci* within the book). The alphabetical ordering of knowledge was not an eighteenth-century invention. Individual encyclopedias arranged alphabetically existed even in the early modern period; the "Pandechion" is not an anomaly in this respect. The aforementioned *Pandectae Locorum Communium*, by John Foxe (1572), was published a few years after Aldrovandi had, by his own account, perfected his method for the "Pandechion." A commonplace book, empty except for the heads, it was arranged alphabetically at the publisher's instigation.[115] Moreover, since the late Middle Ages, the large systematic encyclopedias also always had an alphabetical index. But the rise of the alphabet as the paradigmatic form of organization did not take place until the eighteenth century and was closely connected to the decline of the topical model of knowledge.[116]

The persistence of the topical model of knowledge may be explained first and foremost by the associated notion that knowledge located in categories was more amenable to memorization than knowledge arranged in a different fashion, and that notes were primarily intended to help one recall what one had noted. As Frances Yates concedes in her groundbreaking study on the reception of the ancient *ars memoria*, however, scholars in the sixteenth and seventeenth centuries increasingly expressed criticism of mnemonics using images or location.[117] Aldrovandi aligned himself with these critics, lamenting the fact that the effort required to learn the *loci* system outweighed its benefits.[118] His preferred alphabetical system was, as he wrote, inspired by the way merchants kept their accounts.[119]

What role did early modern commonplace collections and other instruments for the ordering of material play in the persistence of factoids in Renaissance natural history? First, they enabled the naturalist to overcome what contemporaries often perceived or at least lamented as *multitudo librorum*, as *scriptorum abundantia*, or—to use a key term from current research on this phenomenon—as "information overload."[120] They gave them access to more quotations, more case histories, and different textual building blocks for their own discourse. Early modern *loci* or *topoi* were not so much linguistic constituents of text as "inventional" ones.[121] Commonplace books organized the knowledge contained within them in a manner intended to enable the user to locate the relevant morsels of knowledge quickly and easily. The same is true of alphabetically organized collections like Aldrovandi's. We may recall here the above-cited description of the "Pandechion" by Aldrovandi himself: a kind of finding aid.

Tools for ordering knowledge such as the "Pandechion" thus indirectly contributed to morsels of knowledge such as that of the hermaphrodite born in Zurich in 1519 being used frequently by a multitude of authors. In the light of the widespread ideal of the *copia*, the abundance, of one's own discourse, this technology would have appeared all the more useful. Indeed, the concept of *copia* reveals the ambivalence of complaints voiced at the time of an inundation of books: *copia* was both an ideal and a lamentable state, as scholars like Arndt Brendecke have noted. "Multiplicity is an ambivalent phenomenon. In the phase under consideration here, it can, for example, stand for stylistic abundance [*copia verborum*], but also for that which is threateningly insurmountable [*multitudinis librorum, scriptorum abundantia*]."[122] Tools like the "Pandechion" thus enabled early modern scholars both to handle *copia* and to generate it in the first place.[123]

A second effect appears central in the context under discussion here. Commonplace books by learned writers of the late Renaissance typically do not differentiate between "true" and "false" facts.[124] Aldrovandi's "Pandechion Epistemonicon" is no exception. In this respect, Aldrovandi followed his role model, the Roman natural historian Pliny.[125] As the "Pandechion" undoubtedly played a role in the compiling of his published natural history encyclopedia, it may also have influenced the selection of factoids found in the latter. But scholars like Ulisse Aldrovandi were, of course, proficient in the different styles of writing demanded by genres as diverse as the "Pandechion" and published natural history. There are numerous notes in

Aldrovandi's *selva universale* that did not find their way into his printed works; he was quite capable of distinguishing between "true" and "false" facts. However, he was evidently at pains to avoid prematurely excluding from his natural history even those statements and observations that were in dispute. His discussion of the centaur is a good example of this, as we shall see.

One of the most fascinating aspects of the way early modern naturalists organized their notes is the peaceful coexistence, up to a certain point, of notes on observations made by an individual scholar with factoids taken from literature.[126] Indeed, this tool deriving from book-based scholarship, from humanism, could be argued to have supplied the model for the meticulous recording of even the most modest of observations—in natural history and other fields, too. Nevertheless, the more philological, more strictly "learned" entries were not to vanish as quickly from the naturalists' records as the grand narrative of the rise of empiricism in the European early modern period might have us believe. The "Pandechion," the vast majority of whose factoids were taken from literature, is a particularly impressive monument to this fact.

Is something analogous at play in the visual representations of natural history knowledge in Aldrovandi's work? Ulisse Aldrovandi repeatedly stressed the usefulness of drawings that were executed *al vivo*.[127] In the following, I argue that Aldrovandi's approach to visual representations has strong parallels with his approach to textual factoids, and that here, too, on the level of the images, he refers—contrary to his rhetoric—remarkably often to traditional sources. In this respect, too, his working methods resembled those of his Swiss rival Gessner.[128] Aldrovandi did *not* have the majority of his drawings and woodcuts executed "from life"; rather, they were often taken from earlier sources.

A Two-Legged Centaur

Let us turn now to our second example, a woodcut of a two-legged centaur in the *Monstrorum Historia* (fig. 13).[129] It depicts a male centaur that deviates from typical Renaissance representations of such creatures: it has just two legs, and its upper body appears strangely squat. Its arms are crossed across its chest, its pensive gaze turned backward. It seems impervious to gravity.

Some of the particular features of the way it is depicted have a banal explanation. The space available on the square block of wood did not allow the engraver to depict the centaur in a more upright posture. The artist in question here was very probably Cristoforo Lederer. Lederer, occasionally also called Lederlein, was born in Nuremberg in the mid-sixteenth century. But he soon moved to Italy, where

Fig. 13. Ulisse Aldrovandi, "Icon Centauri ex Licosthene," 15 cm × 13 cm; rotated clock-wise by 90 degrees. Ulisse Aldrovandi, *Monstrorum Historia: Cum Paralipomenis Historiæ Omnium Animalium* [. . .]. Paris, Turin, 2002 [1642], 31.

he would eventually become a permanent employee of Aldrovandi, and where his surname was translated literally as Coriolano. From 1587 at the latest, and for more than fifteen years, he produced the woodcuts for Aldrovandi's work. He was thus responsible for the final step in a whole chain of transformations that would create, from a drawing, a printed illustration in one of the volumes of Aldrovandi's natural history. As Giuseppe Olmi has shown, the vast majority of the printing plates in Aldrovandi's possession, whether they were ultimately used, are by Lederlein.[130]

A brief look at the printing block preserved in the Biblioteca Universitaria di Bologna (fig. 14) confirms this hypothesis: with its head and arms, the figure of the centaur extends right up to the top and left-hand edges of the piece of wood that was used. It was therefore necessary to compress parts of the upper body to fit the whole figure on the printing block. Deformations like these are not unusual in early modern woodcuts.[131]

Fig. 14. Printing block for the two-legged centaur, from Aldrovandi's intellectual estate. Biblioteca Universitaria di Bologna: fondo Ulisse Aldrovandi, tavoletta xilografica, inv. 23.

But this does not explain the greater peculiarity of the depiction. What we have here is a two-legged centaur. How and where did it come to Aldrovandi's attention? Let us again turn our attention first to the places in Aldrovandi's oeuvre in which this factoid appears. *De Quadrupedibus Solidipedibus Volumen Integrum*, the volume of Aldrovandi's zoological encyclopedia dealing with odd-toed ungulates, is the earliest work in which the woodcut of the two-legged centaur was used; it was later also printed in the *Monstrorum Historia*.

Let us begin with the latter volume. Within the discussion of monsters in the *Monstrorum Historia*, the centaur—along with the other monstrous races or exotics from the edge of the world such as satyrs, a hermaphrodite race, or the giants believed to exist in America—is discussed under the Aristotelian heading *differentiae* in the first, extensive chapter on humanity in general. The concept of *differentiae* made it possible to discuss the systematic differences between the human inhabitants of different parts of the world, both in terms of their customs (*mores*) and in terms of their form (*figura*).[132] Individual monstrous births that occurred in Europe are addressed elsewhere.

Let us consider this passage in more detail. In it, Aldrovandi subsumes different human-animal hybrids under the category of centaur. He structures this part of his book by first listing all the authors known to him who vouch for the existence of centaurs and citing what they had to say on the matter. He then lists all of the authors known to him who did not believe in their existence. Aldrovandi himself refrains from expressing an opinion on the matter,[133] which may be due to the genre of the *historia*, which as Jean Céard points out allowed the author to assemble facts and ideas on a particular subject without needing to regard any inconsistencies as problematic.[134]

The extensive listing of those for and against centaurs reveals just how essential tools like the "Pandechion Epistemonicon" were for naturalists of the ilk of Ulisse Aldrovandi. Without a well-organized repository of extensive "building material," Aldrovandi would not have been able to put together such a detailed list. Although the notes on centaurs in the "Pandechion" exist only in alphabetical order, here—in spite of Aldrovandi's abstention from any conclusive position—they are arranged in an order that has an underlying objective. The juxtaposition of the conflicting positions recorded in literature represents the first step toward a decision on whether such beings exist.

One is reminded, here, of a statement made by Francis Bacon in his *Novum Organum*, in which he anticipates the criticism that false or unverified observations or experiments might creep into his lists. He accepts this possibility with equanimity, trusting that the individual observations would effectively correct one another or themselves when viewed in the wider context:

> Some will no doubt think when they have read over this same history of ours and the tables of discovery, that there is something in those very experiments which is less than certain or downright wrong, and because of that they may imagine that my discoveries rest on false and doubtful foundations and principles. But this is of no account, for such things necessarily occur when we are starting off. For it is like in writing or printing where if one letter or other be misplaced or wrongly set, it does not generally get in the way of legibility very much, for such errors are easily put right by the context [*ab ibso sensu*].[135]

Was Aldrovandi hoping for a similar effect, that the overall context would enable his readers to decide what to make of the various reports concerning centaurs?

Whatever the case, Ulisse Aldrovandi did not make a clear decision in favor of one of the two points of view—perhaps he was simply not yet sure and believed, like Bacon, that certain imponderables were inevitable at the beginning of a comprehensive survey of the natural history of the world. Rather than aiming for a synthesis of the differing views of the proponents and critics of the category of the centaur, he restricts himself for the time being to changing the boundaries of the category itself. He recounts a series of reports from literature about a variety of beings that are said to combine elements of the form of a human with those of a horse or donkey. He concludes from them that the definition of the term "centaur" should be altered to include all of these variations.[136]

Like most of the textual factoids that make up the passage about centaurs, the woodcut of the two-legged centaur is taken from an earlier publication. The caption reads, "Icon Centauri ex Licosthene." Like the report of the hermaphrodite born in Zurich, then, this woodcut originates from Conrad Lycosthenes's *Prodigiorum ac Ostentorum Chronicon*. The similarities between the illustration in Aldrovandi's text and its counterpart in Lycosthenes's work are undeniable (fig. 15). The latter likewise shows a standing, two-legged centaur whose head is turned back over its front shoulder. The fact that its arms are not folded as closely across its upper body as those of Aldrovandi's centaur, and that its tail does not, like his, describe an arc, is due to the artist's choice of a square wood block that required him to allocate the available space differently.

In fact, almost every one of the more than one hundred woodcuts in Aldrovandi's *Monstrorum Historia* can be traced back to one of the major sixteenth-century publications on monsters, in particular to Konrad Gessner, Ambroise Paré, and—as we have seen—Lycosthenes.[137] Aldrovandi's appropriation and reuse of the latter author's woodcuts was not unusual. Many other reproductions of Lycosthenes's woodcuts can be found in texts from this period. This is significant, because it demonstrates that the standards applied by natural historians to the illustrations in their works did not necessarily differ from those of theologians, philologists, or historians. The woodcuts used by Lycosthenes satisfied the requirements of the natural historian Aldrovandi—at least in those instances where he had not seen the phenomenon in question himself. And as monsters, whether real or fabulous from a modern-day perspective, were by definition rarely sighted, Aldrovandi was forced to rely on other peoples' illustrations particularly frequently where this subject was concerned. These two factors were significant contributors to the observable widespread circulation of the above and other images of monsters at this time.

The caption "Icon centauri ex Licosthene" also helps us to resolve a question that is closely related to that of the source of knowledge. It concerns the epistemic status of the image. As Peter Parshall has shown, it was only after the invention of the printing press that images were furnished with titles in a more systematic fash-

Fig. 15. Conrad Lycosthenes, *Prodigiorum ac Ostentorum Chronicon* [. . .]. Basel, 1557, 12; 7 cm × 7 cm. Staatsbibliothek zu Berlin—Preußischer Kulturbesitz, Department of Early Printed Books, Shelfmark: Bibl. Diez 2° 747.

ion. Regarding their role he writes, "Titles or inscriptions often announce the exact status of the picture as a type of portrayal or representation. Typically, this takes the form of explaining the relation of the subject to its model . . . In this respect prints were an unusually articulate medium in the Renaissance, making them an excellent source to explore when asking questions about what people thought certain kinds of images were meant to be doing."[138] Parshall is referring here in particular to the use of the term *contrafactum*. Sixteenth- and seventeenth-century broadsides telling of monstrous births and similar preternatural phenomena also occasionally make use of this term or its vernacular equivalents. *Contrafactum* came into use in the sixteenth century to describe the epistemological status of the artist as an eyewitness to what is depicted.[139] Likewise, the term *ad vivum* (from life) was used increasingly in natural history from the early sixteenth century on to assure the reader or viewer of the documentary value of an image.[140] However, the phrase did not always imply that an image was based on firsthand observation. It was often used simply to characterize a representation as lifelike and in this sense virtuosic.[141]

Aldrovandi made repeated use of this phrase. In relation to the above woodcut, however, he uses the epistemologically broader term *icon*. This is also the term most frequently used for images in the *Monstrorum Historia*. The subjects of the illustrations thus described could scarcely be more diverse. They depict animal and plant species, parts of the human anatomy, hieroglyphs that in turn depict creatures, and individual monstrous births. Two other terms are often used in this sense in the volume: *effigies* and *figura*. None of these three terms implies that the artist actually witnessed the depicted object.[142]

Let us now turn to the second of Aldrovandi's texts in which the woodcut of the two-legged centaur was used: the volume on odd-toed ungulates. The woodcut is found in the extensive chapter on horses, in the section "Whether centaurs or hippocentaurs are entirely fabulous."[143] There is no caption this time. In the body of the text, Aldrovandi uses the term *icon*, which leaves the epistemological status of the image open: "Finally, I add to these [things] this image [*iconem*] by Conrad Lycosthenes from the book concerning wonders and prodigies."[144] He initially believed, he adds, that the image depicted a hippocentaur (*hippocentauri*), but Lycosthenes himself spoke of Apothami—a term Aldrovandi had not previously encountered. He then goes on to relate Lycosthenes's description of their external appearance: the females are described not as half human, half horse, but as resembling bald and bearded humans. Aldrovandi remarks that this description matches that of the *onocentauri* in the work of Aelian and Philes: four-legged creatures whose face, beard, and upper body with breasts resembled those of humans, and whose lower body resembled that of a donkey.[145]

Here, as in the *Monstrorum Historia*, Aldrovandi's approach was to meticulously, and with details of his sources, lay out for the reader's perusal the entirety of the knowledge he has gathered on the subject. The question raised in the section's heading as to whether centaurs are real or fabulous remains unanswered. Instead, the reader is offered an overview of relevant passages by predominantly ancient authors on the subject. This encyclopedic approach to the literature is typical of Aldrovandi's natural history method in general.[146]

The Apothami: A Chain of Transmission

Unlike many woodcuts of monsters and other rare phenomena on illustrated broadsheets from the sixteenth and early seventeenth centuries, the woodcut of the two-legged centaur used by Aldrovandi is not presented to the viewer as a likeness based on firsthand inspection of the creature itself. Neither a heading to this effect nor the body of the text gives this impression. At first this would seem only natural: after all, the *icon* represents a species and not an individual. Nor is this species intended to be illustrated by a representation of a single specimen; rather, the woodcut is designed to make the viewer aware of the *essential* morphological characteristics of centaurs or of a specific species of centaur. In this respect, it resembles not only the illustrations in many natural history texts of the period but also contemporary drawings and woodcuts of technical equipment, for example. Wolfgang Lefèvre has shown that early modern drawings of machines typically depict *generic* rather than individual ones.[147]

As we have seen, Aldrovandi's illustration is based on a woodcut in Lycosthenes's work. We should not, however, consider the latter to be the original. As many historians have observed in a variety of contexts, woodcuts at this time were often reproduced and reused in other publications—both in the field of natural history and elsewhere. Furthermore, as early as the Middle Ages, a distinct depiction tradition had emerged for the so-called monstrous human races or exotics from the edge of the world—the headless Blemmies, hermaphrodites, and centaurs, for example.[148] It is therefore no coincidence that there is a two-legged centaur very similar to the one found in Aldrovandi's work among the depictions of monsters from the edge of the world in the *Nuremberg Chronicle* (fig. 16).[149]

The similarities between Schedel's woodcut and the one in Lycosthenes's work are so great that it seems reasonable to surmise that the latter is based directly on the former.[150] But Schedel's woodcut is not the first depiction of this species of centaur. It can be better understood as an indication of the general presence of this form of depiction in the pictorial tradition of the late Middle Ages. For as Rudolf Wittkower has shown, the woodcuts in the *Nuremberg Chronicle* are for the most part based on contemporary templates.[151] Other contemporary or earlier depictions of the two-legged centaur are not difficult to find. A 1481 German translation of Jehan de Mandeville's widely read *Travels*, for example, contains a woodcut of a two-legged centaur. It is used to illustrate a passage about the Ӱpotamies, a species living in the kingdom of Walckaria on a peninsula in today's Dutch province of Zeeland, clearly identical to that of the Apothami in Lycosthenes's work.[152] Earlier

Fig. 16. Hartmann Schedel, *Register des Buchs der Croniken und Geschichten mit Figuren und Pildnussen von Anbeginn der Welt bis auf diese unnsere Zeit.* Munich, 1975 [1493], fol. XIIv; approx. 4.2 cm × 4.5 cm.

images again can be found in two illuminated manuscripts of Dante's *Divine Comedy*.[153] And two-legged centaurs are also present in Renaissance ornamentation, as evidenced by tiles on the outer wall of the pavilion in the gardens of the Alcázar in Seville, which depict a two-legged centaur armed with a bow and arrow.[154]

Let us turn again to the woodcuts in the passage of the *Monstrorum Historia* concerned with centaurs. Closer examination reveals significant parallels between Aldrovandi's approach to images and to textual factoids. The woodcut of the two-legged centaur is just one of a total of three woodcuts of centaurs illustrating this passage. None depicts the familiar four-legged version. The two-legged centaur is the first to appear; it has a torso that is half human and half horse. On the following page, a *hippopos* is depicted. It has the head of a man. The entire torso is human. Only its legs are horse's legs. Under the laconic heading "Another Species of Centaur"[155] on the opposite page, a third version of this being is illustrated. Its body corresponds largely to that of a man, but it has donkey's ears on its head, and its right leg is that of a horse. What is more, its right arm ends in two forearms and hands. All in all, then, it can be said that the woodcut of the two-legged centaur, along with the two other images, reflects and underscores the central argument of this passage: that there is, in fact, more than one documented species of centaur.

Here, too, as in the passage about hermaphrodites analyzed above, Aldrovandi assembled all of the knowledge available to him about the subject—including the various illustrations of species that he encountered in his reading and that he believed could be subsumed under the category of centaur. In this case, too, the classification of these findings is complicated by their number and diversity. *Copia* was again more important than separating firsthand observations from knowledge drawn from other sources: all three woodcuts used in this passage come from an earlier work and were copied in order to be reused in Aldrovandi's publications. He found the templates for all three in Lycosthenes's work.[156]

Aldrovandi's version of the woodcut of the two-legged centaur was not only used in the various editions of the *Monstrorum Historia* and *De Quadrupedibus Solidipedibus Volumen Integrum* that were printed in the seventeenth century.[157] It also served as the template for an etching in Gaspar Schott's (1608–1666) *Physica Curiosa*. Schott was a Jesuit and student of Athanasius Kircher, and his *Physica Curiosa* an impressive compendium numbering more than 1,583 pages.[158] The volume was reprinted twice in the early modern period—each time including the etching of the two-legged centaur.[159]

The image of the two-legged centaur thus proved long lived, which was largely due to the fact that naturalists (and other scholars) often reused illustrations from earlier publications. Much like citations in this respect, they were often removed from their original textual environment, copied, and used again in the author's own discourse. But how does this practice tally with the contemporary rhetoric of experience, which is particularly pronounced in Ulisse Aldrovandi's work? This is the question to which we turn our attention.

The Place of Observation in Aldrovandi's Epistemology

In the *Discorso naturale*, Aldrovandi describes his working methods very clearly. His natural history was written trustworthily, he affirms, "and I have not written in it a single thing that I have not seen with my own eyes and touched with my hands and whose outer and also inner parts I have not dissected."[160] It is not easy, however, for historians of science properly to understand the working methods thus described in relation to concrete practices. This is due in large part to the fact that although Aldrovandi repeatedly stresses the central importance of experience in his working methods—at moments like these he seems very much to be speaking the modern language of empiricism—the end products of his naturalist studies have a strongly compilatory character. Historians of science have traditionally perceived the compilatory features of his work as shortcomings. Änne Bäumer-Schleinkofer writes in this connection, "Aldrovandi himself affirms that he always sought directly to inspect the object of his studies . . . Taking as an example the first two books of the *Ornithologia* concerning the eagle, however, Guiseppe Olmi demonstrates that the scientifically relevant data is overlaid with a layer of mere [book-based] scholarship. According to Olmi, this approach—typical of the era—is unfortunately characteristic of Aldrovandi's entire oeuvre."[161] It can therefore come as no surprise that historiographical accounts of his working methods are widely divergent with regard to the relationship between empiricism and book-based scholarship.[162]

A particularly cogent and influential approach to the seemingly contradictory nature of Aldrovandi's working methods relates directly to the comparison of his citation and illustration practices undertaken in this chapter. Let us therefore explore this in a little more detail here. Guiseppe Olmi characterizes Aldrovandi as both backward looking and forward thinking: backward looking in relation to his textual descriptions of objects of natural history, forward thinking in relation to his approach to images. It is typical of this viewpoint that the scientific practice of observation is correlated with the images, whereas textual descriptions are associated with encyclopedism, which is conceived as outdated. Olmi's argumentation makes this plain: "ULISSE ALDROVANDI WAS AMONG THE GREATEST INNOVATORS of the study of nature during the Renaissance, stressing the need for direct personal observation and the value of accurate illustrations in natural history books."[163] In his interpretation, these "accurate" images contrast clearly with his descriptions: "There is no doubt that these illustrations represent the most original and innovative aspect of Aldrovandi's work, for his written descriptions are often obscured under a thick mantle of erudition and encyclopaedism. Driven by the urge to provide the largest possible amount of information, to say every thing about everything, Aldrovandi did not always restrict himself to describing the results of his observations and anatomical dissections; he also reported everything that had been written since antiquity about animals, plants and vegetables."[164] In the light of the comparison undertaken in this chapter, this contrasting of the two appears in need of revision.

An analysis of Aldrovandi's woodcuts and collection of drawings can help us to come to a more nuanced assessment of his approach to images. Many of the thousands of drawings and woodcuts in Aldrovandi's intellectual estate are not based on firsthand inspection of the depicted phenomenon by Aldrovandi himself. Given their intended function, however, this was not at odds with the prominent role of observation in Aldrovandi's epistemology. It was, after all, not a matter of using images to *record* observations. Aldrovandi's drawings (and those of other contemporary naturalists) were rather intended to *enable* naturalists or students to make future observations. The same is true of the woodcuts.[165] Aldrovandi proceeded as encyclopedically with illustrations as he did with words. And with regard to the possibility of images of inexistent beings creeping in, he may have shared Bacon's hope that such errors, like printing errors, would correct themselves in the greater context.

The vast majority of the numerous notes in his "Pandechion Epistemonicon" likewise describe observations by other authors and knowledge that is not based on *experimentum* or *observatio* at all. Like other early modern natural historians, Aldrovandi borrowed both textual and pictorial factoids from previously published works—including, in particular, those concerning preternatural phenomena, the firsthand sighting of which did, after all, require a fair bit of luck. He used them as building blocks for his own texts, thus contributing to the widespread circulation in late sixteenth- and early seventeenth-century natural history of case histories, statements of a theoretical nature, and illustrations of individual preternatural phenomena, as well as of entire classes of phenomena, such as species.

But the parallels evident in Aldrovandi's approach to textual and pictorial factoids go deeper still. Both practically and linguistically, he dealt with descriptions and illustrations in a startlingly similar way.

Text, Images, and Their Material Practices

It is instructive in this context to consider the topographical arrangement of texts and images in Aldrovandi's family palace near Piazza Santo Stefano. Around 1600, Aldrovandi's *museum*—a locus of objects and images—was situated with the two rooms of his library to one side and two further rooms of his *studio*, in which he kept not only further *naturalia* but also drawings and books, to the other.[166] In Maria Cristina Bacchi's convincing interpretation, this physical proximity shows that the *museum* and (other parts of the) *studio* constituted closely connected places of work.[167] The proximity of book collections to cabinets of curiosities was typical of the early modern period. Aldrovandi's is not the only case, then, in which the collecting of books can be placed in the context of collecting typically associated with curiosity cabinets.[168]

An interesting parallel between Aldrovandi's respective approaches to text and images is also evident in his dealings with his assistants. He drew on the help of his artists and amanuenses in comparable ways, as is clear from sources such as this description of his botanical excursions in a letter to Cardinal Maffeo Barberini: "I often, too, wandered over the marshes and mountains, accompanied by my draughtsman and amanuenses, he carrying his pencil, and they their notebooks. The former took a drawing if expedient, the latter noted down to my dictation what occurred to me, and in this way we collected a vast variety of specimens."[169] The parallelism in his characterization of these assistants' activities is no coincidence. This, at least, is what further parallels suggest, for example, on the level of storage.

Both textually and pictorially embodied knowledge had to be stored until such time as it was required for Aldrovandi's own publications. As we have seen, textual factoids were pasted into his "Pandechion Epistemonicon." The printing blocks for the woodcuts in the *Monstrorum Historia* and other works were stored, along with the extensive collection of drawings, in a different epistemic space. It consisted of walls and cabinets rather than paper and was three-dimensional, not two. Significantly, though, it was nevertheless likewise referred to by its owner as a *pandechion* (or *pandechio*). The space in question was his *museum*. In a sense, then, *copia* was operative as much on the level of *naturalia*, *artificialia*, and visual representations[170]—the contents of the three-dimensional *pandechion*—as it was on the level of language.

Paula Findlen has observed that Aldrovandi's use of the term *pandechion*, like other encyclopedic terms, was flexible in the sense that it could refer both to plenitude itself and to its locus. In relation to the second *pandechion*, Aldrovandi's *museum*, she goes on to write, "Not surprisingly, the principle of plenitude was operative in his decision to designate it as an encyclopaedic structure. In a similar fashion, the first cataloguer of Francesco Calzolari's natural history museum in sixteenth-century Verona called it, among other things, a cornucopia. If nature was the 'cornucopian text' which held the interest of the naturalist, then the museum itself was a receptacle of *copia*."[171] In light of the above analysis, it makes sense to extend Findlen's conclusion to include the paper *pandechion*, the "Pandechion Epistemonicon." Aldrovandi was an encyclopedic collector on both levels—of words as well as images and objects—and his twofold *pandechion* served as the loci of this plenitude.

Nel paese del gra Tiburlano si troua cétauri in questa forma da mezze usu fazze e cerpo humano excetto che hano nel loco delle brazze do brazzi curti amodo di rospi nella testa hano doe orechie di cane có tre barbe al uso poi nelli fiáchi hano doi brazze comone e dúa humani el resto sono groppa e pie dide cauallo correno forte e se abbrazzano uno ly fano crepono tanto strégono forte usono solo de elephati sono amici de huomeni er de dene p che 'a essi nó fano molestia alcuna.

Fig. 17. Print of a centaur from Aldrovandi's intellectual estate. Biblioteca Universitaria di Bologna: Ms. Aldrovandi, Tavole vol. 006-2 Animali 057.

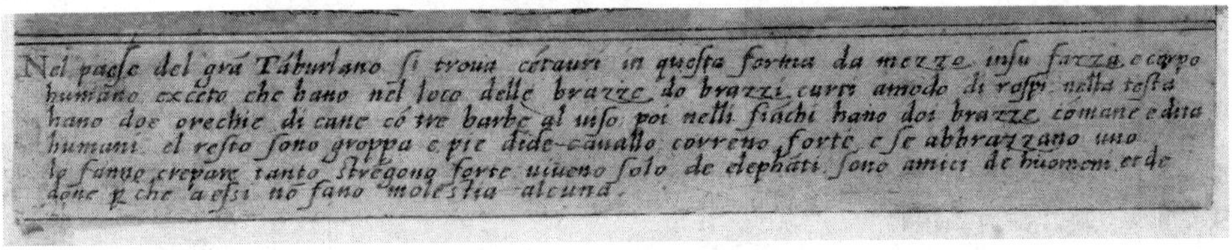

Fig. 18. Detail of fig. 17.

Lastly, let us consider the text/image relationship of the sheets in Aldrovandi's collection of drawings. Most of the surviving sheets have a handwritten caption above or below an illustration. In most cases they merely describe the object portrayed. In some cases, however, the text/image relationship shifts to the point where the pictorial and corresponding textual factoid are found on the same page. This is true, for example, of a hand-colored etching depicting another species of centaur (fig. 17) that Aldrovandi mentions[172] but does not illustrate[173] in both the *Monstrorum Historia* and *De Quadrupedibus Solidipedibus Volumen Integrum*.

The etching was produced by Giovanni Battista de' Cavalieri (1525–1601) and published in 1585 as part of the series *Opera nel a quale vie molti Mostri de tute le parti del mondo antichi et moderni [. . .]*.[174] This copy was carefully cut out, with its border, and glued into one of the volumes of the drawings collection. Here, as in many other instances, Aldrovandi applied the cut-and-paste technique described by Hooke to an image.[175]

An accompanying text in Italian was pasted in with the image (fig. 18). More than merely a caption, it contains all of the knowledge available at the time about the illustrated species. Both the image and the text, like the woodcuts of centaurs used in the *Monstrorum Historia*, originate once again from Lycosthenes's *Chronicon*, where it is written that centaurs of this appearance can be found in the land of a certain Great Tamburlaine.[176] Tamburlaine (also known as Temurlenus, Tamerlane, or Timur) was a Mongol ruler and conqueror of the fourteenth and early fifteenth centuries. His Asian empire was located not far from the latitudes in which the topographical tradition believed the exotics from the edge of the world to exist. During Aldrovandi's lifetime, Tamburlaine was a familiar figure in both historiography and art.[177]

Text and image, reading and observation thus by no means represented opposing media or modes of knowledge production in Ulisse Aldrovandi's work. In the concrete practice of nature study, they were, in fact, inextricably linked. His statements on the central importance of firsthand inspection to his work should not, therefore, be misinterpreted through the lens of a narrow, modern understanding of observation. Angela Fischel is right to highlight the extent to which many of Aldrovandi's drawings were mediated through other media: "Though Aldrovandi, like other late-sixteenth-century scientists, celebrated observation as an innovative epistemic technique, in fact it has more in common with the 'traditional' epistemic technique, the reading of texts, than was—and still is—claimed."[178] We need not assume that claims by actors at the time regarding the central importance of observation to their working methods were false to reconcile their self-presentation with their actual practices. Their concepts of observation and experience were often broader than today's.[179] In a sense, therefore, reading *was* a kind of observation. Let us conclude by recapitulating what we have seen so far.

Compilation *and* Observation

Like his Swiss colleague Konrad Gessner, Ulisse Aldrovandi is a prime example of a proponent of "learned empiricism" as described by Pomata and Siraisi. For him, scholarship and empiricism did not conflict with but rather complemented each another. Laurent Pinon has shown on the basis of Aldrovandi's ornithology, for example, that humanist working methods led to a form of writing that, on the one hand, was characterized by the great significance accorded to citations and references to older works, but, on the other, also allowed for the integration of previously unpublished observations.[180] I can now add that this applies to his images as well as to his writing.

Monsters are a particularly illuminating case study in this connection, not least because they were by definition notoriously rare natural phenomena or, as in the case of the monstrous human races, notoriously remote and therefore inaccessible. A naturalist had to be in the right place at the right time to witness a monstrous plant, a monstrous newborn, or a monstrous apparition in the night sky—to say nothing of centaurs and other exotics at the edge of the world. As we have seen, it was only thanks to Aldrovandi's influential relatives and friends that the Bolognese dragon came into his possession—an unbelievable stroke of luck. In these and other cases where he was in a position to examine the object in question himself, he did. There are many less prominent examples than this dragon, whose particular potency as a divine omen made it more the exception than the rule. Aldrovandi also dissected a four-legged hen, for example, whose skeleton he then preserved and immortalized with a woodcut in the *Monstrorum Historia*.[181]

In the text of the *Monstrorum Historia*, it is repeatedly emphasized that these individual cases were witnessed by Aldrovandi firsthand. The section on the "for-

mation of multiple legs in chicks," for instance, contains the following two passages
on a three-legged hen and a similarly endowed chick; in both cases the author's
firsthand experience is emphasized by the use of the terms "observation" (*observatio*) and "seeing" (*videre*). These examples are found among a whole series of similar cases, some based on Aldrovandi's own experience, some on literature:

> We add to these [aforementioned cases] a three-legged hen from our observa-
> tions [*ex nostris Observationibus*], whose third leg emerged near the anus. Al-
> most the whole body of this hen was gold-yellow in color. Its wings and back
> were flecked with dark brown spots. In the same manner, some dark brown
> feathers hung down from its neck over the base of the wings. In addition, the
> whole tail [was] dark brown, the other [parts] gold-yellow. It is illustrated in
> figure VII along with a sprig of barley.
>
> We were also given the opportunity last year to see [*videre*] a chick with three
> legs that had just hatched from the egg and whose third leg emerged, as it were,
> from its rump: we therefore ensured that this strange variety [*varietatem*] was
> drawn, and offer an illustration of it in figure VIII.[182]

Next to the first of these two paragraphs there is also a printed marginal note that
reads, "Gallino monstrosa observata." In both cases—and this is typical of Aldrovandi's approach to rare natural phenomena—he does not merely emphasize his
firsthand experience but uses it to produce a detailed description of what he had
seen. That he also had drawings made of both animals and then immortalized each
animal in a woodcut demonstrates the importance he attached to documenting
and conveying their precise form.

As comparatively inexpensive farm animals, chickens were commonplace in
early modern households. This explains the disproportionately high number of
monstrous chickens documented. It was presumably not too difficult for a famous
naturalist like Aldrovandi to come by such a chicken, to examine it in detail and
perhaps dissect it. But what alternative did a natural historian with encyclopedic
aspirations like Aldrovandi have in the majority of cases of rarer monsters but to
rely on the earlier records of others—whether pictorial or textual—and to reproduce them in his *Monstrorum Historia*? And as reading and observation were not
necessarily perceived at the time as contradictory, especially by a humanist like
Aldrovandi, this did not present a problem.

༄

As we have seen, Aldrovandi is very difficult for historians of science to categorize.
This is due not least to the fact that the dichotomy between empiricism and book-

based scholarship that is presupposed by the grand narrative of the Scientific Revolution presents an obstacle to understanding Ulisse Aldrovandi's working methods. If we wish to view Aldrovandi from this perspective as a prime example of the increased importance of observation in natural history in the late Renaissance, the compilatory elements in his work—especially those that conflict with our understanding of experience—disturb the picture. If, however, we want to emphasize one-sidedly that he proceeded encyclopedically in gathering the knowledge that was available, his emphasis on the central importance of observation to his working methods does not fit.

But reading and observation were not at odds with each other in Aldrovandi's working practices. He documented things he had seen himself and things he had read in equal measure. Monsters were phenomena that were rare or topographically far removed from the European scholar. It is therefore unsurprising that Aldrovandi often had to refer to literature with regard to monsters specifically. And against the backdrop of the collective empiricism of early modern nature study, this did not present a problem. Indeed, for the humanist Aldrovandi, reading itself seems to have represented a form of empiricism.

Must he not, though, be accused of credulity, as he was by champions of the Enlightenment, for including in his work perfect hermaphrodites, centaurs, and other creatures that, from a late seventeenth- and eighteenth-century perspective, were questionable? On the basis of the above analysis of Aldrovandi's working practices, this question, too, can now be answered more satisfactorily than it has been to date. The gathering of material that he undertook before writing his printed natural history was encyclopedic in conception. Until such time as they were used in print, Aldrovandi stored textual factoids in his "Pandechion Epistemonicon" and pictorial ones in his *museum*. While the "Pandechion" did not distinguish between "true" and "false" factoids, this distinction certainly *was* important in his printed natural history. However, passages like the one on the centaur show that, if in doubt, he reserved judgment. Like Bacon, he perhaps hoped that any errors that might creep into his natural history would be corrected in the long run or the wider context. The reconstruction of Aldrovandi's working methods thus provides no evidence of his "credulity" but rather suggests a different explanation for the presence of fabulous monsters in his work.

Moreover, "credibility" itself is subject to change over time, as Franz Mauelshagen has postulated in his discussion of a similar view, held since the Enlightenment, of Johann Jacob Wick and his *Wickiana*. He therefore calls for research into the cultural history of the belief in wonders "to explore the criteria for what is considered credible and to view them within the complex social and communicative contexts from which they emerged."[183] It is apparent that late Renaissance naturalists worked in an intellectual environment in which *credulitas*, credulity, did not yet constitute a central topos in the discussion of the credibility of authors and witnesses. It is only from the late seventeenth century on that *credulitas* is found in the discourse on monsters in this function, and thus as a discursive weapon of the Enlightenment.[184]

The scholarly practices that were typical at this time—not only of Aldrovandi's approach to images and text—were part of the reason why pictorial and textual factoids about monsters could "travel" so exceptionally well from the work of one author to that of another. The fact that so many were to prove greatly persistent was also due to the widespread use of the practices described.

As the seventeenth century progressed, naturalists' reading and observational practices were to change, and along with them the practice of writing. I investigate these changes in the next chapter with particular reference to the Academia Naturae Curiosorum, which was founded in Schweinfurt in 1652.

CHAPTER THREE

Observing Correctly

On the Ambivalent Relationship of the Academia Naturae Curiosorum to Monsters

By the time they were preparing the third volume of the *Miscellanea Curiosa*, the journal of the Academia Naturae Curiosorum,[1] which had appeared annually since 1670, the *curiosi*, as the members of the academy liked to call themselves, had evidently already lost interest in monsters. In the third volume, Breslau's municipal physician Heinrich Vollgnad (1634–1682) addressed readers directly, setting out the difficulties in the matter. He did so in a scholium, a form of commentary that accompanied many articles and served predominantly to list previous cases from literature that were similar to the phenomena described in the article at hand. Vollgnad, however, had a more fundamental concern:

> It had been decided to delete not only the reports on monsters, but also the images of them from this year's volume of the Ephemerides Curiosae, not so much on account of their seeming lack of relevance in the refinement of medicine or physics, but above all because their causes—namely whether they are of natural or of hidden origin—are, due to the sluggishness and disorder of the human mind, either deliberately suppressed by those who know them, or (completely) ignored by the less educated common people.[2]

As will be shown, Vollgnad was concerned chiefly with human monsters. In the *Miscellanea Curiosa*, however, the word "monstrous" was used, in line with Ulisse Aldrovandi's definition of the word, to describe rare phenomena from all three natural realms.

When it came to the journal of the academy, Vollgnad was not just anyone. Beginning in 1672, he and Johannes Jänisch (1634–1707) were responsible for its publication.[3] There were, then, general doubts about the value of communications concerning monsters. What is more, though, those responsible evidently saw a problem with the approach to the causes of such phenomena. The first point—the benefit to medicine in particular, and therefore indirectly to human well-being—was central to the self-conception and reputation of the academy. The second point—how to deal with speculation that extended beyond what was actually observed, such as to the causes of the observed phenomena—was a question of utmost importance to the "epistemic genre" that underlay the journal, as we shall see. Furthermore, the *curiosi* were not at all pleased by the fact that some authors refused to rule out nonnatural causes for monsters.

Ambivalent feelings about reports on monsters were not restricted to the Academia Naturae Curiosorum. However, the *curiosi*'s publications and correspondence provide a particularly good means to identify them and to examine their causes.[4] If we are to do this here, it is first necessary to clarify what considerations led to the publication of the *Miscellanea Curiosa* and upon which "epistemic genre" they were based.

The founding of the journal marked a significant change in the work program of the young academy. The *curiosi* were responding—belatedly and within a very short period of time—to a change in the concept of authorship within nature study that was taking place particularly among the academies in the seventeenth century. It was less about replacing reading with experience, in line with the grand narrative of the Scientific Revolution (and the rhetoric of many contemporaries), and more about practicing a new kind of experience and a new way of writing about it. For the most part, the *Miscellanea Curiosa* contained authored *single* observations; their authors had to learn to distinguish between experience in the general Aristotelian and scholastic sense and their own observations concerning the specific individual case. Where previously the role of the *curiosi* had been primarily as compilers of encyclopedic monographs, as *collectores*, they now had to learn to observe in a way that had not so far been required of them and, as *auctores*, to write.

I then analyze the articles on monsters in the *Miscellanea Curiosa* and compare them to other sources associated with the academy to better understand the Academia Naturae Curiosorum's ambivalent relationship to monsters. *All* subjects dealt with in its journal required correct observation and description. Because of the commotion their appearance often caused, however, and particularly after the experiences of the Thirty Years' War, monsters were an especially sensitive subject.

Unlike their predecessors in the late sixteenth and early seventeenth centuries, academically educated physicians—as well as large proportions of secular and religious elites—agreed that the divinatory view of monsters that was widespread among the population was dangerous. If "naturalization" is understood in the broad sense to mean the explanation of a phenomenon by its attribution to natural causes, then monsters had already been naturalized by authors in the European Middle Ages.[5] It would seem to make more sense, however, to apply a narrower definition that implies not only a discussion of natural causes but also a rejection of supernatural ones. In this fuller sense, monsters in the Holy Roman Empire were being naturalized only now, after the Thirty Years' War.

Observing Monsters *Correctly*

Despite the concerns mentioned, the *curiosi* eventually decided against excluding monsters from the third volume of the journal. In his scholium, Vollgnad gives the following reasons for this: First, there were, he claimed, successful models for collecting the *historiae* and *icones* of monsters. He was referring here to the large number of monographs on the subject by naturalists that appeared around 1600.[6] "Seeing, though, that people such as Bauhin, Liceti, and Weinrich have filled entire books with similar historiae and exempla, to the subsequent applause of scholars, we did indeed believe that it would not be to our disadvantage to record and preserve for posterity the historiae and icones of those instances of monsters as have occurred in recent and modern times."[7] So the existence of such precursors in scholarly naturalist literature made the collecting and documenting of *historiae* and *icones* on the subject seem legitimate and desirable.[8]

Immediately following this, Vollgnad addresses the question of the benefit of such contributions and at first openly admits that their value to medicine is not immediately obvious. But their value to anatomy was a different matter, and this was the second reason not to exclude them. However, monsters needed to be examined by the *right* men, in the *right* way:

> However much, therefore, similar observed examples pertaining to medicine may be of no value, [medicine being] pursued for the sake of the preservation or restoration of health, they may still afford value to anatomy, one pillar of medi-

cine which you, by no means unjustly, call its second leg, provided that they fell into the hands of those physicians who were either not willing to content themselves with the examination of internal and external defects, or who, despite various obstacles, . . . were always able inasmuch as they were willing to devote themselves to dissections with tenacious industry, or, lastly, those who were not let down by the tools required.[9]

Thus, particularly because of the anatomical insights promised by their examination, monsters were considered a fitting subject for the *Miscellanea Curiosa*. Nevertheless, the right tools were needed and—in terms of the character of the scholar—sustained industry, *industria*, and a tenacious thirst for knowledge.

What Vollgnad merely hints at is set out in more detail in a document that illuminates the observational practices of academies that studied nature. The document in question is an introduction to the *Memoires pour servir à l'histoire des animaux*, by the Académie Royale des Sciences in Paris, written by the anatomist Claude Perrault (1613–1688). The French academics set themselves the aim of achieving, through collective effort, a comparative anatomy of plants and animals. The approach they took was similar to the journal project of the *curiosi*: they carried out the most detailed examinations possible of individual animals or plants and collated the corresponding reports.[10] Published in 1671, the *Memoires pour servir à l'histoire des animaux* was the only volume to emerge from this project.

Perrault's foreword provides a concise summary of both the underlying goal of the publication—a comparative anatomy of animals—and the working methods used. Perrault begins by explaining the genre of *mémoires* chosen by the academics. He differentiates between two distinct methods of writing an *histoire*: One is encyclopedic, bringing together everything that has, at various times, been recorded on a particular subject. The other contents itself with relating the particular facts (*faits particuliers*) of which the author has certain knowledge (*une connoissance certaine*). This second model is called *mémoires*. It covers only a part of the other meaning of *histoire*, he writes, and does not have the same *majesté*. But it does have the advantage of greater certainty (*certitude*) and veracity (*vérité*).

According to Perrault, however, these advantages come into effect only if authors fulfill the necessary requirements for this form of writing: they must be both precise and sincere (*exact, & de bonne foy*). The authors of the other form of *histoire* frequently do not meet these requirements.[11] Later in the foreword, Perrault again summarizes what is required of the observer of particular facts, this time *ex negativo*: most authors of reports on particular things—by which he means authors of travel accounts with no academic background—lack philosophical *esprit*, patience (*patience*), and truthfulness (*fidélité*).[12]

One may well assume that Perrault is referring here to the part played by judgment, *iudicium*, in the act of observing—to the requirement that observers always verify what they perceived with their eyes with the aid of their reason. At least that is how it is put in a statement by Perrault in the context of the academy's anatomy project:

> For although our knowledge of the human body is principally dependent on its inspection and on the precise distinction which is made of all its parts by way of dissection, it is nevertheless true to say that our eyes are not the only guides in this research and that Reason also provides some lights that can guide us there, which do not only serve to clarify the use of the parts that we have found but even the necessity or the probability of those that we hope to discover, and that it can happen quite often that for lack of this advice we search in vain for organs and chanels that Reason makes us judge are not necessary.[13]

From this point of view, *observatio* could *not* be a purely empirical matter, because appearances can deceive. Only observers' power of judgment protects them from misperceptions.

The qualities of the *right* observer listed by Perrault correspond largely with those that Vollgnad had in mind when he wrote his scholia. Both implicitly demand a certain cognitive restraint on the part of observers—that they remain silent on what they have *not* seen. The intention here was also to establish a form of observing and writing that was distinct from the one familiar to us from many of the monographs on monsters published around 1600: it was no longer to be encyclopedic but selective. This requirement took on particular importance when it came to subjects such as monsters, which, considered by some as divine omens, could still cause great unrest wherever they appeared.

For Vollgnad there was a third reason why monsters were a valuable subject after all: they had the effect of strengthening piety, without which medicine would be worthless: "All usefulness in medicine would indeed be lost if strange things were not observed in such examples by physicians who, as is required by Hippocratic Law, should be pious men, so that they may remind others of piety . . . Each man, therefore, should consider that such monsters are shouting at him: BE MERCIFUL, YOU WHO LOOK ON ME, BE PIOUS AND GRATEFUL TO GOD!"[14]

This and the following volumes of the journal contain a series of further accounts of monsters, though their number dropped slightly in comparison to the first two volumes, which were published in 1670 and 1671 by Breslau's municipal physician, Philipp Jacob Sachs von Lewenhaimb (1627–1672). Vollgnad's scholium does not mark the end of monsters in the *Ephemerides*, as the *Miscellanea Curiosa* were also known. Rather, it is an expression of the ambivalent relationship that would characterize the academy's discussion of monsters for decades.

Not only the *Miscellanea Curiosa* but also other journals, such as the *Philosophical Transactions* of the Royal Society of London, Thomas Bartholin's *Acta Medica & Philosophica*, and the *Journal des sçavans*, published articles about monsters—though not as frequently. Probably because of their focus on accounts of a medical nature or of closely related subjects, the *Miscellanea Curiosa* contained a greater than average number of reports on monsters.[15] Similarly frequent subjects were stones in human or animal bodies, and worms or snakes discovered inside the bodies of humans.[16] These subjects are in marked contrast to those of the monographs, the composition of which constituted the principal task of academy members prior to the founding of the journal. However, this was not what most troubled their contemporaries about these publications. Rather, it was the fact that they reflected a concept of authorship in the study of nature—authorship through compilation—that was increasingly contrary to the zeitgeist of the seventeenth century.

Collectores, not *Auctores*: The Original Work Program of the *Curiosi*

A letter dated 1669 to Henry Oldenburg (c. 1615–1677), the secretary of the Royal Society of London, founded in 1660, provides a good introduction to the working methods of the Academia Naturae Curiosorum and its contemporaries' estimation of them prior to the founding of the journal. Oldenburg's correspondent, René-François Sluse (1622–1674), a canon in Liège, describes his visit to the most recent Frankfurt Book Fair, where he was able to look at some of the monographs that had been written *ad normam et formam Academiae Naturae Curiosorum*, that is, according to the statutes of the academy. He was not impressed by them: "At the last Fair there appeared three or four little books by authors calling themselves members of the Investigators' Academy, dealing with wormwood, hematite, the bluestone, chrysocolla, and suchlike, in which you will discover hardly anything except the opinions of ancient authors on these matters, collected together with great care and diligence."[17] He describes the activity of the academy members using the verb *colligere*, to collect or to glean. The term *auctor*, originator, is reserved for the ancient authors they consulted.

The comments of a second correspondent of Oldenburg, the Hamburg physician Matthias Paisen (1643–1670), with whom Oldenburg was in regular contact by letter, were strikingly similar. In return for a package of books lost in transit, he offers books by the members of the Academia Naturae Curiosorum, who, he writes, called themselves the curious but differed sharply from the Royal Society in their studies and theories and were more fittingly classed as collectors (*collectores*) than originators (*auctores*): "I shall see what happened about the books. Should they be lost, I should be no less obliged to you and will take care to repay you for your books by sending others published by the German society, called 'The Society of Investigators,' although they depart widely from the investigations and theories of the English and are rather to be called compilers than authors."[18] Such assessments cannot have contributed to a positive estimation on Oldenburg's part, either—the motto of the Royal Society, *Nullius in verba*, expressing as it did a mistrust of the written record and the desire to supersede it.[19] Paisen had come up with an appropriate exchange for a lost package of books.

Gottfried Wilhelm Leibniz (1646–1714), like Oldenburg a central figure in the European academy movement—but unlike Oldenburg, a mediating authority[20] of the relationship between scholarly knowledge and the "new" knowledge of the so-called Scientific Revolution—appeared unimpressed. Like Sluse and Paisen, he criticized the academy members' monographs in his second plan for a German society (1671); the journal, established a year previously, was more in keeping with his ideas, being founded in experience:

> The Collegium Medicorum Naturæ curiosorum was thus formed that each member should take on a certain materiam Physico-Medicam and elaborate on it; but this institutum, although in itself worthy and not to be disdained, is yet not truly satisfactory, as it merely offers things already available in other books, not those discovered through personal experience. Thus not only has the outside world made little of this Collegio, but there has also been nothing of note occurring within it, until now, since the institutum began less than a year ago to publish from time to time a number of new observationes Medicas, to follow at least somewhat in the footsteps of the English Transactionibus philosophicis, the French Journal des Sçavans, and the Italian Giornale de letterati.[21]

In short, prior to the founding of its journal, the Academia Naturae Curiosorum had not quite caught the naturalist zeitgeist, as far as it can be determined from the academies of the time such as the Royal Society or key figures in the European academy movement such as Oldenburg and Leibniz. This was particularly true in relation to the evolving understanding of the naturalist author as an *originator* of discourse.[22]

As Jacques Roger has described the "new scientific mentality," which the *curiosi* lagged behind: "What characterized the beginning of another age, around 1670, was the broad and rapid diffusion of the new intellectual values. Rejection of the authority of the ancients, scorn for book-learning, and the search for evidence in reasoning and in the certainty of facts: these were the cardinal virtues indispensable to modern scientists, not only in the eyes of the scientists themselves but in those, as well, of an ever-broader public newly enthralled by the young science of its time."[23] An important correlate of this shift on an institutional level was the emergence of learned societies, which at first modeled themselves primarily on the Italian academies.[24] Within the history of science, the histories of the Royal Society in London and the Académie Royale des Sciences in Paris, both founded in the 1660s, and the Akademie der Wissenschaften in Berlin, all are well known. The Academia Naturae Curiosorum, however, was long neglected. This is surprising inasmuch as it was not only the first academy north of the Alps[25] but also, as of 1687, the oldest and only academy with the imperial seal (benefitting from a plethora of rights[26] granted by Leopold I that were unparalleled at any similar institution in the Holy Roman Empire) and as such, is deserving of an important place in the history of science.[27]

The contrast between the "new science" of the so-called Scientific Revolution and older, scholarly knowledge is relativized nowadays on the basis, among other things, of the use of identical or comparable practices of reading and organizing knowledge. One need only think of commonplacing, that is, the ordering of knowledge by *topoi* or *loci*, for example.[28] But in terms of content, too, new didn't necessarily mean new. We would do well, therefore, to take the frequent emphasis on the new by the protagonists of the "new science" with a grain of salt.[29]

Nevertheless, there is no getting around the strong presence of this rhetoric in early modern sources or its positive connotations from, at the latest, the second half of the seventeenth century.[30] And in relation to the referencing structures of the texts, in particular, it was more than mere rhetoric. Helmut Zedelmaier is right to point out that it corresponded to a change in the way that naturalists located themselves, a change in the way that self and time were understood:

> Yet the idea of progress . . . marks a change in attitude. In the search for the perfection of knowledge, the gaze is directed into an open future, not into a past of scholarly knowledge. In *Of the Proficience and Advancement of Learning* (1605), that influential document of empirical science, Bacon prescribes such an openness to future knowledge for knowledge systems, too. "Desiderata," or gaps in the

knowledge system, demand particular attention. The new system of knowledge serves as a program and plan for the advancement of the sciences. Against the "non plus ultra" that, according to ancient lore, was engraved into the Pillars of Hercules, Bacon sets the "plus ultra" that he has printed as a motto on the front page of his "Instauratio magna" (1620).[31]

So what did the academy members' monographs, berated by many contemporaries from the viewpoint of a new concept of authorship in the study of nature, and of the ideal of the "plus ultra," actually look like? What were their underlying aims? As we shall see, they correspond to the second type of *histoire* identified by Perrault, alongside *mémoires*, to the form of *histoire* that gathers all historical knowledge on a given subject. However, as Perrault's preference for *mémoires* shows, the prestige of such comprehensive *summae* in the study of nature had meanwhile greatly diminished.

All Four Corners of the World Explored from the Reading Desk

The four Schweinfurt physicians who founded the Academia Naturae Curiosorum as a private academy on January 1/11, 1652—Johann Laurentius Bausch, who received his doctorate in 1630 in Altdorf and was Schweinfurt's municipal physician from 1636; Johann Michael Fehr (1610–1688), who earned his doctorate in Padua in 1641; Georg Balthasar Wolfahrt (1607–1674), awarded his doctorate at the University of Basel in 1634; and Georg Balthasar Metzger (1623–1687), who received his doctorate in 1650, also in Basel—can be classed as "latecomers to Italian Renaissance humanism."[32] The group comprised all the academically educated physicians living in the free imperial city of Schweinfurt at the time. Not least because of their *peregrinatio academica*, their Grand Tour, they were well acquainted with Italy's intellectual landscape and took inspiration from the academies that, beginning in the fifteenth century, had been founded in almost all Italian cities for the establishment of their own.[33]

In the case of Bausch, in particular, who was regarded even by contemporaries as the *spiritus rector* of the academy's founding,[34] it is worth taking a closer look at the possible impact of this academic background on his own working methods and epistemology. Johann Laurentius Bausch initially attended the Latin school in his Protestant hometown of Schweinfurt. Then, from 1615 to 1621, he attended the Thuringian state school in Schleusingen that, since 1583, had belonged to the Lutheran Ernestine Electorate of Saxony. His school education was followed by private tuition, and, in 1623, Johann Laurentius began his studies in medicine in Jena, in those days considered a center of Lutheran orthodoxy. Like his father before him, he set off in 1628 on his *peregrinatio academica* to Italy and, in particular, to Padua, where he was to study for almost a year. His journey then continued all the way to Naples. The University of Padua was open to both Protestant and Catholic students; it was, as Laetitia Boehm rightly points out, an "interdenominational attraction for physicians from all corners of Europe."[35] Bausch completed his studies at the University of Altdorf, which was founded by the Protestant imperial city of Nuremberg, and it was here he was awarded his doctorate in 1630, in the middle of the Thirty Years' War. In the same year, he entered practice in Schweinfurt where, in 1636, he was eventually to succeed his father, Leonhard (1574–1636), in the office of municipal physician.

Bausch's passion for collecting, to which a funeral oration published in the year of his death testifies, was awakened by his visits to art and natural history cabinets in Italy. While still on his journey, he began to lay the foundations for his own *museum* "of all kinds of beautiful old coins, artistic *Naturalibus*, *artificialibus* and *curiosis exoticis*."[36] During (and as a result of) his *peregrinatio academica*, then, Bausch was already becoming interested in individual objects of nature and art. All in all, one might, like Laetitia Boehm, describe this activity as collecting: "He used the journey to study in Padua as a journey of general educational value, on which to gather information and to observe and collect antiques, rarities, and curiosities, and, not least, for the sake of social contacts and 'famous people.'"[37]

Did Bausch's passion for collecting objects also imply an appreciation of publications on preternatural phenomena or of the *curationes* and *observationes* that were gaining significance during that same period? In the sixteenth and seventeenth centuries, these genres devoted themselves with great enthusiasm to the rare diseases and extraordinary disorders of the human body—and thus also to monstrous births. A glance at the well-documented library of the Schweinfurt municipal physician[38] helps answer this question.

For Johann Laurentius Bausch, collecting books was also "connected with a zeal for creating collections of rare or curious objects."[39] Johann Laurentius did not have to start from scratch. His father had begun as a student to lay the foundations for the scholarly library that his eldest child was to continue. As shown by the owner's markings, over a third of the surviving collection dates back to Leonhard. These works include the monographs described in the founding documents of the Academia Naturae Curiosorum as exemplary for the encyclopedia of remedies to which the academy aspired.[40] Leonhard's books were influential, too, then, in establishing the cornerstones of Johann Laurentius's scholarly medical worldview—even if, admittedly, the possession of a book is no guarantee that its owner has actually read it.[41] In any case, a consequence of Johann Laurentius Bausch's medical worldview was that the original work program of the academy put its focus on a genre that soon after its foundation was already seen as outdated by many actors in the European academy movement.

In 1643, Johann Laurentius added an instruction to his will—in this respect, too, he was entirely his father's son—that made early provision for the continued existence of the library, which thereafter, from the sixteenth to the early eighteenth century, was to remain the property of Schweinfurt's municipal physicians. In 1813 it was donated in trust to Schweinfurt municipal library. Today, parts of the collection have been lost, while others have become separated and scattered. An impressive total of 2,363 volumes with 6,265 titles have been cataloged.[42] Initial viewings show the Bausch library to be a "prime example of the library of a German Renaissance humanist scholar in the confessional age."[43]

As can be seen from the documented collection, the two Bausches undoubtedly took an interest in movements that—in terms of the relationship between ancient authorities and empiricism, philosophy, and medicine—were groundbreaking at the time, even if many of the now firmly established heroes of the so-called Scientific Revolution are not represented in the library.[44] One example is Paduan Aristotelianism.[45] Both Bausches visited Padua during their *peregrinatio academica*. We know that Johann Laurentius was impressed there by the natural philosopher Cesare Cremonini,[46] who, along with Jacopo Zabarella, was one of the most influential thinkers in this form of Aristotelianism.[47] There is, nevertheless, only one text by Cremonini in the Bausch library, and it is not a treatise but a theater play: *Le pompe funebri* from 1599.[48] Neither Zabarella nor Cremonini's pupil Ulisse Aldrovandi feature at all. Fortunio Liceti, however, a late proponent of the Aristotelian Scuola di Padova, does at least have one text in the Bausch library—the second edition of his *De Ortu Animae Humanae* from 1606;[49] there is no sign of his monograph on monsters.

All in all, the Bausches' interest in monsters seems to have been limited. Only two of the monographs published on the subject around 1600 are documented in their library: Gaspard Bauhin's monograph on hermaphrodites and other monsters (1614) and Johann Georg Schenck's *Monstrorum Historia* (1609).[50] This suggests that they were aware of the discourse but certainly does not imply extensive knowledge of it.

The choice of these two specific texts fits the intellectual, social, and geographical profile of the municipal physicians Leonhard and Johann Laurentius Bausch inasmuch as they are primarily historical, collated texts and thus comparatively close to medical practice (rather than theory). Given that Johann Laurentius Bausch was also active as a city chronicler,[51] it is surprising that *Prodigiorum ac Ostentorum Chronicon*, by Conrad Lycosthenes, is missing from the collection. However, this does fit into the picture of a limited interest in preternatural events and phenomena that were seen by some to bear divine messages.

When it comes to the genres of *curationes* and *observationes*, which were to serve as a model for the format of the future journal of the Academia Naturae Curiosorum, the situation is quite different. The Bausch library contains most of the volumes featuring *curationes* and *observationes* published between 1551 and Johann Laurentius Bausch's death in 1665: thirty-eight from a total of around fifty-five.[52] Some of these texts are marked as belonging to Johann Laurentius Bausch, though this does not tell us who originally purchased them.[53] The library's contents in relation to this genre cannot, however, all stem from Leonhard Bausch, as they include volumes that were published after his death (but before that of his son Johann Laurentius).[54] By the time the idea of founding the academy had formed in Johann Laurentius Bausch's mind, the municipal physician's library already contained a comprehensive collection of monographs of this kind. And Johann Laurentius Bausch had a personal interest in *curationes* and *observationes*.

Nevertheless, he and his associates initially decided against this genre of medical publishing for the work program of their academy, though it had, by that time, matured and gained increasing recognition. One reason for this may have been that there was still no guiding model to successfully use the genre by an academy working collectively. Thomas Bartholin's *Acta Medica & Philosophica Hafniensia* was an influential journal that—though not based on the work of a collective—featured articles in the *observationes* format, but it appeared too late to influence the early phase of the Academia Naturae Curiosorum. It was first published in 1673 and very well received by the *curiosi*, as attested by the many cross-references in the *Miscellanea Curiosa*.

The contradictory nature of the academy's work program, as it was originally envisioned, reflects the above positioning of the actors within the history of ideas. It was oriented less toward the increasingly influential genre of medical *curationes* and *observationes* than toward the work of the Italian academies with which the founders were acquainted. It was first set out in the *leges* of the academy—which were first agreed in 1652 and were to undergo several changes even before the end of the seventeenth century[55]—and in the *epistola invitatoria* initiated by Bausch for the purpose of recruiting new members. Monographs written by members at an ambitious rate of two per year, each on one subject of *materia medica*, were to come together to form a systematic description of all subjects of value to medicine.[56]

Originally, as can be concluded from the initial letters of the four subjects the founders chose to work on, it was intended to be an alphabetical series. However, alphabetical order was soon abandoned in favor of personal preference when it came to the choice of topic.[57] The 1662 version of the *leges* also features a second concession to the members introduced since the 1652 version: the topics were no longer allocated by the president of the academy—initially Bausch—but members were allowed instead to choose for themselves.[58] The basic aim of compiling an encyclopedia of remedies,[59] however, remained unchanged.

In view of this aim, it is unsurprising that preternatural phenomena such as monsters play a lesser role in these publications. The total of fifty-three papers known to date to have been published *ad normam et formam Academiae Naturae Curiosorum* by the mid-eighteenth century deal, in line with the work program, with routine occurrences in nature—and not phenomena created *praeter naturam*, which is to say outside the habitual course of nature.[60]

What did these volumes look like? What referencing systems do they use, and what are the sources of knowledge upon which they are based? Like the publishing project as a whole, the individual volumes were—in relation to their respective subjects—encyclopedic in design. The *leges* stipulate their components as follows: "The academy member should handle each topic thoroughly and with the greatest possible care, studying the names, synonyms, origin, occurrence, distinctions, selection [of the subject to be examined], as well as the powers of the whole and of its parts, and the common and chemical medicines, simple and composite, to be prepared from it."[61] Of particular interest here are the knowledge sources used by the authors in the production of these monographs. The *leges* of the academy contain the following stipulation in this regard: "For this purpose, he [the author] will draw on reliable authors, his own observations, credible reports and the perceptions of others: in doing so, he shall not conceal the names of these contributors, but accord them an honorable mention."[62] The author's own observations (*observationes*) and the reports (*relationes*) and perceptions (*animadversiones*) of others, that is to say, personal impressions and material gleaned from literature, were to be incorporated into the texts without distinction. This is a form of empiricism similar to the one we have already encountered in the works of authors such as Fortunio Liceti and Ulisse Aldrovandi.

The texts created in this way are compilations, a fact that was not denied by those involved. In this context, a statement by Sachs von Lewenhaimb, whose 1661 treatise on wine, *Ampelographia*, was—as it says on the front page—the first monograph to be printed *ad normam Collegii Naturae Curiosorum*, is revealing. Sending his manuscript to Schweinfurt on May 15, 1660, as the statutes required, he describes his method as follows: "I roamed through all [four] corners of the world by poring over a great number of writings, naturally from botanists, philologists, historians, physicians, and chemists. I planted my vineyard with vines from every corner, i.e. from books by Italian, Spanish, French, German, Dutch authors, which I had gathered with great diligence and arranged in order, according to what was possible; the sympathetic reader will judge for himself the effort expended on finding a great many attestations by authors, quoted directly from the source, with page and verse duly noted."[63] There is no mention of the author's own observations. Sachs's method was, for the most part, limited to *reading in order to collect and compare* and *writing down in order*; its aim was to make the knowledge gained by such reading as comprehensively accessible as possible. And, indeed, the little octavo volume offers easy access to the inherited knowledge available to Sachs on the subject of wine and its medical applications. This is due not least to its elegant structure of chapters and subchapters and the division of the body of the text into sections. The italicized short titles of the literature used, the alphabetical index of authors at the end of the book that helps to categorize these short titles, and the printed marginalia that briefly describe the adjacent text, all serve to facilitate the *inventio* of the reader.

Thus, in the eyes of a contemporary critic like Matthias Paisen, Sachs's short volume on wine proved him to be an exemplary *collector*, but not an *auctor*. He is not so much the originator as the compiler of the knowledge encountered by the reader in *Ampelographia*. In this respect, the understanding of authorship that underlies this and the other volumes produced *ad normam et formam Academiae Naturae Curiosorum* comes very close to what we have seen, for example, in Aldrovandi's work. Which is not to say that compilation in either Aldrovandi or Sachs resulted in mere patchworks of quotations from earlier authors devoid of new insights. As Anthony Grafton has rightly argued, compilation was much more complex than this.[64]

These authors did not prove their expertise by being the first to address a particular matter or even by using their *testimonium* to vouch for the veracity of a description of that matter. Rather, they demonstrated their skill in evaluating vast numbers of older texts, discovering the relevant factoids within them, then extracting and arranging them in their own texts in a way useful for the reader. Sachs was rightly proud of the effort he went to in compiling his factoids from the texts of numerous authors, but not of their originality.

With regard to Sachs's understanding of authorship, it is revealing that, in the above-cited letter, he describes his reading using the metaphor of roaming the four corners of the world. He saw no contradiction between his own experience and the factoids he gathered through reading: rather, their boundaries were fluid. In fact, the reading *was itself* a form of empiricism. Neither is Sachs's working method inconsistent in this respect with the knowledge sources that the members were to employ in the writing of monographs in accordance with the *leges* of the academy. Here, too, the authors' own observations, reports, and the perceptions of others coexist harmoniously.

Not only in relation to the knowledge sources he used but also in terms of his understanding of experience Sachs still very much conformed to the model of authorship that I have delineated for the writers of naturalist texts around 1600. The gathering of factoids from a wide variety of sources—from the author's own observations to what we now consider to be fictional ancient texts—was quite typical of this model. The writers of these texts saw themselves as authors precisely *because* they gathered factoids in this way, which gave them a share in the authority of the *auctores*. Beginning in the early seventeenth century, however, a different model of authorship increasingly asserted itself in the study of nature that required authors themselves to be the originator of their own discourse. The compiling of factoids from others rapidly lost its prestige.

Ampelographia's Sachs von Lewenhaimb is radically different from the authors we typically encounter in the *Miscellanea Curiosa*, which he was instrumental in founding.[65] In the spirit of the contemporary terminology introduced above, they were to be *auctores* in this new sense and no longer merely *collectores*. Before we can begin to analyze the content and form of the journal, we must first take a look at two contexts that were central to the reform of the academy's work program. The key outcome of these reform efforts was the founding of the journal of the academy.

Context I: The *Miscellanea Curiosa* and the Academy's "Belated Baconian Pedigree"

The Academia Naturae Curiosorum gained international standing only as a result of its journal, the *Miscellanea Curiosa*, published between 1670 and 1706, and again, after a six-year break, under various titles from 1712 on.[66] It is regarded as the first German scholarly journal[67] and the world's first specialist journal,[68] for although it was founded on the model of the *Philosophical Transactions* and the *Journal des sçavans*, its thematic focus was substantially narrower. The *curiosi* sought submissions primarily from medicine and related areas of knowledge.[69]

As Herbert Jaumann rightly points out, both the contents of the journal and the form of reporting reveal the increasing divergence between knowledge concerning the natural world and philological historical knowledge: "The horizon of the reporting and of the topics covered was . . . purely scientific, the older model of the *medicus philologus* is no longer in evidence, and the journal makes clear how far the study of nature, which in this case is closely connected to medicine, could even at this time be separated from the word-centered philological-historical fields."[70] The subject matter restrictions placed on submissions, along with the fact that the journal was not a regular publication, may have helped establish the false view that the *Acta Eruditorum*, published in Leipzig between 1682 and 1782, was the first scholarly journal in the German-speaking world.[71]

What was the relationship between the journal of the Academia Naturae Curiosorum and the monographs produced by academy members? The *Miscellanea Curiosa* were innovative, both in terms of content and epistemology.[72] In contrast to the monographs, they are composed of contributions by named authors usually based on their own observations—observer and author are, for the most part, one and the same person. The contents of the articles are wide ranging, though there is a clear prevalence of rare and astonishing subject matter. The format of these contributions is that of the *observatio*.

The new form of naturalist experience, in the creation of which the Academia Naturae Curiosorum was now involved alongside those academies founded shortly after it—the Royal Society of London and the Académie Royale des Sciences in Paris—represented a fundamental departure from Aristotelian experience, which had centered on what *mostly* occurred.[73] Instead, the new focus was on rare phenomena and individual objects.

As we have seen, the discernable preoccupation of the academies with the particular, and especially with monsters and other natural phenomena defined as rare or preternatural, was not new at all at this point.[74] What was new was the distillation of this attention into a program that was put into practice collectively and with great effort, and whose primary point of reference was the natural philosopher Francis Bacon. His program for the reform of natural history and natural philosophy, developed between 1605 and 1620, marks an important turning point in the study of preternatural phenomena in natural philosophy. Unlike many other natural philosophers engaged in the study of these phenomena, Bacon did not take the categories of conventional, scholastic natural philosophy—which would have implied a definition of extraordinary phenomena as exceptions to the rule—for granted. Rather, he believed that collecting carefully examined and described pre-ternatural phenomena, according to fixed rules, would help to shatter Aristotelian natural philosophy, thus making room for a reformed natural philosophy.[75]

The natural history of the preternatural gained significantly in status thanks to this program. Preternatural phenomena now had two central functions. On the one hand, they were to be turned against existing natural history, giving them a destructive function. As Bacon rejected the view of preternatural phenomena as exceptions to the general rules of nature, they could also be assigned a second, constructive role: the *universal* causes that underpinned both the ordinary and the extraordinary, that it was the task of natural philosophy to seek, lay hidden within them.[76]

In keeping with this dual function, the relationship between natural history and natural philosophy also had to change—in favor of greater autonomy for natural history, and, therefore, with significant consequences for the relationship between observation and its embedding within a line of argument: Bacon insisted on *brief* descriptions of the natural wonders observed. Furthermore, to prevent the falsi-fication of natural history by natural philosophy, he warned against premature theorizing.[77]

Bacon's program had an enormous influence on the "facts" or "truths" of natural philosophy in the late seventeenth century. Academic journals and correspon-dence between naturalists were full of accounts of rare phenomena. However, his reform project was not implemented fully by any of the academies, and the con-structive task it had ascribed to the natural history of the preternatural remained largely unfulfilled. Although the "facts" of the academies—both the Royal Society and the Leopoldina—were largely free from embedding within argument as well, just as Bacon demanded, they became an epistemological end unto themselves and, contrary to his plans, were not incorporated into a methodical search for universal causes.[78] As we shall see later, the *Miscellanea Curiosa*'s reports on monsters largely reflected this pattern. The ordering of the articles in their volumes according to a numbered list is also reminiscent of Bacon, but more on that later.

Lorraine Daston and Katharine Park do not attribute the predominance of "strange facts" in the work of the Royal Society and the Académie Royale solely to the influence of Bacon's reform project. They argue that the absence of attempts to explain the phenomena reported—or at least the meticulous separation of these attempts from the factual reporting—also had to do with the ideals of debate adopted from Italy's private humanist academies. In addition, the requirements of collective empiricism played a role, for disputes about theories and systems risked ending in irreconcilable conflict. The preference for discussing "strange facts," however, resulted in the academies allowing little room for innovation in the theoretical field.[79]

This explanation, although developed in relation to the Royal Society and the Académie Royale, also applies in the case of the Academia Naturae Curiosorum. Humanist educational ideals and the model of the Italian academies were the inspiration for its founding, too.[80] Furthermore, collective empiricism and ideals of debate intended to help avoid conflict had also had a formative influence on the founders of the Schweinfurt academy. An illuminating document in this respect is the letter of invitation Bausch wrote in the academy's founding year, calling on physicians from elsewhere to participate in the academy. It cites the academy's aim as the promotion of medicine and therefore, indirectly, also of the common good.[81] The necessity of collective research is explained by the inexhaustibility of nature itself: "Since God is not only to be admired in his works, but his works in the natural realms of vegetable, animal, and mineral are innumerable [*innumera*], to the point that the lifespan of an individual man is not long enough and sufficient to explore them all with curiosity and to know them precisely, this deficiency, which consists in the short life of a man and in his insufficiency to perceive the infinity of natural things, may be compensated for by many people collaborating collegially on a common work."[82] Collective work requires a framework. Consequently, the fourteen *leges* of the academy that accompany the letter of invitation contain a stipulation on the monographs to be written by members of the academy regulating the form of debate that is strongly reminiscent of corresponding stipulations made by the Royal Society and the Académie Royale: "All this is to be done in a friendly and brotherly manner, without contempt, abuse, envy, or presumption toward others."[83]

In contrast to the case of the Royal Society, Bacon was not the inspiration behind the Academia Naturae Curiosorum at the time of its founding.[84] But he was to gain that title belatedly, so to speak. Shortly after the founding of the journal in 1670, something becomes tangible in the self-presentation of the academy that has been aptly described as the "belated Baconian pedigree" of the Academia Naturae Curiosorum.[85] In fact, even from the early 1660s, correspondence between academy members contains the first indications of efforts to reform the academy in line with the model of the Royal Society. And early correspondence between the two institutions reveals from the beginning, which is to say from around the mid-1660s, the high regard in which the *curiosi* held their English sister academy.[86]

Some of the Academia Naturae Curiosorum founders' core beliefs must have coincided with Bacon's from the outset, however, considering their intellectual background as academically educated physicians, several of whom had received a significant part of their medical training at the medical faculty in Padua: "Long before Bacon formulated his vision for a reformed natural philosophy that took as its basis the ordinary and extraordinary individual matters of natural history in order to discover nature's hidden 'form' or 'laws,' physicians had searched passionately for the hidden traits of rare, extraordinary, and exotic *naturalia* in particular."[87] This influence is significantly more tangible in the academy's journal than in the monographs. It was joined by a second development within the field of medicine itself; namely, the rise of the *historia* in general and of the *observatio* in particular.

Context II: The *Miscellanea Curiosa* and the Rise of *Historia* and *Observatio*

In terms of the format and sources of knowledge used in the articles in the *Miscellanea Curiosa*, the pre-Baconian rise of the factual account, the *historia*, in medicine was of particular significance. This took place—as Arno Seifert showed in his seminal work *Cognitio Historica*—in the context of a comprehensive increase in the importance of the *historia* across the boundaries of academic fields.[88] More recent investigations into the epistemological status of the *historia* in the medical literature of the early modern period have shown that—from a position of marginal importance in scholastic medicine—in the course of the sixteenth century it became a central concept in medical literature. In the same period, *historia* began to appear in the title of numerous medical texts. In short, both in epistemological terms and as a genre, the *historia* experienced an enormous growth in importance in medicine in the sixteenth and seventeenth centuries.[89]

In the context of this study, the epistemological growth in importance of the *historia* in the field of anatomy is of particular significance. In the second half of the sixteenth century, a number of anatomists linked the Aristotelian concept, which had previously had no influence in anatomy, with the Galenic concept of autopsy: *historia* as *autopsia*. Hieronymus Fabricius (c. 1533–1619), for example, who taught "philosophical anatomy" in Padua between 1565 and 1613, lamented that most of his colleagues, Vesalius included, went no further than the *historia*. He viewed it as a preliminary step to examining the *actio* and—as the culmination of the examination—the *usus* of a body part.

Although his most famous pupil, William Harvey, initially accepted his teacher's several-stage model, he later conceded that the dependable *historia*—which he, too, understood as *autopsia*—had a value in itself and the character of knowledge. In the second half of the seventeenth century not only was there, after all, an interested readership for anatomical *historiae*; they were potentially of greater interest even than more extensive anatomical explanations.[90]

The rise of the *historia* was even more noticeable in texts from fields closer to medical practice than it was in anatomical publications.[91] The roots of this development go back a long way. But it was not until the Renaissance that a *curatio* or *observatio* reporting a medical case or medical history gradually took on a value in itself. Gianna Pomata sees the Galenic *historiae* or *curationes* as a model for the early modern case history, which—like these two genres—typically reports successful treatment. Yet she does not deny the great influence of Hippocratic case histories on the text genre of *observationes* that emerged in the sixteenth century, either.[92] If one compares the *curatio* or *observatio* with the older format of the *consilium*, one is struck by the similarity of this genre to Bacon's "matters of fact" described above, and the "facts" of the academies in the second half of the seventeenth century:

> Both the early modern *curatio* and *observatio* as the medieval *consilium* focused on an individual case, but the difference between them is clear: whereas in the *consilium* diagnostic and therapeutic reasoning was tightly anchored to a doctrinal framework, *curatio* and *observatio* give pride of place instead to the precise and detailed description of the single case, abstaining from interpreting it in the light of doctrine (or, if that is done, it is done in a separate part of the text, under the rubric of *scholion*). What is of interest now is precisely the particularity and singularity of each case. The focus on singularity explains the selective attention for whatever is rare, exceptional and unheard-of, that is a constant feature of the late-16th century collections of medical *observationes* and *historiae*. A case seems noteworthy precisely when its singularity is so extreme that it challenges classification: the rare and odd is the epitome of individuality. An instance of the unforeseeable, the odd case proves the complexity and mysteriousness of nature, which defies scholastic systems.[93]

Thus, even before Bacon's reform program or the demands of collective empiricism suggested such formats for the annals of the academies in the second half of the seventeenth century, the *curatio* or *observatio* provided physicians with their own genre for reporting individual observations of rare cases, while excluding theoretical considerations.[94] The Schweinfurt founders would draw on this medical tradition for the *Miscellanea Curiosa*.

The *curiosi* chose the *observatio* as the format for contributions to the *Ephemerides*. With the *epistola invitatoria*, an invitation to all physicians to submit articles for the new journal that was printed in the first volume in 1670, they secured for their journal project a place in the history of the genre. They did this, among other things, by presenting a list of role models that constituted a kind of Who's Who of authors who had so far appeared as collectors of *observationes*.[95]

The fact that the *curiosi* chose the *observatio* format had far-reaching consequences for the working method demanded of contributors. They were required to observe and to separate their own observations meticulously from what they had taken from literature. Like Bacon's "fact," the early modern *observatio* gave much space to the precise and detailed description of a specific case and stopped short of interpreting it in light of a theory. The *Miscellanea Curiosa* used a separate section, the scholium, as a space for anything apropos the case or phenomenon under discussion that had no place in the *observatio* itself.

The separation of the narrative of the case in question from commentary had long been the norm in *observationes*. The founder of the genre, the Jewish physician Amatus Lusitanus, had introduced it in his *Centuriae Curationum*, published between 1551 and 1566, and thus extended to contemporary writers—namely, to himself—the commentary that had, until that point, been customary only for canonical authors: at the top he placed his observation and below it the commentary on it. In his casebook, Lusitanus separated the case history itself, the *curatio*, from its respective scholarly commentary, the *scholium*, and highlighted this distinction typographically: the case history was set in nonitalic roman type and the commentary in italics. Lusitanus's work was read and quoted for centuries; its separation of case-related narrative and commentary was paradigmatic for the genre of *curationes* and *observationes*.[96]

The fact that the individual articles in the *Miscellanea Curiosa* bear the title *observatio* is an important indication, then, of the practices on which they were to be based. The focus was on direct experience, though reading continued to play a significant role, not least in the scholia. As Gianna Pomata highlights, the empirical connotation that the term *observatio* still carries today is by no means deceptive. "In early modern medical language, *observatio* was indeed 'vox empyricorum,' as it is defined in the medical lexicons—the Latin rendition of the Greek *paratéresis* used by the ancient empiricists."[97] However, the *curiosi* were no longer interested in experience in a general, superhistorical sense, but primarily in the experience of their contemporaries.

Collecting and Preserving the Experience of Contemporaries

From the late sixteenth century, academically educated physicians felt the urge to publish their *curationes* and *observationes*. Colleagues were pressed to make available their observations, which they were increasingly documenting by hand, and the prospect of the publication of *observationes* previously only available in manuscript form was eagerly awaited.[98]

The fact that the Academia Naturae Curiosorum had a part in this development, and that its members shared the view of many other physicians that it was important to make the practical experience of each individual accessible to others, was tangible even before the founding of the *Miscellanea Curiosa*. As early as 1668, academy member Georg Hieronymus Welsch (1624–1677), a practicing physician from Augsburg, published his own *observationes* along with those of five colleagues whose notes he had purchased.[99] He revealed the plan for this oeuvre, which he referred to as a "thesaurus of medical experience," in a letter to Johann Michael Fehr, one of the four founders of the academy; on its publication, he sent Fehr an advance copy.[100]

The *epistola invitatoria* printed in the first volume of the *Miscellanea Curiosa*, which cites Welsch as an inspiration, also expresses the same high regard for the practical experience of physicians, and consequently the concern that this wealth of experience might be lost without timely publication. This danger is conveyed using the image of the proverbial prophecies of the ancient oracle Sibyl, whose leaves threaten to be scattered to the winds. If they were merely recorded individually, the physicians' observations might easily be lost. Collected in the volumes of the journal, it was hoped that they had a better chance of standing the test of time.

> Thus, unjust fate erases the writings and *observationes* of many physicians of the ancient [*veteris*] time, as well as of ours, the more salutary *observationes* of which the brilliant Georg Hieronymus Welsch has preceded by the sending of the centuries of his *observationes* in a comprehensive catalogue of profound scholarship and medical knowledge . . . How many learned men . . . would be greatly pleased to communicate discoveries [*inventa*] and observations [*observata*] to subsequent generations or to scholars elsewhere; but by what means? There is neither enough free time available for the compiling of whole treatises, nor does the business of the profession advise it: If these could be gathered into the space of a few pages & would bear distribution, how quickly would the Sibylline leaves become prey to wind and waste. Inspired by these weighty considerations, the *Society of the*

Curious, requesting the help and efforts of all physicians and natural philosophers, has offered annually to collect and compile *Ephemerides Medicas* and finally to distribute them in print.[101]

The practical experience of contemporary physicians was thus "precarious knowledge"[102] in a double sense: collected mostly in manuscript form, it was in great danger of being forever lost to humanity once these physicians passed away. And relatedly, its epistemic status was still relatively low outside of the circles to which the *epistola invitatoria* bows and of which the *curiosi* wanted to become an integral part.

A little further down, the *curiosi* stress how central the experience of the contributors was to the success of the venture. This passage constitutes a fourfold performative speech act that uses rhetoric to endow the experience of the contemporaries with authority: "*Your* [*Vestra*] thoroughness in the examination of hidden things; *your* [*Vestra*] profound knowledge [*scientia*] & reliable experience [*experientia*] in driving out disease, *your* [*Vestra*] very reliable reason and practiced handling of the discovery of new remedies, *your* [*Vestra*] very credible reporting, only *these alone* will be the glory, the ornament, the substance, & the complement of our *Ephemerides*."[103] The fact that the *curiosi* so strongly emphasize the value of their contemporaries' experience must be read against the backdrop of the gradual extension—discussed in chapter 1—of the *auctoritas* of the ancient authorities of the study of nature to include early modern writers. It was not yet a given that the experience of contemporaries was to be awarded the same value as that of the "ancients." Its importance had to be specifically emphasized and, by doing so, established in the first place. Now that the contexts in which the journal was founded and the format chosen for its contributions have been examined, let us turn our attention to the arrangement of articles within the volumes.

Numbered Lists: On the Arrangement of the *Observationes* within the Book Space

With regard to the structuring of knowledge, the step from monograph to scholarly journal seems at first glance greater than it might in fact have been for those involved at the time. Jaumann has pointed out "how close the proximity of the scientific periodical to the usual collective publication of scholarly articles, how instable the contours of the genre"[104] still were at that time. The boundaries between the *Ephemerides* and the collections of *observationes* published in the form of monographs in the sixteenth and seventeenth centuries were fluid.

Likewise, the concrete form of the *Miscellanea Curiosa* as a numbered list of in dividual *observationes* echoes earlier, non-periodical publications. Contemporary collections of *conversationes* or *histories*, for example, often presented their short factoids as numbered lists, as did Erasmus with his *Adagia*, to the extent that we can conclude a general "tendency in the early modern period to an orderly structuring of information about the subject."[105] Thus, readers of the first edition of Peter Lauremberg's *Acerra Philologica* in 1637, for example, were confronted with two hundred German-language histories. These were expanded in later editions to six hundred (1658) and seven hundred (1688 and 1708; now published by Gotthard Heidegger), which by then were written predominantly by other authors.[106] To quote Jaumann again: "From accumulation to series is only a small step here."[107]

Curationes and *observationes*, in particular, were typically published in the form of numbered lists even before the founding of the *Ephemerides*. In fact, many of the collections of this kind published before 1670 bore the numbering in their title. This is true even of the publication that established the genre, Amatus Lusitanus's *Centuriae Curationum*.[108] Later collections of *curationes* or *observationes* were also often presented as *centuriae* or *decuriae*, including those of Georg Hieronymus Welsch.[109]

Astronomical and weather observations were also generally recorded in list form as *ephemerides*. They were kept chronologically, though this was not essential to the term *ephemerides*. In this context, it primarily meant "list."[110] The fact that the *Miscellanea Curiosa* also had the term *ephemerides* in its title, and that scholars at the time liked to use this term to refer to the journal, is a result of the ordering of its observations in list form.

Alongside weather and astronomical observations and the *observatio* and *curatio* genre, there was another important point of reference for this arrangement: the *observationes* of which Bacon speaks in the second volume of *Novum Organum* that occupied a central place in his program for the reform of natural philosophy was also intended to be kept in list form. Bacon's posthumously published *Sylva Sylvarum* conforms to this requirement. The observations and experiments collected in the volume are numbered and grouped into ten centuries.[111]

The arrangement of the *observationes* in the individual volumes of the *Miscellanea Curiosa* followed the order of their submission. This can be seen, for example, by the printing of contributions by the same author in direct succession. A description of the publishing practices of the *Ephemerides* in "Protocollum," a handwritten chronicle of the beginnings of the academy, started by the fourth president, Lucas Schröck (1646–1730), and written retrospectively, confirms this observation.[112]

Even after the founding of the *Miscellanea Curiosa*, in accordance with the revised *leges* of 1671, the production of monographs on matters of medicine was one of the duties of academy members.[113] In addition, they were now expected to contribute reports to the *Ephemerides*. This meant that academy members needed to master comparatively new cognitive practices and forms of writing that were not necessary for penning monographs.

Observatio as a Form of Observing and Writing: *Observationes* about Monsters in the Early Volumes of the *Miscellanea Curiosa*

The monographs of academy members were still based entirely on practices of reading and excerpt making—and of organizing and editing factoids gathered in this way—that were common across the fields of early modern scholarship. The founding of the academy journal meant that the *curiosi* now also had to master the *single* observation, along with the corresponding writing practices, the corresponding "epistemic genre."[114] This takes us another step closer to understanding Vollgnad's above-mentioned demand that it was necessary to observe *correctly*.

In the first instance, it was a matter of conducting and describing an *individual* observation. A single observation or experience was now separated clearly—and linguistically—from experience in the broad sense. This is demonstrated particularly well by a statement by Johann Daniel Major (1634–1693), who had been a member of the academy since 1664.

Major was born in Breslau and studied in Wittenberg, Leipzig, and Padua. In Padua he obtained a doctorate in medicine and in philosophy. After practicing briefly in Wittenberg and Hamburg, he became a professor of medicine in Kiel in 1665, where he was also responsible for the botanical gardens. In 1679, he became personal physician to the Duke of Holstein-Gottorp. Dozens of his Latin and German writings have survived, chiefly on topics related to the study of nature.[115]

Toward the end of the seventeenth century, Major defined the terms *experimentum* and *observatio* as synonymous and distinguished them from *experientia* as follows: "An *experimentum* is the *observatio* of a circumstance or individual thing, perceived with the help of the senses. *Experientia* is a more universal knowledge composed by means of the recollective induction of many *observationes*."[116] In contrast to what we have seen in the case of Johannes Schenck and Johann Georg Schenck von Grafenberg, the term *experimentum* in Major's work refers clearly to a single observation or experience that is authored in the sense that it was undertaken by a named observer. *Experientia*—which for Johannes Schenck was still synonymous with *experimentum*—is now distinguished clearly from the individual experience.

The literary counterpart of this form of individual experience, which in the *Miscellanea Curiosa* and elsewhere was usually labeled *observatio* or *historia*, can be understood, as Gianna Pomata writes, as an "epistemic genre": "Epistemic genres give a literary form to intellectual endeavour, and in so doing they shape and channel the cognitive practice of attention. Some may provide, for instance, a framework for gathering, describing and organizing the raw materials of experience (as was the case of the early modern observationes)."[117] Reports on this kind of authored single observation were what the *Miscellanea Curiosa* was intended to be about.

The fact that learned physicians were not necessarily proficient in this kind of observation and writing can be seen from the laborious way in which the *curiosi* explained to readers the type of journal that the *Ephemerides* was intended to be. Readers were also potential contributors, especially since nonmembers, too, were invited to submit their observations. The *epistola invitatoria* printed in the first volume of the journal in 1670 contains, in addition to the above-mentioned list of exemplary writers of *observationes*, the following statement on the contemporary journals that the *curiosi* took as models and the sources of knowledge on which the articles submitted were to draw: observation (*observatio*), examination (*inquisitio*), and experience (*experientia*). It is likely no coincidence that observation is mentioned first. All three sources of knowledge relate to what physicians experienced directly in their practice. And, as is the case in Major's work, a distinction is made between *observatio*, understood as a single observation, and *experientia*:

> They saw how many advantages the *Ephemerides Anglicæ*, known as Transactiones Philosophicæ, offered daily to devotees of almost every kind of art, how generously the *Gallicæ Ephemerides* made available comprehensive findings on many different things; the *curiosi* considered, therefore, that it would be neither inappropriate nor useless if the *Ephemerides Germanorum* were to be put together only from the field of medicine and its sons and kin, physics, botany, anatomy, pathology, surgery, chemistry [*Physica, Botanologia, Anatomia, Pathologia, Chirurgia, Chymica*], so that whatever the *curiosi* encountered through the gracious communications of the most excellent physicians, or through their own observations [*observationes*], in practice or through thorough examination [*inquisitione*] & experience [*experientiâ*], would be compiled by the *curiosi* in a volume of Ephemerides & published annually.[118]

If one compares this statement with the above-cited passage from the *leges* on the sources of knowledge to be used by academy members in their monographs (and with Sachs von Lewenhaimb's description of his method in drafting *Ampelographia*), it is conspicuous that knowledge gleaned from literature is no longer mentioned.

But did the authors adhere to these stipulations? Were they proficient in the required form of experience, in the cognitive and the literary senses? As mentioned above, the first volume of the *Ephemerides*, published in 1670, already featured seven articles about monsters or singular phenomena categorized as "monstrous" (*monstrosus*). For the most part, they conform to the ideas of the time as to how an *observatio* should be written.

Karl Rayger's *observatio* VII, for example, is, as the title says, a firsthand account of the dissection of a "two-headed monster." As was usual in articles about monstrous newborns, the report itself details the location of the event (his hometown), the date of the birth (January 5, 1669), and the social standing of the parents, or rather, the father's occupation (in this case, townsman and trader). First, Rayger reports on the birth of the monster, a boy with two heads, three arms, and two legs. He writes that he was called "to this spectacle" along with one other learned physician and an unnamed surgeon. At the parents' request he performed a dissection of the dead baby's body. Rayger lists what he discovered (*inveni*) in short, numbered paragraphs. There is no mention of further reflections beyond the dissection report itself, for example, on the cause of this physical malformation.[119]

There are three more *observationes* in the first volume that conform to this pattern.[120] A *monster*, or an organ classified as "monstrous," is described on the basis of firsthand examination and, where possible, dissection. Nonhistorical reflections, for example on the *causae* of these rare formations, are seldom made. There are also some contributions, however, whose authors took less literally the stipulation to submit articles based on their own inspections and to restrict themselves to what they had personally seen. Two are presented in more detail here.[121]

In the lengthy *observatio* XLVIII, Sachs von Lewenhaimb tells of an anthropomorphic beet that he was able to inspect only by means of a painting in the possession of the family of a count—the count himself not actually having seen the beet either. The etching that illustrates the beet (fig. 19) appears to be based on the painting in question.[122] The etchings in the *Ephemerides* sometimes vary noticeably in style. This fact, and the statements about the origins of the respective images that are occasionally interspersed within individual *observationes*, suggest that a draft for the etching was usually either drawn by the author personally or commissioned by him and submitted with the article. The academy then had an etching made on the basis of this draft.[123]

Sachs is not content to restrict himself to a report on this individual object;

Fig. 19. Etching from Philipp Jacob Sachs von Lewenhaimb, "Observatio XLVIII: Rapa Monstrosa Anthropomorpha." *Miscellanea Curiosa Medico-Physica Academiae Naturae Curiosorum sive Ephemeridum Medico-Physicarum Germanicarum Curiosarum* [...]. Annus Primus (1670), bound between pages 138 and 139. Forschungsbibliothek Gotha der Universität Erfurt, Med 4° 00145 (1–2).

rather, he integrates it into a general discussion of nature's tendency to bring forth imitations of the human form. In addition, he cites the examples known to him from literature on the mineral and plant kingdoms and states his position on the question of forgeries of anthropomorphic plants or plant parts. He also touches on the doctrine of signatures. Finally, he refers extensively to a publication by another member of the academy in which much information on monstrous plants could be found.[124]

Observatio CII breaks with the format of the *observatio* even more noticeably, though in a different way and for different reasons than the article on the anthropomorphic beet does. Its author, Johannes Jänisch, who at this point was still an ordinary member of the academy, had, in contrast to Sachs, seen the subject of his observation with his own eyes, and he attached great importance to a precise rendering of what he had seen. Jänisch describes a specimen of a plant called wild bugloss (*Buglossa sylvestris*) that he classifies as "monstrous" because of its unusual growth. The report on the actual observation is concise, at only nineteen lines long. However, the etching (fig. 20) testifies to the importance to Jänisch of conveying what he had seen with precision: The wild bugloss is illustrated in great detail. In addition, a scale included in the etching gives the viewer an impression of the size of the specimen.

The strategy of either producing images in their original size or specifying the ratio of the image size to the actual size of the object was employed particularly frequently in journals such as the *Ephemerides* and *Philosophical Transactions* when unusual natural phenomena such as monsters were involved. It increased the credibility of the report and reinforced the impression of authenticity.[125]

Jänisch also follows up his actual observations with general reflections. He incorporates his report into a systematic discussion of monstrosity in plants and its causes—drawing frequently on both older and more recent literature. He denies right at the outset that nature deliberately makes mistakes in producing *monstra*. Rather, he argues, obstacles force it to take a different route than the one intended.[126] Decades earlier, authors such as Johann Georg Schenck von Grafenberg and Fortunio Liceti had already argued similarly.

Unlike Sachs, Jänisch had an agenda. The wild bugloss had been offered for sale at the market (*in foro publico*), where curiosity and diverse and superstitious speculations led the common people (*Vulgi*) to take it for a prodigy.[127] With his description and the image of the plant "engraved from life," Jänisch hopes to correct this misconception.[128] In keeping with this objective, he is necessarily bound to explain the phenomenon causally and with exclusive reference to natural causes. He blames

Fig. 20. Etching from Johannes Jänisch, "Observatio CII: Buglossum Silvestre Monstrosum." *Miscellanea Curiosa Medico-Physica Academiae Naturae Curiosorum sive Ephemeridum Medico-Physicarum Germanicarum Curiosarum* [. . .]. Annus Primus (1670); bound between pages 232 and 233. Forschungsbibliothek Gotha der Universität Erfurt, Med 4° 00145 (1–2).

the very cold and snowy winter for the formation of the monstrous wild bugloss, having made a similar observation about tulips.[129]

In short, of the impressive total of 160 articles collected in the first volume of the *Ephemerides*, seven deal with monsters or objects classified as "monstrous." The broad spectrum of phenomena classified as "monstrous" shows that, even in the late seventeenth century, the term *monstrum* was still widely applied to preternatural phenomena in all three realms of nature. Only four articles confine themselves—in accordance with the conventions of the *observatio* genre—to reporting an individual observation with no discussion of the *causae* of the object observed or theoretical reflections concerning it. The situation in the second volume is very similar. Once again, monsters and phenomena categorized as "monstrous" are among the most frequently discussed objects. And again, a number of these articles go far beyond the actual observation. Could this have been one of the reasons why those responsible considered excluding such articles from the third volume of the *Miscellanea Curiosa*?

As we have seen, in his above-quoted scholium in the third volume, Vollgnad sets out at length the requirements an observer must meet to undertake a useful *observatio* of a *monstrum*. Clearly, it was also possible to observe *incorrectly*. But what exactly distinguished a successful *observatio* of a *monstrum* from an unsuccessful one? Vollgnad appended his scholium to an article in which he reports on a dead "monstrous" fetus. Thus, by examining the way Vollgnad wrote his own *observatio*, we may get a little closer to answering this question.

The short *observatio* begins with the details of the location, date, and circumstances of the birth that are usual in such reports. It took place in 1668 in the county of Oels (Olsnâ) in Silesia and lasted twenty-four hours. The mother was the wife of a cobbler. Instead of giving a detailed description of the baby's appearance, Vollgnad refers to the accompanying etching, which reproduces it more precisely, he argues, than he was able to in writing.[130] In this case, too, the etching—or the sketch on which it was based—was produced directly from life.

Vollgnad even points out to the reader elements of the picture of particular interest. He names three extraordinary anatomical features of the baby, each labeled on the etching with a small letter for easier identification: a perforation in the ear of one of the two heads, something reminiscent of a chicken claw, and seven toes on this foot. Without any further reference to a letter on the etching, he also mentions that the beginnings of a fourth leg could not be clearly depicted.[131] Indeed, the picture shows only an amorphous mass here.

The portrayal of the baby's outer appearance, undertaken with the help of the etching, is then followed by statements on its inner anatomy. The organs, including the heart, were entirely natural and normal in form, as a certain Johann Wilhelm Agricola, the municipal physician of Oels, saw (*vidit*) on opening up the corpse. In contrast to the external examination, then, Vollgnad did not conduct the dissection himself but cited a source who saw the things described with his own eyes and—as a municipal physician—could be regarded a reliable witness.

The short dissection report brings the *observatio* to an end—with no attempt made at any reflections beyond the detailed and credibly reported observation itself. Vollgnad confined himself, then, to describing and illustrating what was observed of the subject directly. Was this what was required in an *observatio*? And were the frequent transgressions of the genre's boundaries the reason for problematizing submissions concerning monsters? After all, reflections that did not relate directly to the case at hand were meant to be expressed in a scholium, not in the *observatio* itself. Roaming the four corners of the world at the reading desk was to be undertaken separately from the act of observation.

Reading Vollgnad's statements about the *right* physician for a *monstrum* against this backdrop—alongside paratexts such as the *epistola invitatoria* in the first volume of the *Ephemerides*, which explain in detail to readers and potential contributors how submissions to the journal and scholia were to be presented—one cannot avoid the impression that some writers still lacked practice in the epistemic genre of the *observatio*. In other words, the *curiosi* had to insist actively that the texts submitted conform to the rules of the genre. Gianna Pomata has rightly pointed out that epistemic genres influence the perception of naturalists—they shape and channel their perception. Conversely, to write an *observatio* that was acceptable from the viewpoint of those responsible for the *Miscellanea Curiosa*, it was necessary first to learn to observe and write in a certain way. The clear separation of the observation itself from further reflections was one of the keys to a successful *observatio*.

A second aspect is striking here: Vollgnad's experience evidently put him in a good position to assess which aspects of the baby's external form qualified as extraordinary. In turn, the municipal physician who was summoned was able to judge on the basis of his experience that the organs were normally formed. Thus, the specimen fell into the hands of the right physicians, to hark back to Vollgnad's own wording. Observation, as the *curiosi* understood it, was by no means an unconditional activity.

However, these reflections do not adequately explain the problematic relationship of the academy to articles about monsters. After all, they relate to articles on *all* subjects in the *Ephemerides*, not just those on monsters. Why, then, were submissions about monsters, in particular, perceived as problematic?

The Benefits and Joys of Monsters: A Debate Reconstructed

If we are to reconstruct the standards by which an article was accepted or rejected and evaluated, it makes sense to consult the discussions of those responsible for the journal. However, the Academia Naturae Curiosorum was not an academy based on physical presence, as the Royal Society was, for example. Its members were spread throughout the Holy Roman Empire and even beyond and were not therefore able to meet regularly to discuss such questions. Thus, there are no minutes of meetings for us to consult. The above-mentioned "Protocollum" comes closest to this kind of source; however, it makes no mention of the discussion Vollgnad referred to considering the omission of articles on monsters from the *Miscellanea Curiosa*.

The correspondence between leading academy members is more revealing. Some of the surviving letters offer an insight into the criteria applied to the submissions. Five of these letters even mention submissions about monsters explicitly. The first is dated June 18, 1674, the year after the publication of Vollgnad's scholium. In it, the cofounder and then-president of the academy, Johann Michael Fehr, shares his concerns about monsters with academy member Johann Georg Volckamer the Elder (1616–1693), who would succeed Fehr as president in 1686.[132]

Fehr's reservations are based on a concern for sales and for the reputation of the academy's journal. He writes that he does not wish to see so many etchings of monsters and other objects of little importance in the *Ephemerides*, because they only raised the cost of printing the volumes, which, in turn, was unpopular with fellow scholars—the potential readership. He advises the *Ephemerides*' *collectores*—which is to say the editors Heinrich Vollgnad and Johannes Jänisch—to bear this in mind when publishing the fourth volume.

It seems significant, too, that Fehr points out in the same letter that he had already warned the *collectores* many times to satisfy themselves as to the veracity of a thing. In this context, he also mentions a letter from Henry Oldenburg to the *collectores*, two years previously, that contained a similar warning.[133] After the death of Philipp Jacob Sachs von Lewenhaimb, Vollgnad and Jänisch took over responsibility for the *Ephemerides* and, in this capacity, continued Sachs's correspondence with Oldenburg. Tellingly, this warning was issued in the reply to Vollgnad and Jänisch's very first letter.[134]

How concerned the secretary of the Royal Society was to prevent untruthful observations from finding their way into natural history is also illustrated by the fact that, on receiving the first volume of the *Miscellanea Curiosa* from Sachs von Lewenhaimb, he responded in a similar fashion. The Royal Society had entrusted Oldenburg with the task of carefully evaluating the volume. His remarks in his reply to Sachs should be read in the context of this critical reading:

> They [the members of the Royal Society] believe that nothing can better increase the stock of true philosophy than if the learned and skillful of all nations continue to unite their ingenuity and investigations, as now they do, and bring safely into the philosophical granary their rich harvest of observations and experiments properly and honestly performed. And so your most excellent academy, famous Sir, will of its own volition be solicitous that nothing should be insinuated in this philosophical treasury of weak authenticity, or that refuses to be tested by the Lydian stone. That history of nature which we strive to make the foundation for perfecting a more solid and fruitful physics you will wish to see pure and genuine, as we do.[135]

Although both Fehr's and Oldenburg's exhortations to protect this "philosophical granary" from observations that lacked credibility were general and not linked to any specific subject, it was nevertheless clear that reports of preternatural phenomena such as monsters required particular caution. They were, after all, rare by definition and the observations made of them thus scarcely reproducible.[136]

The second letter dates from the year 1679. It, too, was written by Fehr and addressed to Volckamer. In it, Fehr criticizes a report that academy member Ehrenfried Hagendorn (1640–1692), a physician in practice in Görlitz, had offered him for the *Ephemerides*. The subject of the report was a "monstrous" ear of grain (*de spica quaedam monstrosa*). According to Fehr, a similar report had already appeared in the *Ephemerides*. Submissions were only welcome, then, if they dealt with phenomena not previously described in that form. Conversely, of course, this implies that reports on as yet unfeatured kinds of monsters or monstrosities might continue to be included in the *Ephemerides*. But Fehr also expresses fundamental doubts as to the value of such reports: he does not esteem "the monsters highly either . . . , because they bring little benefit or joy [*nutz und lust's*]."[137] Sure enough, Hagendorn's article did not appear in the subsequent issue of the *Ephemerides*.

Similarly fundamental discussions can be found in the correspondence of Fehr and Volckamer from the 1680s. Fehr, who at this point had already taken on responsibility for the publication of the journal along with Volckamer,[138] informed his colleagues on January 25, 1681, that he had received a letter about a terrible *monstrum* from a certain Moritz Hofman (1653–1727).[139] He had no objection to Hofman's description in itself—except perhaps to its excessive length—but he "thought more of other *observationibus,* which would be useful and pleasing to the *lectori."*[140] In addition, the illustrations of such monstrosities ran counter to the purpose of the academy: enough appendices and images had already been accepted to fill the volumes with them, but this practice served only to push the price of the journal up.[141]

It is not only the lack of usefulness of accounts of monsters that Fehr cites in his objections to them; he also laments their lack of entertainment value, and not for the first time. Unlike the Puritan Francis Bacon,[142] the *curiosi* were of the opinion that their observations could, indeed should, be entertaining and enjoyable. In Fehr's view, monsters were not entertaining, which was probably because they were associated more strongly than ever with revolt and war since having been instrumentalized in the Thirty Years' War. This aspect is examined below.

In view of the impressive number of images printed in the individual volumes of the journal, Fehr's price argument was not without basis: The *observationes* on monsters and "monstrous" phenomena were accompanied particularly frequently by etchings,[143] and illustrations made them particularly costly to print. This was true not only of the *Ephemerides.* In the *Philosophical Transactions* during the same period, a third of all reports of monstrous births or unusual pregnancies were accompanied by an etching.[144]

When the *curiosi* were working on the publication of the first volume of the second decury (decade of the journal), Fehr once again took the position—in a letter to Volckamer—that the number of images of monsters should be reduced. He first lists which printer had offered what price for the printing and then notes that they would be financially better off this year anyway, as fewer etchings were planned for the next issue. It would be best to cut as many as possible, he writes, "especially those of monsters (which are not of interest)."[145]

At this time, a certain fatigue with regard to this aspect of nature study became noticeable among other members of the academy, as well, as is shown by a letter from Christian Mentzel (1622–1701) to Volckamer on May 28, 1681. Mentzel was a royal councilor and personal physician to the great elector of Brandenburg. Like Fehr and Volckamer, he had published several *observationes* on monsters or "monstrous" phenomena. Now he wrote that one should not reprint the same old descriptions of monsters time and again—which conversely implied, like Fehr's similar statement, that articles about monsters that deviated significantly from those already described were still welcome.[146]

What can we learn from these letters? For financial reasons, Fehr wanted to reduce the number of etchings in the *Ephemerides* and advocated avoiding or eliminating those that were less useful or entertaining. Reports on monsters were particularly affected by this, as their authors liked them to be accompanied by etchings and submitted visual material accordingly. In addition, articles about monsters were, in Fehr's view, less pleasing per se. Furthermore, both these letters and Vollgnad's scholium tell us that because they had no immediate benefit to medicine, accounts of monsters appeared more dispensable from the outset than, for example, articles about the successful treatment of a rare disease. This problem seems to have worsened as new articles about similar monsters continued to be submitted. Let us now turn our attention to this question of numbers.

Inflationary Rarity

The reservations that were tangible in both Vollgnad's writing and the correspondence of leading *curiosi* regarding (too many) articles on monsters were not limited to just a few individuals. They were to occupy the academy even in later years. In the 1680s, this concern was expressed to potential contributors, as a remark by Michael Bernhard Valentini (1657–1729) in an *observatio* from year three of the *Ephemerides'* second decury attests.

At this point, Valentini was a young physician who had yet to achieve his doctorate and who, after a period in practice, now wished to resume his academic career. After studying in Giessen and finishing with a *licentiatus medicine*, he had first accepted a position as deputy physician with the army garrison in Philippsburg. He did not return to his place of study until 1682, where, from 1687, he held the chair in natural philosophy and later the chair of medicine. The two monsters that were the subject of his *observatio* were, in both cases, conjoined twins recently born in the state of Hesse that he had viewed in person.

Valentini's article begins with the assertion that an excessive *historia* of monsters could not make any great contribution to medical progress, since monsters had no direct connection to medicine. The academy had pointed this out in its admonitions from the year 1684 (*Monita pro Anno 1684*), he writes. Nonetheless, he wished to give a very brief account of the two monsters.[147] Valentini's seemingly contradictory behavior shows that two impulses were at work simultaneously here—a regard for *copia* and *varietas*, as well as doubt as to the benefits of multiplicity, in particular the usefulness of repeated observations of similar rare phenomena.

The two impulses did not exist independently of each other, as the *Breslauische Sammlungen* demonstrate. The founding of this nature study journal represented in itself a reaction to naturalists' long-standing privileged treatment of rare natural phenomena. In terms of their aims, the editors of the *Sammlungen*, a group of Breslau physicians, saw themselves—like the *curiosi*—in the tradition of Bacon and Oldenburg. In the foreword to the first volume in 1718, they name these two figures as points of reference for their journal, alongside the Swiss physician and naturalist Johann Jakob Scheuchzer (1672–1733). They viewed their objectives as in line with those of other learned societies in the Holy Roman Empire, England, the Netherlands, and elsewhere. But they stress that, unlike them, they intended to concentrate on publishing *communia* and *vulgaria*, rather than the ubiquitous rare natural phenomena; the pages of their new journal were to be filled primarily with observations on the weather, epidemics, and growth.[148]

These and other everyday nature observations were, in the view of the editors of the *Breslauische Sammlungen*, more useful than those related to rare natural phenomena, not least because the preternatural had already been discussed countless times before, whereas that which occurred ordinarily was neglected. Accordingly, they devoted no less than three "classes" to them—subject headings (the weather, animal epidemics, and the growth or decline of plants and animals at specific times of year) under which similar articles were grouped.

But even the Breslau physicians did not want to do away entirely with reports on monsters and other rarities of nature. They continue:

> But we also 2) do not deny the rarioribus their publicity / especially those / whose existence habitually causes a noticeable noise or light (*eclat oder bruit*) / and which are mostly accustomed to being noticed first of all; but which we will also mostly think of accompanying with some reflections, for the sake of a more thorough Wissenschaft / or actual use.[149]

The rarities (*rarioribus*) were even given their own subject heading: "Class IV." Monstrous births were often among the cases reported here. Like during the lifetime of Liceti—who had in fact defined the term *monstrum* in this sense—monsters were still making "a noticeable éclat or bruit," to borrow the phrasing used by the editors of the *Breslauische Sammlungen*.

Indeed there are contributions on monsters in almost every volume of the quarterly journal. The "Historia eines Monstri," the tenth article under the heading "Class IV: Of singular reports from nature, July 1717" in the first volume, is just one example of many.[150] The author, who remains anonymous as was the custom of the journal, first recounts how the *monstrum* in question—the stillborn baby of a poor woman from Epperies in upper Hungary, whose head in particular was unusually formed—came into the world and into his custody. This is followed by a detailed description of the baby. In the subsequent second section, the author elaborates further in order to position his observation within the medical tradition. The *curiosi's monita*, mentioned by Valentini in the *Miscellanea Curiosa*, form a key reference point for these reflections, too:

> As this fetus was of unusual formation, it must therefore take up a place in the class of *monstrorum*: whereof many, whether Ulysses Aldrovandus, Joh. Schenkius, Alphonsus de Caranza, Ambrosius Paræus, Martinus Weinrichius, Caspar Bauhinus, Fortunius Licetus, Schotus Wechtlerus, or others, have written extensively and, as God wished it, usefully enough, yet they still have left much for those coming after them to observe; of which, too, Mssrs. *curiosi* have gathered such a large collection throughout all ages that one must also moderate oneself in such great profusion and attend to the extraordinary Monitoria of Anno 1684, as, after all, no great benefit to the Arti Medicæ has been seen, since these have not been subjected to the cure of a Medici.[151]

The admonitions of the Academia Naturae Curiosorum mentioned by Valentini are placed here in the context of the great number of articles on monsters that the *curiosi* had gathered in their journal and made public. Clearly, then, the *monita* issued by the Schweinfurt physicians can be understood only against the backdrop of their "profusion" in the *Ephemerides*.

Admittedly, this author did not allow these reservations to put him off submitting his article any more than Valentini had. But he does feel compelled to justify why his comments deserve to be published: "According to this, I, too, might justifiably have been released from making my comments, were it not so that a number of not entirely useless axioms deserve to be derived from this example."[152]

This makes tangible on a small scale a basic tendency in naturalist discussions of preternatural productions of nature, which were in actual fact defined as intrinsically "rare." Around 1700, after two centuries of intensive collecting of preternatural objects, of illustrating and describing them in curiosity cabinets, notebooks, monographs, and journals, the paradoxical situation increasingly set in that the rare had become extraordinarily common. It was due in large part to this that naturalists gradually lost faith in the central importance of studying such objects, soon turned their backs on them, and even, toward the middle of the eighteenth century, came increasingly to consider it "vulgar" to concern oneself with them.

The above justification by the editors of the *Breslauische Sammlungen* for the orientation of their journal is, in this sense, somewhat polemical. The Breslau physicians considered the long-dominant focus of nature study on the rare and preternatural to be of questionable usefulness. What is more, as it had led to the neglect of what was in their opinion the more useful observation of *communia* and *vulgaria*, they even viewed it as damaging.

At the close of the seventeenth century, the *curiosi* had not yet gone this far. For them, the pull in the opposite direction was stronger. The *curiosi* still generally considered the collection of rare *observationes* to be a valuable exercise in itself. Their understanding of nature emphasized the plenitude and variety of nature's creations, and this explains in no small part the dominance of articles on monsters and other rare phenomena in the academy's journal.

Copia and *Varietas*: The Journal as a Cabinet of Curiosities

The large number of *observationes* on monsters and similar rare and astonishing phenomena published in the *Miscellanea Curiosa* can be explained in part by the Academia Naturae Curiosorum's understanding of the natural world. Like Aldrovandi,[153] it emphasized the plenitude and variety, the *copia* and *varietas*, of nature's creations. In general, intellectual regard for *varietas* contributed to the success of the epistemic genre of *observatio*.[154] This aspect of the academy's understanding of nature was already very much in evidence before the journal's establishment, in the *epistola invitatoria* penned by academy founder Johann Laurentius Bausch in 1652, which is considered the founding document of the Academia Naturae Curiosorum.

In the first section of the letter of invitation, Bausch refers to the ancient personification of nature as a goddess, praising it as ingenious. In his depiction of this personification of nature, the plenitude and variety of nature's productions are reflected in the milk with which she nourishes her creations:

> The ancient Romans and the other pagans did not only imagine nature or the earth as a goddess and dedicate temples to her, they also imagined her—if not in a Christian sense, nevertheless ingeniously [*ingeniose*]—as a woman who was characterized by quite a number of breasts; as a woman, namely, because Titan or Sol (he was said to be her husband) were, as it were, constantly at work in the matter contained in her for the creation of animals and metals, stones and plants, and she, as a woman, was incited, as it were, through the warmth of the sun [or Sol] to create and, as it were, took into herself seeds containing the concentrated

potency of all elements: her countless breasts, however, were meant only to show the marvelous plenitude [*copia*] and, especially, the variety [*varietas*] of that milk with which she suckled everything that her womb had brought forth, according to its needs.[155]

In the second section of the letter, Bausch translates this pagan understanding of nature into his Christian worldview. In doing so, he preserves the notion of *varietas*, adapting it to his Christian understanding of nature: "We, who have been taught Christianity to a greater extent through Holy Scripture, know that nature itself is not a goddess, but was created by the true, triune God, through whose mercy it brings forth such a countless variety [*varietatem*] of things."[156] For Bausch, the "countless" variety of the things in nature reflects divine intent. What his Christian understanding of nature has in common with the pagan view that he praises, then, is the variety brought forth by nature and, therefore, of the subject matter of medicine and its kin.

According to the letter of invitation, it was, as we have seen, this variety—too immense for the individual scholar to cope with—that made the *collective* study of nature necessary in the first place. And the need for binding rules for the academy's work was also an indirect result of the founders' understanding of nature as characterized by *copia* and *varietas*. This, at least, is what the letter of invitation's reasoning suggests. The connection is made explicit in the passage of the *epistola invitatoria* that introduces the *leges*:

> But as we do not only notice with our eyes those [things] that cross our path daily and are stored in great number in the pharmacies, but rather the natural world, abounding luxuriantly in fecundity [*foecunde luxurians*], also brings every day before our eyes new and as yet unknown, and for the ancients hardly imaginable [things] to be examined, that have been brought forth on and in the earth, in the sea and in the air, whose qualities are hidden from us and not sufficiently known; therefore any *Academicus Naturae Curiosus* who wishes to join our society shall adhere to the following [rules], which have the force of law.[157]

The *copia* and *varietas* of nature feature throughout later academy publications, too, particularly those of primary interest to us here—the volumes of the *Miscellanea Curiosa*. The dedication in the first volume of the journal to Emperor Leopold I— to whom the academy was also to dedicate the subsequent volumes—expends just one sentence on characterizing its contents. It can be assumed, then, that the words of the sentence were carefully chosen. So it is surely no coincidence that the adjective *varius* is used three times, and highlighted each time in italics; only the rare and astounding nature of the contents of the volume is given similar emphasis: "Therefore they [the members of the academy] dare to present *astonishing* jokes of nature, reproduced in beautiful etchings: *various* monsters deviating from nature, to Hercules, the vanquisher of monsters [meaning the emperor]; *various* cases of rare diseases, *rare* medicines, *various* new inventions by the most distinguished minds, all bound together in one bouquet, and offered not to the ruined Temple of Asclepius, but to the IMPERIAL AUGUSTAN SHRINE."[158] Only in combination with the other adjectives employed here, *mirus* and *rarus*, does the association of *varietas* with the valuable, the collector's item, become clear. The imperial shrine to which the academy symbolically presents its contributions in this dedication is a (paper) treasure chest or cabinet of curiosities.[159] The volume, with its numerous articles, is portrayed as a bouquet of astonishing, rare, and varied natural objects and products of human ingenuity.

The parallel drawn between the vanquisher of monsters, Hercules, and the dedication's addressee is also noteworthy. This is an allusion to the twelve labors that Eurystheus imposed on the Greek hero, in the course of which he had to overcome numerous dangerous beasts, such as the Nemean lions. The comparison was intended to flatter Leopold—a common gesture in the genre of dedication texts. But the choice of this specific comparison is revealing in several respects. It shows, once again, the familiarity of the *curiosi* with the texts of ancient Greece and thus their humanist scholarship. It is also further proof that monsters played a central part in their project. They were an essential element in—if not the embodiment of—the *copia* and *varietas* of nature.

Significantly, Ulisse Aldrovandi's *Monstrorum Historia*, which is characterized by an understanding of nature similar to that of the *Miscellanea Curiosa*, also features a dedication comparing its addressee to a vanquisher of monsters in ancient Greece. The brief dedication in the first edition of *Monstrorum Historia* is addressed to Ferdinand II de' Medici, Grand Duke of Tuscany. It is not written by Aldrovandi himself, as he died before Ferdinand was born. It must have been written by Bartolomeo Ambrosini, who published the volume. As *Monstrorum Historia* first appeared in 1642, the paratext dates from a few decades before the dedication in volume 1 of the *Ephemerides*. The same applies to the text on the page immediately following the dedication. Here, Ambrosini honors the addressee of the dedication with a four-verse poem that he has penned himself.[160] Its subject matter is interesting, as it illuminates how dedicating a volume on monsters to the grand duke might help to increase his reputation.

The poem begins with a question: "What does the entire genus of monsters that was dedicated to you show?" The next line gives the answer: it shows that the Grand Duke tames wild monsters. The statement is then explained. Using the term *magnus* (great, grand), which features in both the honorific title of Alexander the Great and the title of the Grand Duke, draws a rhetorical parallel between the two rulers. Alexander the Great tamed terrible monsters, and what that great man could do—now Ferdinand is addressed directly—this grand one can do, too. Thus, Ambrosini also draws a parallel between the addressee of his dedication and a vanquisher of monsters from the ancient world—and by implication between the monsters of antiquity and those of the early modern present, too.

In an equally prominent position, and one that was similarly defining for the public image of the academy, the etched frontispieces of the *Miscellanea Curiosa* also point to an understanding of nature centered on plenitude and variety. The volumes from years one to seven of the first *decuria* all use the same frontispiece (fig. 21), which presents itself to the viewer as follows: The center of the image shows allegories of the three realms of nature. They are recognizable as such first by the symbols on their robes, and they are labeled accordingly. Kneeling and facing away from the viewer, they hold aloft a piece of parchment. Above them, on a sphere on which is written "Consilio et Industria" and that is itself set on an altar, the imperial eagle sits enthroned. It bears the imperial insignia. The text on the parchment urges it to accept the dedication. Thus, the frontispiece reenacts the symbolic presentation of the work to the emperor that is undertaken in the dedication.

The far left of the image is revealing with regard to the *curiosi*'s understanding of nature. At the outer edge of the scene stands a statue that corresponds to Diana of Ephesus, the ancient personification of nature whom Bausch introduces in the letter of invitation. In the *Miscellanea Curiosa* frontispiece, she is wearing a crown and a mantle and displays the obligatory multitude of breasts. The Greek letters at the base of the statue identify her as Physis ("ΦΥΣΙΣ") and thus as an allegorical representation of nature. Opposite her, on the far right of the image, the scene is bordered by another statue—also the figure of a woman, wrapped in a toga and holding a chalice in her hands. The inscription "ΥΓΕΙΑ" marks her as the personification of health, *sanitas*. She represents the *curiosi*'s primary aim in their engagement with nature: through knowledge thus gained about mineralogical, zoological, and botanical objects wholly or in part suitable for use as remedies, they wish to contribute to human health.

The portrayal of Diana used by the *curiosi* was exceptionally widespread in the art of the early modern period as an allegory for nature. An original creation of

Fig. 21. Frontispiece, *Miscellanea Curiosa Medico-Physica Academiae Naturae Curiosorum sive Ephemeridum Medico-Physicarum Germanicarum Curiosarum* [...]. Annus Primus (1670). Forschungsbibliothek Gotha der Universität Erfurt, Med 4° 00145 (1–2).

humanism, even its beginnings are linked to the study of nature. The oldest known depiction of this kind originated in Naples in the 1470s, when a Roman scholar and a miniaturist were working together on a manuscript of Pliny's *Naturalis historia* commissioned by Cardinal Juan de Aragón.[161] The qualities of nature that artists and authors of the early modern period associated with this allegory included benevolence and great fecundity. These qualities contradict the prevailing ideas about nature in late antiquity and the Middle Ages, which tended to emphasize scarcity.[162]

From the eighth volume of the first *decuria* onward, the above frontispiece was superseded by a new one crowded with floral ornamentation (fig. 22). This frontispiece begins the remaining volumes of the first and all of the second *decuria* of the *Ephemerides*. The opulence of this image in itself expresses the understanding of nature, one of great fecundity, present here. Among many other visual elements representing or characterizing the academy and its journal, the etching features, on both the lower left and right, a cornucopia filled with exemplifications of the natural realms and individual subject areas to which the *Ephemerides* were devoted. The cornucopias represent the profusion and boundlessness of nature's productions, and thus also of the (anatomical, chemical, and botanical) subjects of the *observationes* in the *Miscellanea Curiosa*.

What is evident here on an iconographical level continues in the rhetoric and epistemology of the articles in the *Ephemerides*. Generally speaking, the authors shared the understanding of nature outlined here: one of plenitude and variety. In this respect, they, too, followed in the tradition of Aldrovandi. His *Monstrorum Historia* is very much present in the *Ephemerides*; it is referenced time and again when a current *observatio* is set in the context of similar accounts from literature. The large number of examples that the book offers to the reader meant that it was ideally suited to this purpose. The author of an article about a monstrous lycopod in the second volume of the first *decuria* of the *Ephemerides*, for example, refers the reader to plants in Aldrovandi's work that were also shaped like cornucopias and could be attributed to the same cause.[163]

A similar approach is taken in an article about a monstrous two-headed calf in the volume covering the fifth and sixth year of the first *decuria*. The observation itself is followed by a list of similar cases from literature. The author mentions Aldrovandi first as a source of various (*varias*) images and descriptions of two-headed calves, one very similar to the case described here.[164] The *curiosi* and the other contributors to the *Ephemerides* were mostly in agreement with Aldrovandi, then, that nature was characterized by the *copia* and *varietas* it brought forth.

Fig. 22. Frontispiece, *Miscellanea Curiosa, sive Ephemeridum Medico-Physicarum Germanicarum Academiae Naturae Curiosorum* [...]. Annus Octavus (1678). Forschungsbibliothek Gotha der Universität Erfurt, Med 4° 00147/01 (8).

And, like Aldrovandi, they attributed great importance to these categories on an epistemological level, too. Clearly, it made sense to assume that such boundless nature would best be described by a collaboratively realized, comprehensive collection of equally varied *observationes*. In a Baconian sense, these observations were to concern nature's rare and extraordinary creations first and foremost. Accordingly, the *epistola invitatoria* printed in the first volume of the *Ephemerides* announced the aim to be the compilation of various (*varias*) very rare but credible *observationes* from medicine and nature study:

> Prompted by these important considerations, the society of the curiosi, requesting the help and effort of all physicians and natural philosophers, has offered annually to collect & compile, and finally to disseminate in print, the *Ephemerides Medicas*, in which they have decided to distribute various [*varias*] very rare [*rariores*] but credible [*fide dignas*] natural philosophical & medical *observationes* from practical medicine and surgery, new experiments in natural philosophy and anatomy, new discoveries from the mineral, animal, and plant realms, most select remedies & the finest chemistry, whose unique benefit is recognized, which were either communicated by letter or, as it were, drawn from the rarest books, often [written] in foreign languages, as in a garden adorned with the varied [*multifariis*] charms of flowers.[165]

The extent to which this stipulation was adhered to through all the years of the *Ephemerides* can be inferred from a comment that introduces the fourth *observatio* from the volume covering the fifth and sixth years of the third *decuria*, published in 1700. In it, Gustav Casimir Garliep von der Mühlen (1630–1713) addresses the case of a "monstrous" bone that protruded from a woman's eye. He begins by remarking that the *Ephemerides* had already shown their curious readers a great variety (*varios*) of monstrosities (*abortus*) produced by nature deviating from its proper path. This, however, did not stop Garliep from presenting his observation, too.[166]

With continually new *observationes* on rarities, the *curiosi* created—as the *epistola invitatoria* states—a "garden adorned with the varied charms of flowers." But there was also a trend in the opposite direction. Alongside their regard for *copia* and *varietas*, the *curiosi* showed the beginnings of the same impulse that was to find expression a few decades later in the founding of the *Breslauische Sammlungen*. This manifested itself, not least, in the Schweinfurt scholars' cautious approach to the emotion that had given their academy its name: *curiositas*.

Ambivalence toward *Curiositas*

When, as in Vollgnad's scholium, the question of usefulness was discussed, it came with more associations for the members of the Academia Naturae Curiosorum than just the immediate benefit to medicine, the advancement of which the academy had made its objective. An important subtext in such statements was an ambivalence toward curiosity as an emotion in the context of the study of nature. Even in the late seventeenth century, not all naturalists regarded *curiositas* as conducive to the production of knowledge. It was therefore of central importance to the Academia Naturae Curiosorum to avoid any appearance of indulging in a useless, vain, and frivolous curiosity. As a preemptive justification in anticipation of such criticism, as it were, the founders emphasized in the *epistola invitatoria* of 1652 that their form of curiosity was comparable to tenacious dedication, or *diligentia*, and was to be considered a virtue.[167]

This assessment corresponds in two respects with the results of Neil Kenny's extensive study on the ways *curiositas* was used in early modern France and Germany. The Academia Naturae Curiosorum was not alone among the academies in making positive use of curiosity: "Among the several academies and learned societies that were founded in the second half of the seventeenth century, several shaped nature and art into curiosities and investigators of them into curious people. These new institutions tended to be far more enthusiastic than universities or churches about shaping knowledge as a metaphorical collection of curiosities."[168] By invoking curiosity, then, the *curiosi* were signaling, first, that they had dedicated themselves to a European knowledge-gathering project that transcended national and confessional boundaries. *Curiositas* stood for membership in a community focused on this task of collecting. It is no coincidence in this regard that the academy was founded so soon after the end of the Thirty Years' War: the concern was also to reestablish a "tolerant civility" among scholars.[169]

Second, the *curiosi* and their contemporaries would still have been aware of *diligentia* as the usual meaning of the word *curiositas* in antiquity.[170] This was one of the reasons both the *Miscellanea Curiosa* and many of the academy members' carefully worked encyclopedic monographs bore the adjective *curiosus* in their title, despite by no means dealing with curiosities. Here, *curiosus* refers to the thorough exploration of the material and, consequently, the high quality of the discourse— not to the rarity of the object in question.[171]

Nevertheless, the *curiosi* were under the impression that they should specifically emphasize their rejection of useless, frivolous curiosity to preempt criticism.

Kenny speaks, in view of the openness and polyvalence of *curiositas*, of a "semantic swamp."[172] Even the motto of the academy, *Nunquam otiosus*—never idle—was part of this self-presentation that avoided any semblance of idleness or frivolous curiosity and emphasized the usefulness of academy members' research.[173] Because of their ambivalence toward *curiositas*, they wanted at all costs to avoid the impression of gathering *observationes* of monsters or other rarities merely for their own sake.

Contrary to what one might assume at first glance, the *Miscellanea Curiosa* were not by any means an indiscriminate hodgepodge of disparate articles. The practices of observation and writing that the *curiosi* demanded of their contributors had implications for the knowledge sources on which texts about monsters were principally to be based from then on, as well as on the circulation of the corresponding factoids. We can identify two limiting effects of this change on the scholarly discourse on monsters: through their choice of the epistemic genre of the *observatio* on the one hand, and their criticism of credulity on the other, the *curiosi* decreased both the number of monstrous factoids and the scope available to authors in discussing the causes of monstrosity in the cases addressed. The circulation of older factoids was brought almost to a complete halt, as shown below.

The Limiting Effect of the Genre

The trend toward authored observation had significant consequences for the discourse on monsters among naturalists. Journals devoted to nature study such as the *Miscellanea Curiosa* favored authored observations vouched for by an author over examples taken from earlier literature that had been so popular in previous decades. This indirectly limited the discourse on monsters in terms of the number of factoids in circulation.

The older examples were still available in *summae* such as that of Johann Georg Schenck von Grafenberg. And such compilations continued to be consulted. But the examples were no longer constantly being reproduced. By contrast, *new* factoids based on an author's observations were in great demand. Closely connected to this was the fact that the writers of such articles quickly developed a conception of themselves as *auctores*, as originators of their discourse. This is illustrated clearly in the *Ephemerides* and in the discussions that accompanied their publication.

The Academia Naturae Curiosorum's call for *observationes* that were rare and based on direct experience led to the submission of many contributions about monsters. Generally speaking, the articles themselves dealt with current as yet unpublished cases. Only the scholia referred to older cases, and typically only in synoptic form, whereas the current case was presented in detail. We must remember in this context that to divide the *observatio* in this way was to turn the concept of authorship veritably on its head: the observation of a *contemporary* author—not a passage of text from an ancient authority—now had center stage, while older cases from literature were relegated to the *commentary*. The scholia were clearly separated from the actual observation typographically and written as a commentary on it. The parallel cases cited in a scholium served primarily to enable a comparison of the current case with the historical knowledge already available, thus increasing the credibility of the observation.

The shift toward contemporary observations vouched for by their author is also clearly reflected in the illustrations. Unlike in Aldrovandi's or Johann Georg Schenck von Grafenberg's *Monstrorum Historia*, older images of monsters were now very rarely reproduced in the *Ephemerides*. Moreover, the etchings that accompanied individual *observationes* were generally based on the author's own inspection of the subject in question, portraying it in all its individuality. And this was explicitly encouraged.

A good example of this is the etching that accompanies *observatio* XX in the eighth volume of the second decury. Gustav Casimir Garliep, whom we know from his *observatio* on the "monstrous" bone protruding from a woman's eye, reports here on a specimen of "monstrous" fleabane. In many respects, it conformed to what might be expected of fleabane in terms of leaf shape, color, and with regard to the soil in which it grew, for instance. Its unusually large number of stems and the bushiness of the plant were enough for Garliep to classify it as "monstrous."[174]

Garliep was not the lucky finder of this monster. Rather, the East Pomeranian state councilor (*regiminis electoralis*) sent it to him, as he wished to have the Academia Naturae Curiosorum's opinion as to its genus and name. The article delivers the requested expertise. At the same time, the last section of the article places the specimen, along with all the various (*varia*) similar cases previously published in the *Ephemerides*, in the category of "jokes of nature" in plants.[175] For our purposes, it is interesting to note the obvious effort that Garliep expended—and was expected to expend—on conveying as precisely as possible the individuality of this specimen.

This effort is clearly visible in the etching, with its three figures (fig. 23). The first illustration shows the entire plant, upright and including its roots, with letters marking its parts. The style of the image matches that of earlier illustrations of monstrous plants in the *Miscellanea Curiosa* but is exceptionally detailed. The two illustrations on the right show the plant's leaves and are also labeled with (in this case Greek) letters or numbers marking individual parts. These images, too, are

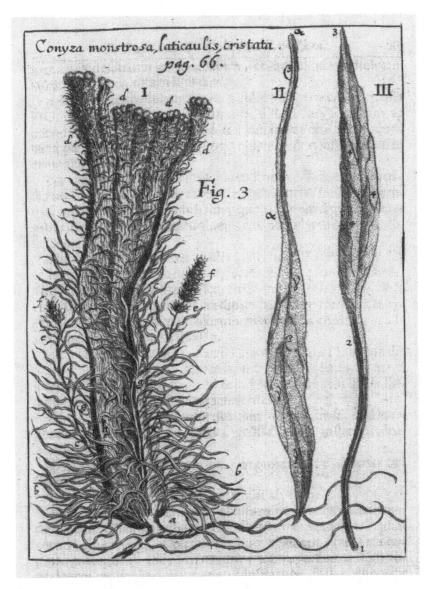

Fig. 23. Etching from Gustav Casimir Garliep, "Observatio XX: De Conyza Monstrosa, Laticauli, Cristata," *Miscellanea Curiosa Medico-Physica Academiae Naturae Curiosorum sive Ephemeridum Medico-Physicarum Germanicarum Curiosarum* [. . .]. Decuriae II. Annus Octavus (1690), bound between pages 66 and 67. Forschungsbibliothek Gotha der Universität Erfurt, Med 4° 00148–149 (2,8).

especially rich in detail and highly individualized—although the leaves depicted are intended to exemplify all the plant's leaves. The letters and numbers in the images are explained in the legend entitled "FIGURARUM." This not only names the parts of the plant but also describes some of them in detail.

Garliep's remarks on the making of the etching are also illuminating. He and Christian Mentzel had examined the monstrous fleabane plant together. As the plant seemed too large to be dried whole and sent to Schröck, the publisher of the *Ephemerides*, Mentzel then entrusted Garliep with the task of drawing its parts. And Garliep took the job very seriously. He recounts in detail over several lines how he produced the drawings—meticulously and with patient dedication—but does not fail to mention the difficulties his aging eyes caused him. At this point, Garliep was already around sixty years old. He admits that reproducing the plant's numerous leaves was a job too great for his eyes and brush, particularly just above the root, where countless leaves surrounded the stem like a crown.

Garliep made every effort, then, to depict the plant in all its individuality and considered this to be in keeping with the academy's wishes. He had every reason to give it his all: Mentzel had been a member of the academy for years and had authored numerous *observationes* in the *Miscellanea Curiosa*. But Garliep himself was only to be accepted into the academy on March 15, 1690. According to the statement of provenance printed at the bottom of the article, his *observatio* was received in Nuremberg on July 8, 1689, and thus predated his acceptance.[176] We can assume, therefore, that at the point when he was producing the article and the drawing, Garliep was keen to improve his membership prospects by means of valuable contributions to the academy's journal.

His efforts to present the subject of his *observatio* on the basis of *autopsia* and with the greatest possible care and attention to detail can reasonably be interpreted—with all due caution—to mean that this was precisely what the academy expected from its contributors. The etchings and descriptions were to be based on *autopsia* and record their subject in all its particularity. Secondhand images or those of uncertain origin, such as the depiction of the anthropomorphic beet reproduced above, are rare even in the first volume of the *Ephemerides*. In later volumes, they appear even less frequently. The practice of reproducing older images of monsters, still common around 1600, was now out of the question for journals such as the *Miscellanea Curiosa*.

The shift in focus toward direct observation had an impact on the self-perception of the authors. They increasingly saw themselves as *auctores*—originators—of their texts. One manifestation of this was the fact that several authors approached the academy requesting permission to republish in monographic form the *observatio-nes* they had printed in the *Ephemerides*. Some even did this without first asking permission. Such endeavors ran contrary, of course, to the interests of the academy and the publishers of its journal.[177]

A further indication that contributors were developing an understanding of themselves as originators of their discourse was the indignation with which many reacted when the academy rejected one of their articles. Lucas Schröck tells of such reactions in a letter to Leopold I responding to an admonition the emperor had given the *curiosi* in 1700, as Oldenburg had decades earlier, to print only reliable reports. This admonition concerned him personally: after all, Lucas Schröck had been responsible for publishing the *Ephemerides* since 1685. He refers to the in-dignation of authors whose submissions are rejected to prove how carefully the *curiosi*—not least himself—already sought to distinguish the reliable from the un-reliable without fear of the rejected authors' reactions.[178]

But it seems that those responsible did not always act quite so fearlessly. In the "Protocollum," Schröck describes a controversy that he, longtime academy member Günther Christoph Schelhammer (1649–1716), and then-president of the academy Johann Georg Volckamer the Elder, dealt with in 1691. Schröck's account reveals that, for fear of putting off potential future contributors, they rejected submissions of *observationes* less frequently than actually seemed warranted in view of their poor quality.

This hesitant rejection practice came at a double cost. First, the *Ephemerides* continued to receive contributions that failed to meet the standards of the Aca-demia Naturae Curiosorum—and of prominent readers such as Oldenburg and Leopold I. In addition, the volumes became increasingly bulky, which increased their price and thus impaired sales. Schröck's final position, approved by the acad-emy's president—that in the future the journal's editors would select more strictly from submissions but at the same time be willing to overlook shortcomings here and there to avoid any resentment toward the academy on the part of the authors—suggests that there was little change in the rejection practice even after 1691.[179]

The authors of already published *observationes* also reacted extremely sensi-tively to criticism at times—another clear indication of their growing understand-ing of themselves as originators of their discourse. In this respect, the scholia were a particularly delicate matter. Until his death in 1672, they were mostly composed by Sachs von Lewenhaimb on behalf of the Academia Naturae Curiosorum. Later, these commentaries on individual *observationes* were written either by other mem-bers of the academy or by the contributors themselves. From the outset, the *curiosi* were aware of the sensitivity of these sections of the journal. In a paratext entitled "Benevolo Lectori Salutem & Officia" in the first volume of the *Ephemerides*, they explain the intention behind the scholia, aiming particularly to alleviate any concern potential contributors might have that their *observationes* would expose them to criticism from the academy. According to the *curiosi*, the scholia were to be historical (*historica*), not critical (*critica*).[180] They expressly emphasize that they did not presume to claim the censor's rod (*censoriam virgulam*). Rather, the scholia were intended to provide those physicians who had less well-stocked libraries with *historiae* from literature that had parallels with the rare observation in question and thus help corroborate it.[181] Similar cases from literature were cited with the aim of increasing the credibility of rare and astonishing *observationes* in particular, not of criticizing their authors.

But it was to no avail. In 1688, longstanding member Gabriel Clauder (1633–1691) complained bitterly that Daniel Crüger (1639–1711), also a *curiosus*, had, in his rejection of remedies containing cinnabar in the most recent volume of the *Ephemerides*, adopted a position that Clauder considered an attack. Clauder also took the opportunity to criticize the institution of the scholia. He did not see himself as alone in this criticism. Many, he claimed, were dissatisfied.

Once again, Schröck coordinated his answer with the then-president of the academy, as he had in the discussion with Schelhammer. In view of the academy's aims, Schröck wrote to Volckamer, it was neither possible nor desirable to avoid controversies such as that between Crüger and Clauder. Furthermore, their exclusion from the *Ephemerides* might deter potential contributors from submitting their reports. They had already had such experiences, he argued. In reference to Clauder's remarks on the scholia, Schröck indicated that he could not accept the opinion expressed from time to time that they represented an instrument of censorship. The argument was untenable. In any case, a scholium was only ever added to an article with the consent of the author.[182]

This appears to have been the end of the discussion. But it was not the first of its kind: Schröck adds in the "Protocollum" that it was evident from a letter in his possession written by Philipp Jacob Sachs von Lewenhaimb to Georg Hieronymus Welsch on June 29, 1670, that the scholia had come under criticism before. He goes on to quote this letter at length, which picks up on the argument made in the letter of invitation in the first volume of the *Ephemerides* and expands on it:

> I hear that some do not like the articles to be accompanied by scholia. They only accompany a few, and also deliver no judgment, which is far from our intention.

They are historical, not critical, and illuminate an unusual object by means of related reports from rare texts. This is, above all, because our *Ephemerides* are not written solely for physicians with well-furnished libraries at their disposal, but also for others: researchers, persons of standing, clergy, and other educated people. And the learned scholia added by Your Honour to some contributions were an impetus to continue on the laudable path chosen.[183]

Schröck concludes by summarizing the outcome of the repeated discussion of the scholia: "Thus President Helianthus also came to the verdict that the scholia, contrary to the view of certain harsh critics, were not to be omitted, especially since they met with the great approval of the most supreme emperor, who alone could and must outweigh a thousand who thought the opposite."[184] For all the trouble that the new understanding of authorship caused the academy, it is nevertheless to be regarded as a correlate of the kind of observation and writing demanded of the contributors. Thus, disputes such as those documented above are, in a sense, evidence of the *curiosi*'s success. The authors of the *observationes* really did regard their reports as *their* observations and were reluctant to accept any interference.

The Limiting Effect of the Criticism of *Credulitas*

The form of observation and writing required by the academies had a second limiting effect on the monstrous factoids that were in circulation, which had less to do with the way reports and images circulated than with approaches to the causes of the phenomenon in question. Even in the late seventeenth century, an echo of the old understanding of monsters as divine omens lingered whenever the subject arose. The contributors to the *Ephemerides* are remarkably united in their opinion of such interpretations. Like Jänisch in the *observatio* discussed above concerning the monstrous wild bugloss on sale at the market in his hometown—which was seen by many as a prodigy—they distance themselves almost without exception from the notion of monsters as divine portents or warnings.

They are emphatic in their criticism of the credulity, the *credulitas*, of the "common man," the *vulgus*, in relation to this possible interpretation. They do not question the phenomena themselves; the debate concerns their interpretation. This particular feature of the objects categorized as "monstrous"—their divinatory legacy—is a principal reason for the ambivalence with which the *curiosi* so often received submissions that dealt with monsters.

Recall that, while assembling the third volume of the *Ephemerides*, the *curiosi*, according to Vollgnad, considered banishing monsters from the journal entirely, primarily because of the problems in how to approach their causes. The uneducated populace in particular, Vollgnad claimed, disregarded the true causes of monsters. It seems that even more so than monstrous plants, monstrous newborns were still frequently regarded as prodigies. Accordingly, they were the primary target of Vollgnad's scholia and of the numerous critical remarks in the academy's correspondence.

The divinatory legacy of monsters made it particularly important that their observation was conducted by the *right* people—those who were not credulous—and in the *right* way, that is, not in the manner of the common person. This is revealed particularly clearly by one of the *observationes* in the volume of the *Ephemerides* that covers the fifth year of the second decury. Its author, Philipp Jakob Hartmann (1648–1707), academy member and professor of history and medicine at the University of Königsberg, had the rare opportunity to examine and perform a dissection on a newborn baby categorized as a *monstrum*. All his effort is aimed at disproving this assessment and combating the credulity of common folks.

His article begins with the usual information about the place (Lacke, on the outskirts of Königsberg) and date of the recent birth, and about the baby's parents. The third in a set of triplets, the baby was stillborn and taken for a *monstrum* by the midwives and other women, meaning that they interpreted it as a divine omen. This idea then spread among the common people (*vulgus*). No wonder, the Protestant Hartmann seems to have thought—the inhabitants of the region were predominantly Catholic.[185]

After reporting, furthermore, that there had been no unusual occurrences during the pregnancy that could have influenced the baby's development, he proceeds to the results of his examination of the body. On the basis of his external examination, Hartmann was already certain that this was not a *monstrum*. The baby had wrongly been said to have the appearance of a satyr, he writes.[186] Hartmann's detailed description of the baby aims at disproving this erroneous assessment. Thus, he does not conceal the baby's large ears, but he asserts that what the common people had wrongly taken for a second pair of ears were nothing but enlarged, shriveled eyelids. In a similar fashion, he interprets the baby's supposed horns as eyebrows, the hair of which was drawn together in an unusual way.[187]

The dissection, too, aimed to disprove the view of the common person, which Hartmann judged to be wrong. He begins his dissection report with the statement that the dissection proved that the baby was not a *monstrum*. Rather, all the baby's cartilage had detached itself from the bones, and the bones themselves were disarranged. Here, the author reveals something of the mechanical explanation for the baby's unusual formation that he expresses later in the text. A list of the individual observations made in the course of the dissection then follows, during the course of which Hartmann stresses repeatedly that almost everything belonging to the human anatomy was present. There were, however, a few exceptions: he found no brain, for example, nor any eyes.[188]

After the subheading "Usus ex Anatome Monstrosi Crediti Fœtus," there follows a list of eleven valuable insights provided by the dissection. For the most part, they serve to explain the anatomical particularities of the baby on the basis of "natural" causes and therefore to indirectly criticize the erroneous assessment of the common folk. In point 2, for example, the author attributes the baby's deformities (*vitia*) to a lack of space in the uterus due to the multiple pregnancy. The baby probably died, he writes, of an attendant tearing of the skin. Point 4 gives the position of the baby in the uterus as the explanation for the clumps formed by the hair of the eyebrows, among other things. In point 5, the pulled-up eyelids and the large ears are also attributed to the position of the baby and to dehydration.

Hartmann's first insight stands out a little from the others, insofar as its *explanans* is situated on a different level. It explains how it was even possible for the baby to be wrongly understood as a divine omen in the first place: the common, credulous man (*vulgus credulum*) veritably devours *portenta* and *monstra*.[189] In his view, then, the common people tended to see divine omens in all sorts of things. The etching later included in the addenda of the tenth volume of the second decury of the *Ephemerides* (fig. 24) shows an illustration of the baby that is entirely in line with these insights. The eyebrows are clearly recognizable as such, there is no second pair of ears to be seen, and in every other respect, the baby has very little resemblance to a satyr. The image has a clear message: one must be inclined to credulity to see a satyr-like *monstrum* in this deformed fetus.

Hartmann was, as mentioned, by no means an isolated case. Where did this unanimity among the authors of *observationes* about monsters in the *Ephemerides* come from? As shown in the first chapter, when faced with rare natural phenomena such as monsters, physicians and other naturalists already tended around 1600—and indeed significantly earlier—to concentrate on the investigation of "natural" causes. Fortunio Liceti, for example, states expressly in *De Monstrorum Natura, Caussis, et Differentiis* (1616) that monsters could come into being not only in "natural" but also in "supernatural" and "subnatural" ways. However, he sees his task to be the discussion of the first variety only. In *De Ortu Monstrorum Commentarius* (1595), Martin Weinrich argues similarly in this respect: as a professor of physics in Breslau, he believed that theologians and physicians had differing aims and arrived

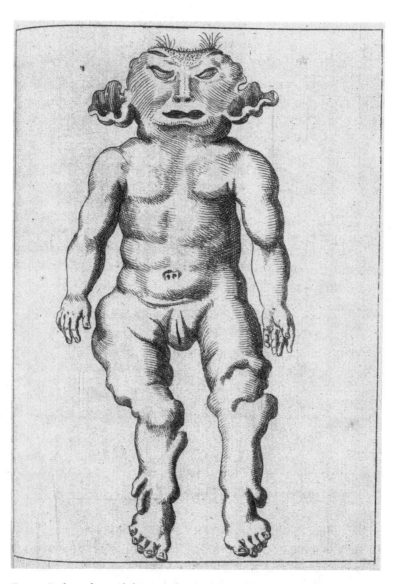

Fig. 24. Etching from Philipp Jakob Hartmann, "Observatio LXXVI, Anatome Monstrosi Crediti Fœtûs," *Miscellanea Curiosa Medico-Physica Academiae Naturae Curiosorum sive Ephemeridum Medico-Physicarum Germanicarum Curiosarum* [. . .]. Decuriae II. Annus Quintus (1687), bound between pages 418 and 419 in *Miscellanea Curiosa Medico-Physica Academiae Naturae Curiosorum sive Ephemeridum Medico-Physicarum Germanicarum Curiosarum* [. . .]. Decuriae II. Annus Decimus (1692). Forschungsbibliothek Gotha der Universität Erfurt, Med 4° 00148–149 (2,10).

at their conclusions using different principles. Soaring to the heavens, the theologians' knowledge was based on divine revelation, whereas physicians based theirs on reason alone and sought explanations that were closer at hand.[190]

There was also already a difference at this point between the ways monsters were received among scholars and among their less learned contemporaries, inasmuch as religious significance often played a more central role for the latter group.[191] What was new in the late seventeenth century was the vehemence with which the *curiosi* and other authors in the *Ephemerides* criticized the *credulitas* of commoners. Liceti and these authors are separated by the experience of the Thirty Years' War, which constituted an important context for the founding of the Academia Naturae Curiosorum. As Lorraine Daston and Katherine Park—following Robin B. Barnes— have shown, the interpretation of monsters and other rare occurrences and natural phenomena was particularly popular in the Holy Roman Empire in times of inter-confessional conflict and war. Thus, it flourished in the unrest that followed the Reformation, for instance, and during the Schmalkaldic War (1546–1547) and— closer still in time to the founding of the academy—in the years from 1618 to 1630, the worst phase of the Thirty Years' War.[192] The *curiosi* had learned their lesson: the interpretation of rare events as a presumed divine message was foreign to them.

A comparison of the Holy Roman Empire with other European countries shows how significant the political and religious circumstances were to the rise and fall of belief in prodigies. In the territories that would become present-day Italy, the belief gradually started to become less prevalent after the Italian Wars. During the French Wars of Religion, significantly, Michel de Montaigne, in the context of a case of conjoined twins he had personally examined, critiqued the notion that monstrous newborns constituted prodigies.[193] In England, belief in prodigies waned at the beginning of the seventeenth century, only to be rekindled by the political and religious unrest of the 1640s and the Civil War. It was not until after about 1670 that it disappeared for good.[194] At the end of the seventeenth century, European physicians, natural philosophers, and theologians were increasingly united in rejecting the interpretation of rare phenomena as prodigies. After the confessionally and politically motivated unrest of the early modern period, prodigies appeared to be beyond the control of politics or religion and therefore dangerous.[195]

Monsters could also still cause great turmoil—even up to the beginning of the eighteenth century, as we have learned from the publishers of the *Breslauische Sammlungen*. This was not in the interests of the *curiosi*, many of whom held public positions as municipal physicians, royal physicians, or councilors. Neither could it be in the interests of the ecclesiastical or secular rulers whose protection they sought—in the case of Emperor Leopold I with some success. In 1677, he approved the academy, and in 1687/1688 he granted it extensive privileges.[196]

⌒

The *curiosi's* relationship to monsters was characterized by a series of ambivalences. On the one hand, *observationes* on monsters were welcome in the *Miscellanea Curiosa* for their value to the study of anatomy, as long as the rules of the "epistemic genre" of the *observatio* were adhered to. Moreover, as rare and preternatural phenomena, they lay at the heart of what the journal of the Academia Naturae Curiosorum—in contrast to its monographs—was supposed, in the *curiosi's* view, to concern itself with. On the other hand, those responsible repeatedly expressed reservations about such submissions. These objections can partly be attributed to the fact that the elusive monster was now, after many decades of intensive study by naturalists, in a sense no longer rare, and that ongoing submissions concerning them must therefore have seemed increasingly less valuable or desirable.

In addition, newborns categorized as monstrous—but also, for instance, "monstrous" plants—continued frequently to be classed as divine omens by large parts of the population. In Europe, and particularly in the Holy Roman Empire, the memory of the devastations of the Thirty Years' War was still very much alive. This was, in no small part, also a war of propaganda over the interpretation of divine omens in pamphlets and other media, which explains the vehement criticism of the *vulgus* by the usually university-educated physicians whose *observationes* on the subject were published in the *Miscellanea Curiosa*. A *monstrum* interpreted as a divine omen had, in their view, not only been observed "incorrectly" but was veritably dangerous. On this point, the *curiosi* agreed with most of the European naturalists and theologians of their time. The interconfessional conflicts of the early modern period had undermined the authority of both the two major confessions and the secular rulers, and in the face of this numerous scholars in both fields worked to repress belief in prodigies (without, of course, fundamentally questioning God's ability to create such omens).[197]

We can thus understand why the *curiosi*, unlike their predecessors in the late sixteenth and early seventeenth centuries, no longer left potential preternatural causes of monstrous births to the theologians but, largely in harmony with the theologians, ruled them out from the start. In the Holy Roman Empire, then, the naturalization of monsters did not take place until the late seventeenth century. It was not the result of a "scientific spirit" that floated above political events and was obliged to assert itself over "religion." Rather, it was intrinsically linked to historical events and unwelcome neither to ecclesiastical nor to secular rulers.

The epistemic genre of *observationes* chosen by the academy for its journal was, like the genre of *mémoires* described by Perrault in the context of the Parisian sister academy, responsible for a further important effect upon referencing structures of the naturalist discourse on monsters: the circulation of older monstrous factoids was brought to a halt by the forms of observation and writing required; they were now reproduced only in the scholium section, if there was one. The perceived value of the older model of naturalist authorship (based on the collecting of factoids) having decreased in the seventeenth century, scholarly merit could better be earned by providing a new, authored observation.

In the following decades, too, naturalists would continue to engage intensively with the subject of monsters. Their suspicion of the uneducated population's approach to them was to become even stronger. As we shall see in the next and final chapter, in the mid-eighteenth century, European naturalists were fond of criticizing the tireless curiosity of common people with regard to monsters and other exotic subjects. In keeping with the ethos of the Enlightenment, they furthermore tried to propagate their views on such subjects among the general population, too.

In the course of these processes, monsters underwent a further transformation: naturalists increasingly considered it self-evident that all monsters occurred naturally, without any divine intervention. Their criticism of the *credulitas* of common people was therefore increasingly aimed at the pleasure with which they consumed accounts of monsters, rather than at their belief in prodigies.[198] Numerous canonical elements in the early modern discourse on monsters were now singled out with greater determination than before and banished to the realm of "fables."

CHAPTER FOUR

A Centaur in London

The residents of London were used to all sorts of "wonders" on exhibit in the city. But the phenomenon announced in March 1751[1] in a slim volume (fig. 25) distributed by Cooper's, a bookseller specializing in pamphlets on Pater-Noster-Row,[2] near St Paul's Cathedral, promised to be "the Greatest Wonder Produced by Nature in these 3000 Years"[3] in more than just its title. The public would shortly see a live centaur, in the flesh, at Charing Cross. Those who spoke a little German may have been struck by the centaur's name as it appeared on the front page: "Mr. Jehan-Paul-Ernest Christian Lodovick Manpferdt"[4] (*Pferd* being the German word for "horse"). His surname was clearly a fabrication. Was its bearer, then, too? To anyone pausing to consider for a moment the date of the advertised spectacle, "the First of Next Month,"[5] April 1, it would have been immediately clear that the announcement was meant tongue in cheek. April Fool's jokes were already a familiar concept in eighteenth-century England.[6] The March 1751 edition of the *Gentleman's Magazine* offers further insight into the contemporary reception of the pamphlet, which it lists among recent publications. The brief description of its contents is favorable: "Contains a short narrative of the Centaur's birth, adventures and peculiarities, with an answer to some objections against shewing him in publick, and is

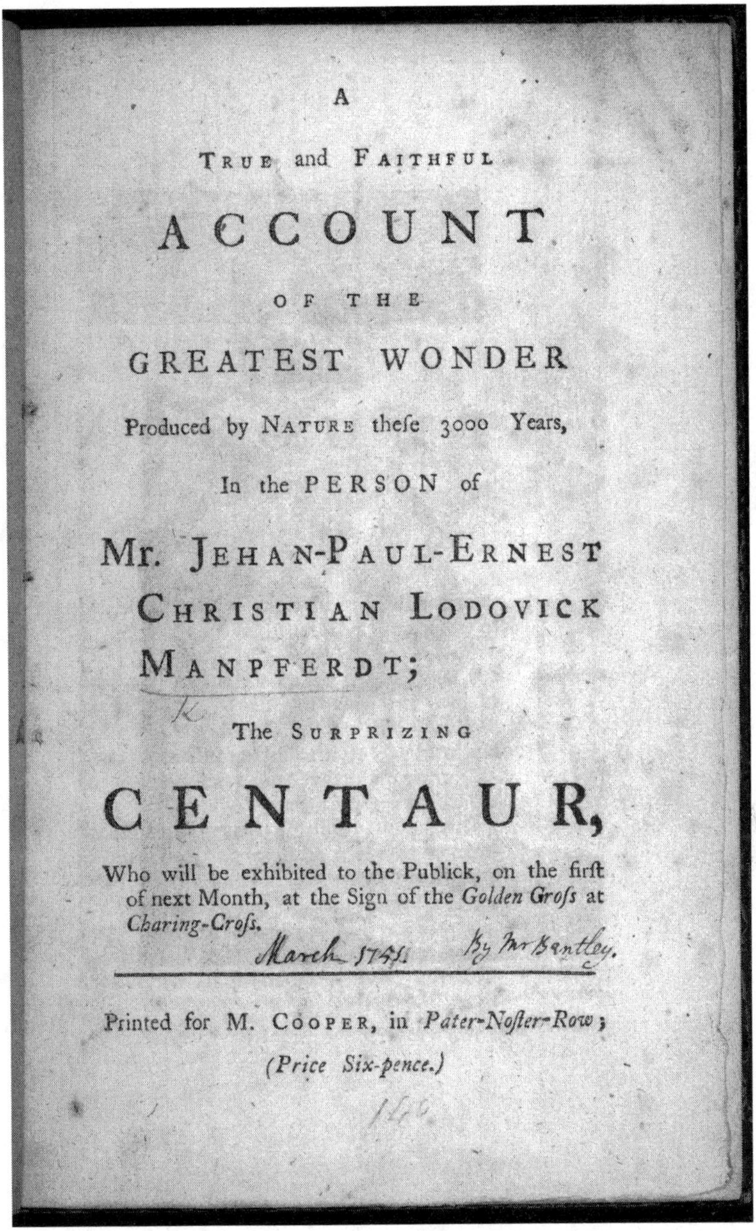

A

TRUE and FAITHFUL

ACCOUNT

OF THE

GREATEST WONDER

Produced by NATURE thefe 3000 Years,

In the PERSON of

Mr. JEHAN-PAUL-ERNEST CHRISTIAN LODOVICK MANPFERDT;

The SURPRIZING

CENTAUR,

Who will be exhibited to the Publick, on the firft of next Month, at the Sign of the *Golden Grofs* at *Charing-Crofs.*

March 1751. *by Mr Bentley.*

Printed for M. COOPER, in *Pater-Nofter-Row*;

(Price Six-pence.)

Fig. 25. Front page of Richard Bentley, *A True and Faithful Account of the Greatest Wonder Produced by Nature in these 3000 Years, in the Person of Mr. Jehan-Paul-Ernest Christian Lodovick Manpferdt; the Surprising Centaur, who Will Be Exhibited to the Publick, on the First of Next Month, at the Sign of the Golden Cross at Charing Cross*, London 1751. The British Library Board: 12330.g.26.

not destitute of humour or moral."[7] The author of this anonymous summary[8] thus viewed the booklet as a successful satire—a witty text with a moral.

The April edition of the magazine again mentions the Manpferdt pamphlet. This time it is reviewed in detail, and very positively. In addition, the image of the centaur from the brochure is reproduced (fig. 26). The first sentence of the review gives us an indication of how the satire was read at the time, as a criticism of credulity: "OF the many attempts to expose the ridiculous credulity of the idle part of this city, that of the wonderful Centaur, lately proposed to be exhibited to public view, on the first of April seems to demand the preference."[9]

"Idle" as it is used here does not simply mean "lazy." Rather, it refers to more general moral shortcomings in the individuals described, and thus to their vulgarity. The phrase "the idle part of this city" refers to those Londoners who concerned themselves with fruitless things[10]—in other words, with matters that did not deserve the attention of the distinguished readers of the *Gentleman's Magazine*.

The author of the satirical text, which was by no means his only contribution to this genre, was the notoriously indebted writer and illustrator Richard Bentley (1708–1782). His Manpferdt satire is a commentary on the exhibition and consumption of wonders in the London of his time, and it also addresses the scholarly discourse on monsters. Bentley's critique is astute, due no doubt in part to the fact that he was the son of the critic and librarian Dr. Richard Bentley Sr.—who was central in the development of historical philology—and had enjoyed a solid academic education in the liberal arts at Cambridge. He completed an MA and was for a time a fellow of Trinity College.[11]

An analysis of his text affords some insight into the economy of curiosity that was expressed in the cultural practices that he satirized. Bentley's text also sheds light on the critical view taken by contemporary intellectuals of the tireless curiosity of the unlearned, one of the objects of which was frequently monsters. Closely connected to this criticism was the need these authors felt to distance themselves from such individuals by virtue of taste.

In the following, I examine this text along with the influential polemic *A mechanical and critical inquiry into the nature of hermaphrodites* (1741) by a member of the Royal Society, James Parsons. It provides significant insights into the premises on which the naturalist discussion of this subject matter was based at the time. The lessons learned from this will place, in a second step, the starting point of this study—the rumor of a centaur mentioned by Albrecht von Haller, and thus Haller's discussion of the naturalist subject of monsters as a whole—in its proper context.

Fig. 26. "An authentick Account of the surprising CENTAUR, the greatest Wonder produced by Nature these 3000 Years, lately proposed to be exhibited to public View, &c." *Gentleman's Magazine* 21 (1751), 153, British Library Board: 249.c.21.

Haller held his Göttingen lectures on forensic medicine (*Vorlesungen über die gerichtliche Arzneiwissenschaft*), in which he mentioned the rumor of the London centaur, in the summer semester of the year Bentley's centaur satire appeared. In this chapter, we return to this passage, which is familiar to us from the introduction, to lastly clarify why, in the eighteenth century, it was no longer possible from a naturalist's point of view for centaurs to exist—not in London, and not in distant parts of the world or in earlier times.

As we shall see, the disappearance of the centaur from "the true,"[12] which stands here as a pars pro toto for fundamental changes in the object of the *monstrum*, does not express a victory of observation over reading, a victory of empiricism over book-based scholarship. Albrecht von Haller was no less studious a reader than the scholars of earlier generations described in the preceding chapters. But the scholarly practices employed in the study of nature had evolved. Observation became firmly established as a scholarly practice and an epistemic category.[13] Nevertheless, observation and reading continued to be inextricably linked. Generally speaking, a naturalist now read about monsters with a view to *critical selection* rather than *collecting*. The same applied to observation: wherever possible, observations—one's own as well as those of other observers—were to be repeated. Not because they were fundamentally questionable, but because it was believed impossible to prevent an error creeping in. The now-prevailing scholarly practices were no longer focused on the *copia* and *varietas* of one's own discourse.

These changes in reading and observation practices had major implications for the way in which the object of the *monstrum* was defined. These implications were the greater because of a new approach to intellectual time unique to the Enlightenment: from the perspective of an "enlightened" intellectual, the observations and opinions of earlier authors were no more reliable than the views and perceptions of unenlightened contemporaries. After all, earlier generations of scholars had not had the benefit of the blessings of the Enlightenment. How, then, could one give unreserved credence to the observations of, for example, a Renaissance naturalist? This approach to intellectual time placed the observations of earlier scholars, along with the perceptions of unenlightened contemporaries, under general suspicion. Both groups lived in unenlightened times, as it were, and from this perspective appeared susceptible to credulity (*credulitas*).

Consequently, the monsters written about by naturalists such as James Parsons and Albrecht von Haller correspond only partly with the monsters of their predecessors in the sixteenth and seventeenth centuries. This relates not so much to their place in specific academic fields or realms of nature—like Fortunio Liceti and others, Haller not only discusses anatomical and physiological monsters from the field of zoology but also makes passing mention of monstrous plants—as to the fact that certain kinds of monsters had disappeared, or were on the verge of disappearing, from "the true." While for Renaissance naturalists, collecting monstrous factoids with a view to a *scientia monstrorum* represented a value in its own right, enlightened scholars like Haller and Parsons were concerned with a critical survey of recorded cases—even if this only ever involved the highlighting of individual cases and never a systematic review.[14] Accumulating a great wealth and variety of cases was no longer considered desirable per se.

How a Centaur Came to Early Modern London

Let us now return to the question raised in the introduction of how a centaur could have found its way to early modern London. Bentley's satire provides specific information about this. It presents itself to the reader in the guise of an announcement in which the fictional authors, the "Proprietors of Mr. *Jehan-Paul-Ernest-Christian-Lodovick Manpferdt*," address themselves directly to "the Publick,"[15] and particularly to "the Curious,"[16] to inform them of the forthcoming exhibition of the centaur in London.

Manpferdt's proprietors announce two objectives. First, they expect the actual exhibition on April 1, 1751, to attract too great a crowd of all ranks to be able satisfactorily to answer all questions at the event.[17] Their brochure therefore provides some answers in advance. Second, they wish to counter "the many fabulous Accounts of the Centaur which have been, it seems, industriously spread, by certain Persons, gifted with so unaccountable a turn of Mind, as to find greater Satisfaction in their own idle Interventions than from any Information of Truth."[18] These alleged adversaries of Manpferdt's proprietors play an important role in the satire. They are said to be rival showmen, the proprietors of various localities in London with such expressive names as "the Panopticon," "Colossus," "Sea-Lioness," and "*Chien Sçavant*."[19] Evidently anxious about their own earnings, they seek to hinder Manpferdt's successful exhibition. The false information circulated by them is only hinted at; it seems to involve the suggestion that Manpferdt's proprietors, having announced a centaur, are now unable to keep their word.[20]

These rivals were not the only parties placing obstacles in the path of the booklet's fictional authors. Transporting Manpferdt to London proved to be exceptionally difficult for a number of reasons. To give one example, when they wanted to ship the Swiss-born centaur[21] to London from a Dutch port, an embargo was placed on the proprietors at the instigation of a "powerful prince" who was said to be determined "to force Mr. *Manpferdt* to serve in his Army as a Trooper."[22]

Is this an allusion to the practice of the "Soldier King"—the Prussian king Frederick William I—of enlisting from Prussia and abroad for his regiment of "tall lads"? It seems likely. During Frederick William I's lifetime, the minimum height for the members of these three battalions, who were recruited starting in 1718 even beyond Prussia's borders using the coarse and sometimes violent methods typical of the time, was six Prussian feet, or 1.88 meters.[23] It was therefore entirely possible that contemporaries might have made the connection to giants, and thus indirectly to other canonical monsters such as centaurs—especially in early modern London, where it was not uncommon for giants to be exhibited for a fee.[24] Certainly the English writer Horace Walpole (1717–1797), who in the course of his reading made numerous marks in the copy of Bentley's satire preserved in the British Library, did not hesitate to note "K. of Prussia" next to this passage.[25]

Manpferdt was brought to London, then, overcoming these and other hurdles, because his proprietors justifiably hoped to make money exhibiting him there. Besides entertaining the reader—and even today, the slim volume certainly makes for an amusing read—Bentley's main intention was to comment on certain basic traits of the "public" of London at the time, particularly of those sections he describes as "the Curious," with sometimes sympathetic, sometimes caustic humor. These commentaries are often couched in the language of sin.

The Economy of Curiosity

Bentley's commentary on the London public begins immediately after the account of Manpferdt's fraught passage to England. On arriving home, his proprietors allegedly found a letter waiting for them from a certain Mr. Whitfield, imploring them "by every thing they hold sacred"[26] to abandon the plan to exhibit their centaur. This spectacle "will be giving an Encouragement to the most horrid Vices, among the fine Gentlemen of the Age, to which (he says) they are already only too much addicted; if not a Means of introducing new ones, to which they may be the proner, from an Appearance of Utility. He owns his Apprehensions are exceedingly heightened by the near Approach of *Newmarket* Meeting."[27] Whitfield feared that this spectacle might exacerbate the sinful behavior already prevalent among the "fine Gentlemen of the Age" and perhaps even incite them to further sin.

Both points would likely have been self-explanatory to the contemporary reader, but today they require clarification. The sin to which the "fine Gentlemen of the Age" had supposedly already succumbed is clear from the intention of the letter's addressees to put their centaur on display. The reference is to the curiosity with which the continually new wonders that could be marveled at, for a fee, in eighteenth-century London were met by some inhabitants of (and visitors to) the city.

It is true that the status of curiosity among scholars had risen over the course of the early modern period, but as we have seen with the *curiosi*—the members of the Academia Naturae Curiosorum—curiosity did not shed its negative associations entirely in the process.[28] For one thing, older positions that condemned *curiositas* lock, stock, and barrel never fell completely silent. And for another, critical contemporaries were able to distinguish between the selective and tenacious curiosity of the scholars and the curiosity for novel marvels of the common folk.

The famous lexicographer and critic Samuel Johnson (1709–1784), who lived in London from 1737 on, was one such staunch advocate of scholarly curiosity. He published a number of essays on the subject in his periodical the *Rambler*, which was issued twice weekly between 1750 and 1752. In the final issue, he stresses once again how important it was to him to distinguish between "good," scholarly curiosity and curiosity that was "bad" because it was focused on novelty: "I have never complied with temporary curiosity, nor enabled my readers to discuss the topick of the day."[29] As Barbara M. Benedict has shown, Johnson advocated an "elite curiosity that concerns high matters of learning and beauty,"[30] that is, a form of curiosity very different both in subject matter and method from the "low wonder mongering" of the general, unlearned public from whom he sought to distance himself.[31] Johnson's pejorative use of the early modern coinage "wonder mongering" described individuals who contributed to the circulation of wonders, whether by sharing accounts of them or by earning money through their exhibition or sale.[32]

For Johnson, the objects of these two forms of curiosity differed in ontological status, too. In his view, desirable curiosity engaged only with verified objects—unlike its base counterpart: "Noble curiosity explores the real, whereas ignoble curiosity explores the rumored, marvellous or fantastic."[33]

What Whitfield held to be the widespread sin of "fine Gentlemen" can properly be understood only in the context of this contemporary distinction of Johnson's. Mr. Whitfield emphasizes the pervasive "addiction" to marvels among his contemporaries. The exhibition of a centaur must have seemed eminently suited to fanning the flames of this passion for perpetually new—and often unreliably verified—marvels.

The new sin Whitfield's letter alleges was of a sexual nature and—in the early modern understanding of nature—unnatural, which no doubt explains at least in part why it is only hinted at. The reference to the upcoming "Newmarket Meeting" provides the first indication of what is meant. Bentley is alluding here to a horse race held annually since the seventeenth century in Newmarket, in the county of Suffolk, scarcely more than one hundred kilometers northeast of London.[34] This sin's purported "utility" provides the second clue: Mr. Whitfield is alluding to sodomy, in this case the mixing of man and horse, and is insinuating that the "fine Gentlemen" might decide to breed centaurs for commercial horse racing.

The response by Manpferdt's proprietors is a miniature masterpiece of satire. It rebuts Whitfield's concerns with a series of rhetorical questions—passing an even more pessimistic judgment on the sinfulness of London society:

> The Proprietors only beg the Favour of Mr. *Whitfield* seriously to ask himself if he really thinks the present Mode of Vice can actually be encreased? or, if any Kind of it can be introduc'd new? and, whether the *Appearance of Utility*, as he is pleased to call it (by which he must mean, if he means any thing, the establishing of a Breed of Centaurs here, as of Mules abroad) would not of itself be enough to deter any a fine Gentleman from a Crime? So that they have all the Reason in the World to flatter themselves, that if their Shew were of Consequence enough to bring about any Change in the present Taste, it is impossible it should not be for the better; and that so far from being chargeable with giving a helping Hand to the Depravity of their Country, they should deserve Thanks for having diminished it or at least turn'd it to the publick Advantage.[35]

Whitfield's concerns are unfounded, they say, because the sinfulness of his contemporaries could scarcely be increased, and because, far from encouraging a "fine Gentleman" to commit a crime, the "Appearance of Utility" would in fact be a deterrent.

It was not only the consumption of the showmen's spectacles, however, that appeared questionable to contemporary intellectuals but also showmanship in general. This was thanks to a widely held view—one shared by many scholars and naturalists—that participation in commerce had a negative effect on an individual's virtue. Adrian Johns has shown, for example, that early modern natural philosophers often had an ambivalent relationship to the book trade. They depended on it to reach a wider circle of readers, but as gentlemen they distanced themselves from any notion that their writing should be turned to financial account—especially for third parties.[36]

This tension between commerce and virtue was not unique to the early modern period. It can be traced back to the ancient Greek belief that physical labor was harmful to body and mind and impaired the ability to exercise virtue. This view was widely accepted in the Renaissance, and the commercial aspect of physical labor was central to its devaluation. Commercial enterprises acquired a collective reputation for being morally questionable. Henry Peacham's (1576?–1643?) guide to living as a gentleman, for example, contains the following list of inadmissible activities: "Whosoever labour for their livelihood and gaine have no share at all in Nobilitie or Gentry. As painters, stageplayers, Tumblers, ordinary Fiddlers, Inne-keepers, Fencers, Iugglers, Dancers, Mountebancks, Bearewards and the like . . . The reason is, because their bodies are spent with laboure and travaile."[37] As can be deduced from Peacham's list, the devaluation of physical labor and profit seeking cast showmanship in a doubly negative light. From this perspective, traveling showmen were not virtuous, and as Mr. Whitfield's letter to Manpferdt's proprietors shows, neither was the consumption of their labor.

Bentley's pamphlet thus indeed has a "moral," as the anonymous author in the *Gentleman's Magazine* puts it. It relates primarily to the curiosity and credulity of many contemporaries concerning questionable subjects such as "fabulous" monsters. But the widely held belief that a member of the social elite should eschew commercial things is also expressed in the brochure.

In addition to a general critique of contemporary society, Bentley's booklet also contains a whole series of specific allusions to happenings in London that show how well informed the author was about the cultural life of the metropolis. They contribute to the text's satirical effect, that is, its suitability for interpretation as a commentary on real events and conditions. One of these references is found in the explanation given by Manpferdt's proprietors for their disinclination to make him available to the public right away: "So much for Mr. *Whitfield*; notwithstanding whose Remonstrance, the Proprietors would not have so long deferr'd producing their Centaur to the Publick, . . . if they had not thought it incumbent upon themselves to postpone it, out of Respect to a Troop of Persons of Distinction, who thought proper to make a Shew of themselves at that Time."[38] This is a reference to a performance of *Othello* held at the Theatre Royal on Drury Lane on March 7, 1751.[39] The brochure's fictional authors felt obliged, it is suggested, to wait until after this performance lest the two shows find themselves competing for an audience.

The explanation of their motives that follows is, at the same time, an astute commentary on the economy of curiosity in the middle of the eighteenth century: "They own besides, their own Interest directed them not to let any thing so extraordinary interfere with what they had to produce, as the first Run of Curiosity and Surprise, is the only Harvest they have to reckon upon from the Town, having constantly observ'd that very little Time is requisite to familiarise it to the greatest Monsters in Nature."[40] Unlike the persistent, tenacious scholarly *curiositas*, the curiosity of London commoners was, in the view of its critics, itinerant, and their attention span brief. They soon had their fill of any *monstrum*, though in the long term their insatiable hunger for new curiosities remained unappeased.

The social criticism that Bentley ascribes to Whitfield—and, indirectly, to Manpferdt's proprietors—was, as Johnson's example shows, not an isolated position. Many intellectuals, and naturalists in particular, took a very critical view of the consumption of wonders by the London public (and urbanites elsewhere). As with Johnson, the credibility of the wonders on display was often a focus, as I demonstrate by way of a prominent example. First, however, we must ascertain who and where the audiences of such spectacles were.

The Monsters of London and Their Audience

The satire of the London centaur makes a revealing reference to the extratextual reality of the city in the location of the exhibit: "at the *Golden Cross* at *Charing-Cross*."[41] At this time, Charing Cross—which in the mid-eighteenth century was already an important crossroads in the heart of the city—was indeed an arena for numerous spectacles of this kind.

Human and animal monsters, exotic animals and "savages" from the colonies, and curiosities such as dancing bears and monkeys performing tricks were exhibited throughout the year all over London. Like Manpferdt, they were often procured from distant lands specifically for this purpose. They were to be encountered not only at fairs but also out on the streets in many parts of the city, where they could be viewed for a small fee. Traveling showmen frequently lodged near public houses. And their shows often took place in the taverns themselves or in their courtyards and stables.

In the eighteenth century, the public houses around Charing Cross were established venues for such entertainment.[42] The Golden Cross at Charing Cross, in the vicinity of which the centaur in Bentley's satire was to be exhibited, was one such inn. First mentioned in 1643, it survived into the nineteenth century, when it was demolished for the construction of Trafalgar Square.[43]

Where did the clientele for these spectacles come from? The population of London in the eighteenth century was already approaching the one-million mark. Potential customers were mostly members of the urban working class: they did not read, or only a little, but they had enough spare cash to indulge from time to time in shows like these. Then there were the inhabitants of the region surrounding London—at this time England's population was concentrated chiefly in the South, whereas the Midlands and the North were comparatively sparsely inhabited. London was the country's only metropolis, so it was the obvious choice for anyone wanting to take a trip to the big city.[44]

But the exhibitors' spectacles were not consumed only by the common people. In addition to workers, who were probably the largest group in numbers, the curiosities showcased in public houses and at fairs also attracted noble virtuosi and scholars. This latter group, however, often mingled with the other onlookers for reasons entirely of their own: "Although most came because their innate relish for the sensational, the mysterious, and the grotesque was titillated by stridently announced new importations, some—the educated minority—came out of genuinely scientific motives, to amplify and verify the descriptions they had read in learned treatises."[45] The showmen were aware of this distinction and accommodated it. An undated handbill preserved in the British Library advertising a hermaphrodite on display in London in 1741 divides the audience in two on the basis of language. It bears the heading "Hermaphrodite (Lately brought over from Angola)" and begins with a general description of the hermaphrodite in English. The second half of the text, meanwhile, is written in Latin. It describes the anatomy of the individual[46]—a subject that, first, would have been of greater interest to scholars than to the rest of the audience, and the details of which, second, decency demanded be made available to educated (and male) readers only. The advertisements for this spectacle that were placed in various newspapers throughout 1741 were likewise bilingual.[47]

Advertising copy for shows with monsters often referred to the writings of contemporary scholars. This practice is reflected in Richard Bentley's satire, too: Manpferdt's proprietors venture the supposition that they are expected "to give some Account of what the learned Professor *Zeiglerus* has already publish'd upon this Head in Two large Vols. folio, at *Basle*, 1744."[48] To meet this expectation, they continue, they have gone to great lengths to obtain one of the few copies of this work available in Great Britain.[49]

The reproduction that then follows of the table of contents from the first book of Zeiglerus's treatise on centaurs attests to the fact that Richard Bentley was familiar with at least some of the scholarly literature on this subject. More importantly, he understood how this discourse operated. He uses the chapter headings to comment on scholarship in general, particularly on the way subjects such as centaurs were discussed by contemporary naturalists. This is especially striking in the case of the sixteenth chapter. Its title serves as a comment on the central importance typically ascribed to direct observation in eighteenth-century nature study:

> XVI.
> How cautious we ought to be how we charge Antiquity with Fable, for relating Things, the like of which have not happen'd in our Days, as if our own narrow Observation was the Test of all that had been, or could be in the World.[50]

Zeiglerus appears here as an advocate of an outdated model of scholarship, so to speak, on the basis of which he criticizes what—from this perspective—is the exaggerated and one-sided contemporary belief in enlightened observation and the overhasty demotion of ancient knowledge.

Fittingly, Zeiglerus himself is indirectly characterized as an example of a scholar who has lost touch with the spirit of the times. This characterization is achieved first by means of his name: it is a play on the German verb *zeigen*—"to demonstrate or show"—and thus indirectly characterizes its bearer as an old-fashioned, professorial type of scholar. Second, Bentley has him hail from Switzerland. Contemporaries considered the Swiss Confederation backward compared to the centers of European scholarship.[51] In short, Bentley has the proponent of an outdated form of scholarship from the intellectual periphery wield his knowledge and his way of knowing against the views and working methods of contemporary naturalists and intellectuals. In so doing, he reveals the contours of the contemporary discourse on monsters.

The function of this appeal to the scholar Zeiglerus is to lend greater authority to the words of Manpferdt's proprietors[52] and to extend their narrative to include the creature's background. The Swiss scholar confirms the existence of the centaur and tells the story of his birth in Switzerland and flight from his home country. The centaur's owners present this account in the form of a long English quotation incorporated into their discourse.[53]

In addition to the differing expectations among those consuming showmen's spectacles that already existed in the seventeenth century, new developments emerged in the middle decades of the eighteenth century that related to the social composition of the audience. Members of the upper classes, particularly the nobility, increasingly considered it an affront to their sense of decency to contemplate monsters and other curiosities *together* with the common people. The traveling showmen accommodated this desire for distinction, too.[54]

The Angolan hermaphrodite was therefore not only viewable daily, like Manpferdt, at the Golden Cross inn near Charing Cross—from 9 a.m. to 3 p.m. and then again from 4 p.m. to 10 p.m.—but interested parties were also able to arrange private viewings on their own premises.[55] Such offers were geared toward members of the upper classes and as such made provision for their desire for distinction, as is again clearly evident in Bentley's satire: Manpferdt could, his owners stated, be viewed daily from 10 a.m. to 4 p.m. In addition, "ladies" could arrange private conversations and rides.[56] Moreover, access to the centaur in a private setting had been granted free of charge to sections of London society in the two weeks prior to the brochure's publication, and thus before the official commencement of the public viewing. These private audiences were made up of "many of the Nobility and Virtuosi," in other words, members of the nobility and affluent nature study enthusiasts.[57]

Among intellectuals, this desire for distinction was often expressed as criticism of the audience. Alongside their curiosity, it was particularly the presumed superstition and lack of enlightenment of the uneducated—or poorly educated—audience of such shows that were depicted as objectionable. Even in the seventeenth century, the ways monsters were viewed by naturalists and by the *vulgus* differed significantly. Initially, however, naturalists discussed among themselves what they deemed the objectionable views of the commoners for the most part, primarily their tendency to interpret monsters as prodigies.[58]

In the eighteenth century, against the backdrop of belief in the human capacity and need for improvement, these differences manifested in new ways. Intellectuals now increasingly sought to address the commoners directly in order to disabuse them of erroneous views. And it was less often their belief in prodigies (which was meanwhile less pronounced) with which they found fault but rather increasingly their willingness to believe in the existence of certain monsters at all.[59] This is demonstrated in the following, with the Angolan hermaphrodite as a case in point.

The Hermaphrodite at the Golden Cross

It was, primarily, the exhibition in 1741 of the "Angolan hermaphrodite" that prompted James Parsons (1705–1770) to publish a polemic that same year.[60] Like Manpferdt, the hermaphrodite was displayed at the Golden Cross near Charing Cross—and for at least eight consecutive months, which testifies to the show's extraordinary popularity.[61] Parsons had studied medicine in Paris before going on to acquire extensive anatomical knowledge and experience in obstetrics in London and in 1741, before the publication of his polemic, became a fellow of the Royal Society.[62] Although just a single edition was published, it enjoyed wide circulation.[63]

And this can only have been as Parsons wished, because his aim was, after all, to establish what he considered an enlightened view of the subject.

James Parsons advocated an uncompromising version of the thesis that was increasingly to become dominant over the course of the rest of the eighteenth century, according to which hermaphrodites did not, indeed could not, exist. For all their differences with regard to details, the critics of this category agreed that supposed hermaphrodites merely appeared to be of dual sex. The marks of the sexes were, in their view, only superficially combined in such individuals.[64] Thus Parsons reasons in his polemic "that no hermaphrodital Nature can exist in human Bodies; and, in fine, that those Subjects hitherto so accounted, were only Females in all Respects, superstitiously, and through Ignorance, mistaken for those Kind of Creatures, or for Men."[65]

For the sake of brevity and decency, Parsons describes the women who in his view were wrongly regarded as hermaphrodites as *macroclitorideæ*. Parsons coined this term himself.[66] Ultimately, though, it corresponds with the older category of the tribade, a woman with an enlarged clitoris. And the fear of sexual acts between women that, as early as 1600, characterized the discussions about hermaphrodites and tribades that raged, particularly in France, is present in his text, too. It was wise, Parsons writes, that clitoridectomies were performed in certain regions of Asia and Africa where there was a high incidence of women with enlarged clitorises—not only because of the difficulties it posed for sexual intercourse between a man and a woman, but also because it could provoke sexual acts between women.[67]

Of greater importance to our analysis than the individual theses put forward by Parsons is the fact that he used them to call for a different *perception* of the individuals in question. The occasion for his text—the exhibition of the (purported) hermaphrodite at the Golden Cross—in itself shows that he was concerned to set himself apart from the "unenlightened" and "superstitious," and that he saw his intervention as an act of edification and enlightenment. His rhetoric points in the same direction.

In addition to Parsons's use of the imagery of light that was so typical of the Enlightenment,[68] the role of anatomy in his rhetoric is striking. Anatomical knowledge and monsters had long been intimately connected. The physicians describing deceased monstrous newborns in seventeenth-century scholarly journals always sought to obtain their bodies for dissection. And as early as the sixteenth century, anatomists like Realdo Colombo engaged intensively with rare anatomical phenomena, including hermaphrodites and other elements of the discourse on monsters that were canonical at the time.

Knowledge of human anatomy has an important function in Parsons's argumentation, too. Among other factors, he holds the anatomical ignorance of his predecessors responsible for their erroneous belief in the existence of human hermaphrodites: "It will not be very difficult to account in some Measure, for the rise of such erroneous Imaginations, if we only consider how ignorant the World was in former Ages of the animal Structure, and even of those that understood ought of it, how few there were, who (from the Obscurity of the Clitoris in Females in a natural State) knew that any such Part existed."[69] Anatomy lent itself superbly to Enlightenment imagery. As a field of study that lays open the interior of the body, which is initially hidden from view, anatomy easily came to epitomize the working method of the enlightened naturalist. Just like an anatomist, the enlightened naturalist had to lay bare the truth obscured by ignorance and superstition. In his introduction, Parsons characterizes his objectives in these terms: "to hunt out the Truth, which is often very intricately environed round with dark Veils of Ignorance and Superstition."[70]

The predilection for the rare and unusual that naturalists, at the end of the seventeenth century, still shared with large sections of society was now proof of a lack of enlightenment, in scholars and common folk alike.[71] It was for this reason that the exhibition of the "Angolan hermaphrodite" prompted Parsons's treatise, particularly as it had already inspired several of his colleagues to publish or to include illustrations of the sex organs of the (purported) hermaphrodite in their writings. And not all of them had shown the same profound skepticism as Parsons.[72] James Parsons's text was, as Palmira Fontes da Costa observes, a veritable " 'Enlightenment guide' to what should be seen at Golden Cross."[73]

It was not unusual in the eighteenth century for an intellectual to try to tell sections of the population how to view a rare phenomenon. At Reform universities like Halle and Göttingen, the venue of Haller's lectures on forensic medicine, lectures in which the educated were given instruction in what they were to make of a certain phenomenon were a veritable genre in their own right. Questions of taste also came into play inasmuch as certain views of wonders had come to be regarded as "vulgar."[74]

On March 24, 1716, for example, professor of philosophy Christian Wolff (1679–1754) gave a public lecture to an unusually large audience at the University of Halle about a sighting of the northern lights in Halle and elsewhere a few days previously, on the evening of March 17. Many "inexperienced in the understanding of nature," Wolff said, were "filled with consternation,"[75] because they believed that it could be a divine omen. He took the opportunity to challenge the widespread and, in his view, superstitious,[76] belief that such optical phenomena were omens sent by God. Superstition, he said, often invented things that were not there,[77] which was why northern lights were often erroneously described as armies battling in the skies, crosses, or the like. The lecture appeared in print that same year. The fact that the lecture was held and published in German shows that, unlike the *curiosi* in their discussion of belief in prodigies in the late seventeenth century, Wolff no longer wanted to preach solely to the choir.[78]

It was also an eighteenth-century development that naturalists sought to dissociate themselves from public spectacles of the preternatural such as the exhibition of the Angolan hermaphrodite. This was a product of the devaluation of wonder and of curiosity about marvels. Having formerly been considered emotions that could, under certain conditions, be conducive to understanding, they were now reduced to badges of the vulgar, credulous general populace.[79] Just a few decades earlier, naturalists were included as a matter of course in the segment of the public referred to in Richard Bentley's centaur satire as "the Curious," who readily paid to see monsters and other curiosities.[80]

Ample evidence that seventeenth-century scholars consumed the wonders of the fairground can be found, for instance, in the diary of the gentleman, virtuoso, and Royal Society founding member John Evelyn (1620–1706). At a fair in Rotterdam in August 1641, for example, he observed with great interest a hen with two large spurs growing out of its side, and a four-legged rooster with two "rumps or vents." He judged the rooster "most strange" and made note of the information given by the exhibitors that, from the surplus hindquarters growing out of its breast, it "likewise voyded dongue, as they assur'd us."[81] On September 15, 1657, he traveled in company to London, where he not only admired a famous tightrope performer named "the Turk" but also spoke at length with a "Hairy Maid" from Augsburg. He made notes on her outer appearance as well as her assertion that neither her child nor any other member of her family exhibited the same anomaly.[82] At "St. *Margarites* faire" in Southwark on September 13, 1660, he observed, alongside trained monkeys and an Italian acrobat, a severely deformed one-year-old child and a preserved specimen of a "monstrous birth of Twinns, both female & perfectly shaped, save that they were joyn'd breast to breast, & incorporated at the *navil*."[83]

These activities did not conflict with Evelyn's involvement in the Royal Society but fit seamlessly in with its work. This is clear from his report of a meeting of the society in the latter half of April 1705. "Nothing remarkable this Weeke," his journal entry begins. Evelyn then reports that he visited the Royal Society again for the first time in a long while. Topics of discussion there included the self-portrait of a giant, a soldier in the service of Elector Palatine Johann Wilhelm.[84]

Parsons sees the lack of distance from traveling showmen and their exhibits that was maintained by earlier naturalists as one reason many of them believed in the existence of hermaphrodites. He speculates, for example, that Thomas Allen—who in 1667 published a report in the *Philosophical Transactions* on a hermaphrodite exhibited in London that year and mentioned by Evelyn—had relied on statements made by the exhibitor. This, he believes, led Allen mistakenly to conclude that the individual in question really was a hermaphrodite.[85] Parsons goes to great lengths to refute Allen's impression that Anna Wilde was a hermaphrodite, maintaining that she was simply a woman.[86] He concludes by stressing "how little credit ought to be given to the Tales of Shew-men, by the Learned."[87]

Parsons's monograph can be described as remarkably learned—not only in view of its second chapter, in which the established theories on the genesis of hermaphrodites are discussed at length. It testifies to his intimate acquaintance with the relevant literature, which he frequently interrogates with philological finesse. Nevertheless, Parsons overemphasizes the importance of observation for his work. He repeatedly stresses that he accepts nothing but his own direct observations and those of trustworthy witnesses as sources of knowledge. And this is the case not only in relation to the individual exhibited near Charing Cross. He is, for instance, willing to acknowledge the existence of hermaphrodites in a small number of lower animals only because, first, he has had the opportunity on a number of occasions to assure himself of this fact with his own eyes and, second, he has "good testimony":

> Whatever the Necessity might be for the Creation of certain of the Reptiles of this Nature [i.e., the hermaphrodite "nature"], such as the Garden shell'd Snail, and the large Earth-worm, both of which are certainly so, which I can affirm from my own Knowledge, having often drawn both these asunder when in Coition, and observ'd them; as well as from so good Testimony, as Mr *Bradley* in his Philosophical Account of the Works of Nature, where he has several curious Observations on these Animals, and a Figure of the Parts of Generation of a Snail, done as they appeared in a Microscope.[88]

The example of the snails and worms that Parsons explicitly claims to have observed on multiple occasions shows that he believed that observations should be repeated—as was customary in the study of nature at the time. The call for not only more cumulative but also more repetitive observation began in astronomy and gradually prevailed in all areas of nature study in the early modern period. By the mid-eighteenth century, naturalists almost without exception held the view that scientific observation should ideally be practiced repeatedly and continuously. Its purpose was to take note of any details that may have been overlooked on first observation and to identify any errors that crept into an observation.

In this manner naturalists repeated not only their own observations but also those of other observers.[89] The repetition of others' observations did not normally imply any particular mistrust of one's fellow observers. It was a matter of routine, just as it was with observations of one's own.[90] Parsons's criticism of the *wrong* perception of hermaphrodites—which applied to contemporary naturalists, too— was an exception in this respect. As we have seen, this was linked not least to the object itself.

There was a second instance in which James Parsons felt compelled to intervene in the commerce of curiosity. In the case of at least one other monster on display, likewise at Charing Cross, he stepped in determinedly, and a showman was banished from city at his instigation. The man presented in a glass case a mermaid allegedly caught on the coast of Acapulco, which on closer and—in Parsons's view— *correct* observation proved to be a monstrous fetus.[91]

In other words, it was not only the category of the (human) hermaphrodite for which the air was getting thin in the naturalist discourse of the eighteenth century. A veritable shake-up of the discourse on the monstrous was underway, in the course of which the boundaries of the object were drawn anew, and increasingly narrowly. Those monstrosities that were still held to be real were, like Parsons's *macroclitorideæ*, increasingly considered in the context of embryological and anatomical questions.

The Epistemic Promise of Monsters

By the eighteenth century, in contrast to the late seventeenth, monsters were no longer discussed primarily as isolated phenomena detached from any systematic investigation of their causes. What gained currency instead was the view—already in evidence in Bacon's work—that the study of rare natural phenomena could provide general insights into hidden causes in nature. Monsters were now of particular interest because they promised insight into "normal" anatomy and the development of the embryo. The epistemic promise that monsters embodied in these two fields was a key factor in ensuring that, rather than dying down, the naturalist discussion of monsters in the early eighteenth century in fact generally intensified.[92]

Bernard le Bovier de Fontenelle (1657–1757) spelled out in programmatic terms the ends that the study of monsters was expected to serve. In 1709, in his capacity as *secrétaire perpétuel* of the Académie des Sciences in Paris, he defended his academy's comparative anatomy of both animals and monsters, pointing out that it furthered our knowledge about the normal human body.[93] The volumes of the academy annals in the following years implemented his agenda. And anatomists elsewhere, too, discussed the monsters they examined in relation to normal anatomy.[94]

Monsters were also viewed as offering insights for the fields of physiology and embryology. With regard to the concept of procreation—*generatio*—the life sciences in the late seventeenth and early eighteenth centuries were "largely dominated by theories that postulated the preexistence of preformed germs."[95] But the differences between the two main approaches available then, preformation and epigenesis, were less substantial than is often assumed, as Hans-Jörg Rheinberger and Staffan Müller-Wille explain: "Preformation, as opposed to epigenesis, meant simply that a germ capable of development is not an undifferentiated mass, but already possesses a given structure in which the future being, though not depicted in its entirety on a reduced scale, must nevertheless be sketched out in a certain manner. This is entirely compatible with the notion that the preformed germ is produced in the body of the parents, or indeed only at conception."[96]

From a physiological standpoint, the subject of monsters was of great interest in the middle decades of the eighteenth century, primarily because monsters appeared to enable observations that promised to resolve the bitter dispute between the supporters of these two main models of *generatio*. Proof that a conformation classed as monstrous could or could not have existed in an incipient form from the beginning but only arose from accidents during pregnancy would point to epigenesis or vice versa. In their debate on the topic, Albrecht von Haller and Caspar Friedrich Wolff invoked monsters in this way to support their respective positions. In the early eighteenth century, Benignus Winslow and Louis Leméry conducted their debate concerning divine versus accidental causes of monsters similarly.[97]

These disputes did not end in any consensus as to the theoretical explanation for monstrous births. Even the definition and classification of monsters were contentious. But the perspective from which monsters were studied changed radically—and with it the way they were discussed:

> What united the early eighteenth-century anatomies of monsters was not any particular theory, but the general framework of inquiry. Although each report began with an account of the historical particulars that could have been taken

from one of the more laconic broadsides, the emphasis throughout was upon matching form to function—a normal function either disrupted by malformation (emphasized by those who, like Leméry, believed monsters to have accidental causes), or served by extraordinary means (the favourite examples of Winslow and his allies, who insisted on divinely guided preformationism). Anatomists no longer exclaimed over the rarity of a malformation, but rather over its perverse functionalism.[98]

Thus, the *monstrum* now no longer embodied the product of a playful nature, a *lusus naturae*—that is, an exception from the usual course of nature—a concept rejected by Fontenelle and others with reference to contemporary philosophers. Indeed, the *monstrum* was used by anatomists to demonstrate that nature, even when it appeared to be disporting itself, in fact obeyed fixed and universal rules.[99]

Albrecht von Haller and the Rumor of the London Centaur

In the year in which Bentley's satire was published, Albrecht von Haller also mentioned a London centaur to his students. But he described the news of its sighting as an unconfirmed rumor: "In the forties of the century in which we live, the rumor also circulated that there was a true, live centaur to be seen in London. However, this legend was never confirmed."[100] What epistemic status did this "rumor" have for Haller? And was the centaur of which Haller claimed to have heard perhaps in reality Mr. Jehan-Paul-Ernest Christian Lodovick Manpferdt?

The words of the Swiss physiologist and polymath were preserved in a late eighteenth-century German edition of the lectures on forensic medicine for his Göttingen students in the summer semester of 1751. At this point, Haller was professor of physiology, anatomy, and surgery at the recently founded University of Göttingen, and an influential one at that. His influence on anatomy and physiology, in particular, was considered defining by his contemporaries, and the same was true of embryology.[101]

The translation of the lectures is based on a Latin transcript by Haller's eldest son, Gottlieb Emanuel (1735–1786).[102] His original transcript has, to our current knowledge, not survived.[103] We are therefore unable to ascertain the Latin term rendered by the translator as "Gerücht" (rumor). Nevertheless, I endeavor to reconstruct the status held by the rumor in Albrecht von Haller's epistemology.

From an early modern point of view, a rumor was by no means always harmless. A rumor meant disquiet—not only literally but figuratively, in relation to social order. The entry on *Gerücht* in Johann Heinrich Zedler's *Universal-Lexicon* gives a number of Latin and French synonyms—"Fama, Rumor, Renommée, Bruit"—and defines it as "talk or report as is spread abroad either to honor or to insult, and can therefore be a good and ill report."[104] What the four synonyms given have in common is that, in their basic meaning, they each point to a sound, in particular noise caused by chatter or cries. Thus, for *fama*, the Georges Latin-German dictionary gives the basic meaning "Gerede" (talk, gossip, rumor, or report), and for *rumor* the meaning "Geräusch" (a rushing, rustling, or murmuring sound).[105] The meanings of the German term *Gerücht* also included "loud cries or shouting, noise, bustle, and bluster."[106] And the French *bruit* has this acoustic aspect of meaning, too.

In contemporary usage, there was also another form of disquiet involved. Monsters, according to the editors of the *Breslauische Sammlungen*, regularly provoked "a perceptible éclat or bruit."[107] Both literally and—at a societal level—metaphorically, they could cause unrest. Given the fact that monsters had often been regarded by common people as divine omens and therefore held the potential to stir up the populace, Haller's choice of words—whether he used the term *fama* in his lecture or *rumor*—may not have been coincidental.

Another central meaning of the term *Gerücht* is evident from Haller's statement: rumors were notoriously uncertain, which made them problematic for jurists and naturalists alike. Rumors could cause harm beyond the epistemological: their circulation could result in personal harm as well. The entry on *Gerücht* in Zedler's encyclopedia draws particular attention to this latter potential: of the two possibilities indicated—that a rumor could be good or ill—it expands only on the second. Ironically, the Breslau-born, Protestant Zedler uses the Inquisition as an example of the consequences of an ill report: "An ill report, if it is clamorous and without cease, and some few other allegations are added to it, can easily bring one before the Inquisition. But in this case a judge must investigate the true cause of such a rumor, whether it came about through lawful, rational causes, and from honest people, and indeed also consider the previous lifestyle of the person about whom the ill rumor circulates."[108] This brings us to another central, if not *the* central characteristic of rumors, one that was of great significance not only in courts of law but in the study of nature: the origins of rumors were (and are) notoriously unclear. A rumor was, in a sense, the opposite of an "authored observation," that is, an observation undertaken by a named observer and documented in writing.

A further complication was the fact that, since the Middle Ages, rumors had been associated with the common people. As "clamor, vulgar chatter, hearsay, gossip circulating among the common people,"[109] they were associated with precisely that (large) part of the population that was believed by eighteenth-century scholars still to be in need of enlightenment. The views and observations of this section of the population were accordingly treated with great reservation. We may recall that, according to Samuel Johnson, the "bad" curiosity of the common people was devoted to what was known merely through rumor and to other unverified phenomena ("the rumored, marvellous or fantastic"), whereas the "good" curiosity of the scholar concerned itself solely with verified objects ("the real").

In times when the self-conception of naturalists meant that they were willing to accept little other than their own direct observations or those of trustworthy colleagues known to them by name and that they had ideally repeated themselves, they inevitably took a critical view of rumors. If they were worth the trouble at all, they had to be verified. Albrecht von Haller, for his part, remained skeptical about the rumor of the London centaur—not least because it had, to his knowledge, never been confirmed, which is to say there was no credible, authored observation to support it. It is therefore only fitting that he concludes by characterizing the "rumor" as a *Sage*—a legend or tale. This term likewise referred, as it still does today, to a statement of obscure origin and uncertain veracity.[110]

But where did the rumor originate? Could Haller, in some circuitous way, have heard about Richard Bentley's Manpferdt? In other words, in the context of the satire's reception, could the rumor have taken hold that a live centaur had been displayed in London? It is not out of the question, even though Haller dates his rumor to the 1740s, whereas Bentley's brochure was published just a few weeks or months before the lecture.

Albrecht von Haller had close ties to England. His *peregrinatio academica*, after obtaining his doctorate in Leiden in 1727, took him not only to Paris, Strasbourg, and Basel but also to London and Oxford.[111] He had numerous correspondents in the British Isles[112] and was accepted into the Royal Society in 1739.[113] Haller spoke English, and numerous letters addressed to him—a total of 14.5 percent of all his extensive surviving correspondence—were written in English.[114] He also took note of at least some of the English-language literature of his day. In his lecture on forensic medicine, for instance, Haller mentions the Irish satirist Jonathan Swift, whom he regarded as "one of the wittiest minds ever to live."[115] And his knowledge of the *monstrum* that he mentions in his *De Monstris Libri II.* (1768) came from the *Gentleman's Magazine*,[116] of which he was a regular reader. So he certainly could have heard or read of Richard Bentley Junior's Manpferdt brochure, though this cannot currently be proven.[117] For the time being, the origin of the centaur rumor cited by Albrecht von Haller remains, as befits a rumor, obscure.

What can, however, be clarified is the place of the centaur in Haller's forensic lectures and in his extensive studies on monsters. It is evident that he held centaurs and other human-animal hybrids to be impossible. This was partly because, from the perspective of eighteenth-century physiology, the mixing of different species was possible only within very narrow limits. It was also because, for Haller and his contemporaries, the emphasis continued to shift toward authored observation by contemporary naturalists and away from the former ancient authorities of nature study and other sources of knowledge and groups of people who, even in the late Renaissance, were still viewed as authoritative. This meant that particular recorded cases, and even entire classes of phenomena, no longer fell within the realm of verified knowledge about monsters.

Haller's Critical Review of Knowledge about Monsters

In what context did Haller mention the rumor of the London centaur? And what was it doing in his lecture on forensic medicine anyway? Let us begin by clarifying when and on what textual basis the lecture took place. Haller lectured on forensic medicine on several occasions during his time as a professor at the University of Göttingen. In the *Göttingische Zeitungen von gelehrten Sachen*, a journal edited by Haller himself, a lecture on forensic medicine to be given by him is announced for the summer semesters of 1748, 1749, and 1751. The twenty-seventh volume of the journal, dated March 28, 1751, includes the following notice: "Court Councilor von Haller will . . . , at 1 o'clock, . . . present those things from the field of medicine that have a bearing on the law, in accordance with Teichmeyer's Compendii."[118]

Haller's lecture was based on the textbook *Institutiones Medicinae Legalis vel Forensis*, written by his father-in-law Hermann Friedrich Teichmeyer (1684–1744) and first published in 1723.[119] But as Thomas René Rohrbach rightly points out in his medicohistorical dissertation on the printed version of the lecture, Teichmeyer's text—it seems to have been the 1740 edition that Haller used[120]—provided no more than a general outline for the lecture. Haller followed the chapter structure of the textbook for the most part, but after relating what Teichmeyer had to say on a topic, he frequently made comments of his own. They were commonly longer than the passage taken from Teichmeyer and often contradicted it. Rohrbach contends that Haller's comments were off-the-cuff remarks, citing in support of this hypothesis the lack of a lecture manuscript in Haller's intellectual estate.[121] The fact that Haller's comments are only sometimes anecdotal—many are in fact coherently and systematically structured—casts doubt on this hypothesis.

The existing edition of the lecture notes taken by Albrecht von Haller's eldest son was published by the German physician Friedrich August Weber (1753–1806)[122] and features a third voice in the form of Weber's interpolations and additions.[123] Weber did the reader—and particularly the historian—the favor of clearly marking the different textual levels typographically. With the exception of one chapter, the work keeps strictly to the chapter arrangement of Teichmeyer's textbook.[124] Haller's comments are identified throughout with quotation marks.[125] In addition, the table of contents indicates the provenance of individual chapters, subchapters, and sections.[126] This allows us, for instance, to attribute the above-cited statement concerning the London centaur unambiguously to Haller.[127]

Albrecht von Haller's remark about the rumor of the London centaur is found in chapter 13 of the *Lectures on Forensic Medicine,* which is entitled "Of the Deformed Creatures" ("Von den Missgeburten"). Weber compares the transcript in his possession with Teichmeyer's textbook and adopts the structure of the latter for his publication.[128] Teichmeyer's thirteenth chapter is entitled "De Monstris" and provides the foundation for the extensive discussion of monsters, or "deformed creatures," in Haller's lecture.

Haller's chapter begins with a section on the "connection of this teaching to forensic medicine," based on Teichmeyer.[129] Teichmeyer begins by explaining why medical knowledge about monsters is of relevance to forensic medicine. "The spiritual and temporal authorities are urged, for more than one reason, to seek knowledge of this learning from the physicians," states Haller's text—on the basis of Teichmeyer's.[130] He then goes on to list a total of seven—mostly legal—problem areas that could be affected by a monstrous birth.[131] These include, for example, the question of how to determine whether a "deformed" newborn was a son or a daughter. Answering this question was important from a legal viewpoint, we are told, not least because it could have implications for the right of inheritance or succession and because it was necessary generally to determine whether a "deformed creature" was entitled to the legal privileges that fell to male descendants. As another legal problem area on which the "deformed creatures" had a bearing, he addresses "sodomitry," the mixing of human and animal, by which monsters appearing to have human and animal traits were traditionally explained. In short, as so-called deformed creatures tended to raise many legal and theological problems in early modern society and it appeared that the expertise of the physician could, in certain cases, contribute to their solution, teachings on monsters fell within the domain of the forensic physician. Up to this point, Haller followed Teichmeyer.

Immediately following this section is a comment by Haller on the question of the field to which teachings on monsters belonged. He postulates that monsters are, in the first instance, a physiological subject: "Strictly speaking, this entire chapter appertains more to physiology than forensic medicine. And yet this teaching is not to be disregarded, because it elucidates and helps to answer various questions of law, including that of whether a monster may be regarded as a legitimate birth."[132] It is unsurprising that Haller considered monsters to be primarily a problem of physiology—after all, they were at the center of intense physiological and embryological debate, in which he himself was involved.

In this introductory section it is conspicuous that neither Teichmeyer nor Haller so much as mentions the possibility of supernatural causes. There is also, therefore, no discussion of the role of the physician, which had been so important in the sixteenth and seventeenth century, as an expert in drawing a boundary between the supernatural and the preternatural. The fact that European naturalists were so concerned with monsters from the late sixteenth century on was due in part to the relative novelty of the category of the preternatural at the time. Academically educated physicians like Johann Georg Schenck, and philosophically trained colleagues like Fortunio Liceti even more so, seized the opportunity to make their mark in this new field and to present themselves as experts in the demarcation of the two categories.[133] In the eighteenth century, however, the expert opinion of the forensic physician was no longer focused on determining whether a birth could be explained by natural causes. Instead, it was now for the most part taken for granted that monsters were natural.

The defining of the term that follows this section points in the same direction and likewise initially follows Teichmeyer's textbook. Though this part of the chapter commences by acknowledging the contribution to disambiguation made by the "Latin scholars," Teichmeyer—like many other contemporaries—had no appreciation for their "subtlety." "The Latin scholars took great pleasure in revealing the subtlety of their intellect in the development of this term. They created a distinction between monstrum, ostentum or portentum, and prodigium, and defined and distinguished in this regard so marvelously, as we can read . . . in the work of *Zachias*."[134] The only "Latin scholar" named, who thus appears as an example of this outdated subtlety, is Paul Zacchias (1584–1659), the Italian personal physician to Pope Innocent X considered the founder of forensic medicine. This shows that in the view of eighteenth-century scholars, authors from the first half of the seventeenth century were already outdated.

Hermann Friedrich Teichmeyer adopts the far less subtle classification of his contemporary Jacob Friedrich Ludovici (1671–1723).[135] The Halle law professor operated with just three terms, *monstrum, ostentum,* and *portentum*:

> The late *Ludovici* sought to untangle what these brooders confused, and reduces everything to monstrum and ostentum or portentum. The first, he says, is a fetus without a human head. Here, there exist two principal cases: The first is when a baby is born with no head at all . . . The second case is when in place of a human head there is one that resembles that of an animal . . . An ostentum is, according to the aforementioned Halle teacher, everything that is indeed human, but that deviates to a greater or lesser degree from the usual form of individual parts of the body.[136]

The passage in Ludovici's writing to which Teichmeyer refers reflects the state of the debate among legal scholars at the time as to what a *monstrum* was and what rights were to be accorded such a creature.

The distinction between *monstra* on the one hand and *ostenta* or *portenta* on the other was of particular legal relevance inasmuch as assigning a newborn baby to one category or the other determined whether it was entitled to the rights of a human being. Unlike *monstra, ostenta* and *portenta* were, for instance, to receive baptism, since they were classified as belonging to the human race.[137] The divinatory context of the genesis of the terms *monstrum, ostentum,* and *portentum,* however, was no longer of any significance to Ludovici (or to Teichmeyer or Haller).

As a physiologist, Albrecht von Haller was unable to reconcile himself to this schematic legal definition. Following on from the definition taken from Teichmeyer, he gives a much longer and more nuanced commentary. It is far more than a mere addition. Rather, Haller wanted to impress upon his students the difficulty of defining the object of the *monstrum*. In so doing, he provides a sample of the "critical" thinking that characterized nature study at the time—with its limiting effect on the monsters that were deemed existent.

The commentary opens by immediately setting what Haller held to be the correct—which is to say critical—tone: "One cannot answer the question: what is a deformed creature [*Misgeburt*]? as precisely as it is believed."[138] Haller begins by questioning the notion that it is easy to draw a clear line between a "deformed creature" and an individual with a physical form that does not deviate, or only minimally, from the norm: "If one says: it is an animal whose natural form is lacking in something, it is surely an inaccurate definition that one hereby provides."[139] He proceeds to give the example of a highwayman who undergoes an autopsy after his death and "in whom the position of the heart was found to be completely inverted."[140] He could no more be called a "deformed creature" than a person who had four or six fingers on one hand but was in all other respects "normal," Haller argues.[141]

Ludovici's definition, according to which a person without a human head was a *monstrum*, also failed to satisfy Haller. It was "no better, because the head may have the most regular form in the world, and yet deformities may be found in the rest of the body that make of the baby a deformed creature."[142] After all, he argues, the regularly formed head might sit atop a body with four feet.[143]

Haller thus questions the validity of Ludovici's—and, indirectly, Teichmeyer's—first "principal case" of monstrosity. Babies with no head at all were indeed to be denied human status. Nor were they to be ranked as animals. But, according to Haller, "this characteristic is not yet sufficient to derive from it the name of a deformed creature."[144] After all, it was not possible to turn the proposition around and conclude that a baby was not a *monstrum* if the head was present and human.

Haller then turns to the second "principal case" of monstrosity in Ludovici's definition: the newborn whose head "resembles that of an animal." He fundamentally questions the existence of such babies, never having seen one himself. It is, he writes, "almost to be doubted whether such a fetus has in fact been found; at least I on my travels have, although I have carefully inspected all natural history cabinets and other collections, not encountered the like in a single one."[145]

Haller did not, then, wish to accept the existence of newborns with the heads of animals, even though they were as much canonical components of early modern tracts and summae about monsters as the reverse case of animals with human heads. His criticism of this type of monster is based on his own observation, which he considered more trustworthy than the perceptions of others. Thus, he had "admittedly seen a baby that was considered an ape, but it was fully human, and merely had a nose that was flattened more than usual, and was very hirsute on its body."[146]

Haller was skeptical of the perceptions of earlier naturalists, too. He considered the cases they reported of humans with the heads of animals to be implausible. One such example was Gerard Blasius (1626–1692), who edited the third edition of Fortunio Liceti's *De Monstrorum Natura, Caussis, et Differentiis*. Blasius had, in Haller's words, "taken the trouble to compose an entire book about such supposed deformed creatures. Most of them, however, are blatant and vulgar fictions, or one must strain one's imagination to the utmost to discover the monstrous."[147] In fundamentally questioning the credibility of the monstrous factoids that were gathered or even personally attested by earlier generations of scholars, Haller was, in the middle of the eighteenth century, in good company. The reasons for this are considered in detail further below.

Some central features of the critical review of earlier wonders by eighteenth-century naturalists are particularly evident in the third case cited by Albrecht von Haller: that of a baby whose head allegedly exhibited nonhuman traits. The object in question was the fluid-preserved specimen of a supposedly monstrous newborn that was only able to be divested of its monstrosity by dissection. The facts of the matter, Haller contended, were the same with the alleged monsters of Gerard Blasius as they were in the case of the "alleged little devil [*cacodaemon*] that Friedrich Ruysch preserved in spirits of wine, and that, alongside other collections made by this industrious dissector, Peter the Great purchased and presented as a gift to the Saint Petersburg Academy of Sciences."[148] Here, too, as in the case of the baby examined by Haller himself, it was firsthand inspection by suitable observers that in his view brought the truth to light. In this case it was even possible to perform a dissection, since Peter the Great had permitted the use of Ruysch's fluid-preserved specimens for the purposes of study: "This little devil required examination by an anatomist in order to exchange its devilry [*diabolicitas*] for humanity: for the anatomist found nothing but a fluid discharged in the cellular tissue, an olecranon [the bony portion of the elbow] extending somewhat over the usual length, and also other bone ends and attached conformations that were somewhat too long, and that were marketed to the ignorant as claims to devilry."[149]

This case is reminiscent of the dead newborn in Lacke near Königsberg who was thought to be a *monstrum* because the baby appeared to exhibit some of the features of a satyr, described at such length by Philipp Jakob Hartmann in the fifth year of the second decury of the *Miscellanea Curiosa*. Hartmann also employed dissection to prove that, contrary to the view of common people, the baby in question was not a *monstrum*.[150] What was required in both cases was the right observer—namely, someone versed in anatomy—and dissection as an antidote to the credulity of those viewers less suited to the task.

Haller then returns to a detailed discussion of babies born without a head. He had fundamental problems with this category of Ludovici's, too, because in his view no headless babies existed whose bodies did not also exhibit other deviations from the usual anatomy. He was aware from literature of a dozen or so cases of babies born without a head, and in each of these cases "other considerable deviations from the natural constitution of the body"[151] were found. After listing some of these deformations, which concerned the internal organs as well as the bone structure and the skin, he concludes that such babies must "assuredly forgo the privileges of humanity,"[152] which is to say that they were to be viewed as monsters.

In a further step, Albrecht von Haller seeks to distinguish those babies with no head at all from babies "who are born with an uncovered head"[153]—in other words, those who are missing the cranium or at least the roof of the skull. Such newborns were likewise to be denied the privileges of humanity, as they had no brain and no will, were like machines, and, moreover, lacked vitality. For this reason, they were also not to be admitted to baptism. Unlike the headless babies, however, they could—Haller believed—be traced back to an accidental cause. Their deformation was pathological, rather than congenital. It was most commonly a case of hydrocephalus, or "water on the brain."[154]

The final argument that Haller makes against Ludovici's overly schematic definition of *monstrum* is that there are also defects in the formation of the head that lie below the threshold of monstrosity. As examples he cites the cleft palate and the case of a face resembling that of an animal. The latter occurred "when the cellular tissue of the face is so stretched and swollen that the countenance thus appears very broad and more bestial than human."[155] Haller's description of the second example shows once again that he did not believe it possible for human newborns to have truly animal traits. What had been identified as bestial would, in his view, inevitably prove to be no more than *seemingly* bestial. And such a baby was thus only *seemingly* a *monstrum*.

Following his lengthy discussion of the first category in Ludovici's definition, Haller turns to its second category, *ostenta* or *portenta*. He follows Ludovici's and, indirectly, Teichmeyer's use of the terms and likewise uses them for babies whose bodies exhibit deviations from the usual anatomy from the head down. Specifically, he mentions babies with a third leg or four feet. He also believed that these deviations lay below the threshold of monstrosity inasmuch as the babies in question "possess a will, hence personality, and with as much justification as the hydrocephalic babies, indeed with even greater, have legal capacity to inherit."[156]

But this part of his commentary, too, includes a criticism of what Haller saw as a widespread, inaccurate perception of (supposed) monsters. *Ostenta* or *portenta* were, according to the Göttingen professor, commonly referred to as changelings (*Wechselbälger*). He traces this usage back to an error made by Luther: "The honorable Doctor Luther believed that such births were conceived with the aid of the Devil; an error for which he may be easily forgiven, since anatomy was not his affair."[157] But this criticism does not apply solely to early modern scholars, like

Luther, who were unversed in the study of nature. After all, naturalists of the same period likewise ascribed to the devil or demons the ability to create monsters, or at least to convince people of their existence.[158]

A central theme in this passage in Haller's *Lectures on Forensic Medicine*, then, is the mistrust of the perceptions of (supposed) monsters by others, be they contemporaries or members of earlier generations of scholars, like Martin Luther or Friedrich Ruysch. As an antidote to such errors, Haller invokes anatomy, whose function is revealed particularly clearly in his cautious criticism of Luther. But what, in Haller's view, did these disparate observers whose perceptions he questioned have in common? What was the source of their error?

For Albrecht von Haller and others, mistrust of the perceptions of others centered on the imagination, against which eighteenth-century intellectuals believed one had to be on guard. This is clearly evident in Haller's discussion of one classic *topos* of scholarly literature on monsters: animals purportedly born to women. He addresses this subject, too, in his commentary on Jacob Friedrich Ludovici's concept of monsters, and has a simple explanation for such accounts: "We also read in the observers that people sometimes believed themselves to have encountered deformed creatures that were taken for lions, toads, etc., and were valiantly slain under these names . . . Such pseudolions and the like are actually nothing but sacs filled with fluid that have a protuberance here and there, refashioned by the busy imagination into bestial limbs."[159] Haller attributes all human newborns with an allegedly animal appearance to the imagination. What is meant here is no longer the imagination of the parents, which in the wake of an influential treatise by Thomas Feyens (or Fienus, 1567–1631) had all but become the universal cause of monsters in the seventeenth century.[160] Rather, he is employing the *topos* of the "pathological imagination,"[161] to which we now turn our attention.

Imaginary Monsters

The late seventeenth-century physicians who published articles on monsters in the *Miscellanea Curiosa* often attributed deformations to the imagination of the mother. If, for example, a pregnant woman took fright at the sight of something, it was said that this could affect the anatomy of the embryo. Although, for these authors, *imaginatio* thus constituted a kind of universal explanation for *actual* physical effects, it became in the eighteenth century a no less universally applicable explanation for *false* accounts of monsters and other wonders.[162]

Even in antiquity, the power of the imagination was sometimes viewed critically—particularly by the Stoic philosophers, who contrasted it with reason.[163] But it was not until the late seventeenth century that this *topos* was systematically deployed against monsters and other natural wonders.[164] This change brought with it new ideas about the way the imagination worked and, more importantly, the scope of its effects. Unlike Thomas Feyens and other early modern naturalists, eighteenth-century naturalists viewed the imagination as limited in its effect on the body of the individual in question.[165]

The change did not take place abruptly and radically, however, but gradually. Albrecht von Haller himself serves as an example here. Haller was among the critics of the notion that the power of the maternal imagination had an influence on the anatomy of an embryo in utero. This critique is given a section of its own in chapter 7 of the *Lectures on Forensic Medicine*, which Weber identifies in the table of contents as the work of Haller.[166]

Haller's stance on this question could scarcely have been more categorical. He begins his remarks with an apodictic rejection of the theory, which he viewed as archaic:

> We dismiss this figment of earlier physicians and scholars of nature, for the following reasons:
> 1) No mutual connection between the soul of the mother and the baby occurs;
> 2) the passage of a local defect [*vitium topicum*] from the mother to the baby is physically impossible;
> 3) the effect of the fright is far less severe and general than one believes; and
> 4) the birthmarks and imperfections in the structure of the body can be explained by pathological causes, which are much more evident, and it follows that we must not resort to causes that are so unlikely.[167]

For the reasons given, which are then explained in detail, it was more than certain, he argued, "that the mere imagination of the mother could not have an effect on the baby."[168]

The fact that Haller gives such a number of reasons for his hostile stance can, however, be read as an indication that the opposite position was still a tenable one. The theory of the influence of the maternal imagination on the form of an embryo had not yet disappeared from "the true." Even years after Haller's lecture, in the eighth volume of Denis Diderot and Jean le Rond d'Alembert's *Encyclopédie*, published in 1765, the article by Voltaire under the lemma "Imagination, Imaginer" supported the position that a fetus could be shaped by the power of the mother's imagination.[169]

In his textbook on forensic medicine, Hermann Friedrich Teichmeyer also took the opposing position.[170] Friedrich August Weber's explanation of his decision to reproduce Haller's remarks in the body of the text and to document Teichmeyer's statements in the appendix of the volume is also illuminating in this respect: "Here, Haller's teaching has deliberately been set in the place of Teichmeyer's. But since, as of 1751, not all patrons of the theory of maternal imagination (*Versehen*) have died out, the section from the Latin original of our textbook has, in order to oblige them, been transplanted to here."[171] Even when the transcript of the lecture was published in the late eighteenth century, then, isolated scholars still supported the obsolete theory.

Having initially done no more than list the reasons for his critical position, Albrecht von Haller then goes on to explain them. It is striking here that, once again, it is primarily anatomical knowledge, or physiological knowledge based upon it, that points the way to true insight: in the absence of a nerve connection between the mother and the baby, the imagination *cannot* shape the embryo. Experience, he argues, confirms this.[172]

As far as Haller was concerned, then, the matter was clear. And he also had an explanation to offer as to how this—in his view—false theory had arisen. The mothers were to blame:

> To say in two words, then, what the entire matter is about, women, when they have given birth to a baby with deviations from the natural form, ruminate in their thoughts on everything that they may have encountered throughout the approximate time of their pregnancy. And then it never fails to happen that some circumstance should come to their mind on which they hereafter cast all the blame. Had they not, however, noted anything particular in the form of their baby, they would surely not have thought of such circumstances.[173]

The fact that Haller looked to women as the origin of a theory that was often put forward by naturalists, rather than to the naturalists themselves, may at first seem surprising. But as many Enlightenment scholars assumed that the female intellect was inferior to that of men, and women were, in their view, also notoriously vulnerable to the power of the imagination in the new sense of the word, Haller's explanation was only logical.

The inconsistent approach to imagination in the *Encyclopédie* provides further evidence that both perspectives on the imagination were still current in the middle decades of the eighteenth century. Following immediately after Voltaire's "Imagination, Imaginer" is a second article with no author's initials devoted entirely to disproving the theory of the power of the maternal imagination. It attributes its supposed effects to the imagination of the observer: "All effects, which, if they depend on the imagination, must much more reasonably be attributed to that of the persons who believe to perceive them, than to that of the mother, who has not really, nor is likely to have any power of this kind."[174]

Zedler's *Universallexicon* deals very similarly with the subject; it is surely no coincidence that the term *Einbildung* (imagination) or *imaginatio* is given not one but two entries by Zedler, each reflecting one of the two schools of thought. The shorter article is concerned with the formative power of the parental *imaginatio* over the embryo: "Imaginatio, in German Einbildung; what powers this has in the formation of the offspring of man and beast, *Fienus* and others have described in entire books, and illuminated with many examples."[175] This idea is picked up in the article on *Mißgeburt* (deformed creature, or monster) in the same reference work and explained using the common wax metaphor. We read there that a "deformed creature" comes into being "according to most common opinion for the most part from the false imagination of the mother, which impresses on the tender body very abhorrent forms and images, just as a stamp seal does to wax."[176]

By contrast, the article on *Einbildung*, which is several times longer, discusses the effects of the power of the imagination on the human mind itself, and thus on perception. It distinguishes four effects of imaginative power or forms of imagination. First, it states, the imagination enables a person to visualize memories vividly. The images the imagination paints in the mind of things that are real but not present were therefore called *ideae imaginativae*.

The second form of imagination did entirely without a basis in reality. It was, the article states, an effect of the *ingenium*, which is to say mental talent or fantasy. The classical early modern origin of such imaginings was novel reading. Such "ingenious fantasies" existed "when the ingenium imagines in the mind its own suppositions to be present in such measure that it is amused or saddened by them. People of much ingenio and vivid imagination experience this in the reading of novels."[177] While *ideae imaginativae* were classed as fundamentally harmless, this second effect of the imagination was viewed ambivalently at best, even before the eighteenth century. In *Don Quijote de la Mancha*, Miguel de Cervantes—at the beginning of the seventeenth century—created a monument to the novel's dangers for readers with a vivid imagination.

The third effect of the imagination relates to *iudicium*, judgment, and likewise seems to have been a source of unease to the author of the entry in Zedler's *Universallexicon*. It is the anticipation—by means of the imagination—of an outcome already regarded as certain: "Thirdly, they are judicious fantasies when the thoughts of the *iudicium*, recognized as true, occupy the mind in the same way. For example, when one imagines the fortunate and unfortunate outcome of a matter about which the *iudicium* makes a judgment to be present already."[178]

The fourth and final effect of the imagination is the one that Haller had in mind when, in his Göttingen lecture, he attributed lions and toads born of women to the power of the observer's imagination: "Otherwise, one tends to use the word imagination in the sense that it is a prejudice. What this means, namely, is that someone only imagines in his thoughts a matter to be so, which in fact, and outside his thoughts, is entirely different."[179] It is evident from its classification as a prejudice—a term that had clearly negative connotations in Enlightenment philosophy—that this effect of the imagination could have been viewed only negatively by an eighteenth-century intellectual.

Not all effects created by the power of the imagination were regarded as dangerous, then. And they were also said to vary depending on how pronounced the imagination, *ingenium*, and *iudicium* of the individual in question were. Whether a person saw—and considered real—products solely of fantasy depended above all on the mental faculties of the individual. This idea is summed up concisely in the entry on *Einbildungs-Krafft* (power of the imagination) in Zedler, which states "that people of great imagination and poor intellect could bring forth astounding things: which, however, has not yet been sufficiently confirmed by experience."[180]

Conversely, it follows from such statements that a "good" intellect might offer protection from the pitfalls of the pathological imagination. This also applied to two pathologies of perception that, in the view of contemporary intellectuals, were no less dangerous—namely, superstition and enthusiasm. At this point, superstition and enthusiasm had also been subject to criticism for some decades, because they had the potential to undermine social order. In contrast to the criticism of the imagination, however, this had a confessional connection, as Daston and Park point out: "[The] association of prodigies with the religious manipulation of the folk had been forged in the learned critique of enthusiasm, superstition, and imagination published by theologians, physicians, and philosophers in the late seventeenth century. Although these treatises were inflected by nationality and confession—'enthusiasm' (*Schwärmerei* in German) being the preferred target of Protestant writers and 'superstition' that of their Catholic counterparts—all shared an anxious preoccupation with the nefarious role played by popular fear and wonder in religious and political subversion."[181] From the point of view of eighteenth-century intellectuals, intellect and will offered protection from the whisperings of imagination, enthusiasm, and superstition. But as intellect and will were unevenly distributed, different groups of people were considered to be at risk to differing degrees. Most defenseless against the threefold danger were "women, the very young, the very old, primitive peoples, and the uneducated masses, a motley group collectively designated as 'the vulgar.' "[182]

Depending on disposition, then, the power of the imagination could deceive the observer into seeing monsters where there were none. In view of such pitfalls of human perception, it seemed to eighteenth-century naturalists that particular caution should be exercised, specifically with regard to reports of (purported) monsters. In this sense, Albrecht von Haller's remarks on monsters in his *Lectures on Forensic Medicine* conform to the Enlightenment zeitgeist.

In view of this conspicuous skepticism of the observations of others, one might wonder what sources an enlightened naturalist such as Haller believed he could base his knowledge about monsters on, and how, in his view, they were to be dealt with. This is the question to which we now turn our attention.

The Observations of Others

It is not without good reason that Albrecht von Haller has been characterized as "the most celebrated scientific observer of the Enlightenment."[183] The notebooks preserved in his intellectual estate in which he recorded his observations testify to the fact that Haller was a diligent and disciplined observer. With regard to contentious questions, in particular, such as that of the development of the embryo, he undertook numerous observations and repeated them in accordance with the stipulations of nature study at the time. For instance, he repeated Marcello Malpighi's (1628–1694) investigation into the development of the chick inside the egg of a hen—without calling Malpighi's observations fundamentally into question by doing so.[184] In this sense, it stands entirely to reason that a mere rumor such as that of the London centaur was not, in his view, a suitable starting point for the generation of reliable knowledge about nature.

Even so distinguished an exponent of the art of observation could not get by without recourse to the knowledge—and, in particular, the observations—of others. The *respublica literaria medica* remained characterized by the collective empiricism that emerged in the early modern period. But Haller's working methods differed from those of earlier generations of scholars, specifically from those of academy members in the second half of the seventeenth century, in one significant respect: he frequently questioned the veracity of accounts of monsters written by naturalists.

It was not a new development that accounts of monsters (and all other objects of nature study, too) had to come from a named observer to be considered usable by naturalists. Authored observations were already called for at the close of the seventeenth century.[185] Even in the seventeenth century, then, mere hearsay was problematic. The shake-up of the discourse on monsters was due in part to the increasingly strict verification standards applied by naturalists to accounts of monsters and other rare phenomena, and to changes in ideas about which sources of knowledge were dependable in such cases.

Until the early eighteenth century, it was extremely rare for the credibility of the observations published in naturalist journals to be questioned. And this was true even though that credibility was, for the most part, based solely on the testimony of the author (and sometimes additional witnesses mentioned by name in the text). Incredulity was the exception. There was a number of reasons for this. First, it was considered insulting, in England especially, "to gainsay the word of a gentleman—and most correspondents of the Royal Society counted themselves as gentlemen."[186] Second, the collective empiricism of the academies necessitated the collaboration of many people, including volunteers. There was no alternative—particularly in view of the rarity of preternatural phenomena—but to place trust in them. Third, as we know, Bacon's reform program had recently made a virtue of the description of unheard-of phenomena. And then there were also "metaphysical grounds for lowering the threshold of belief for strange facts. Although phrases like 'the laws of nature' had become common currency in late-seventeenth-century natural philosophy, such laws were seldom taken to imply strict, much less mathematical, regularity throughout nature. Interlocking 'municipal' and 'catholic' laws could create as much variability in nature as they did within the polity whence the metaphor was borrowed."[187]

By the mid-eighteenth century, the situation was very different. There were changes in the understanding of nature and in scholarly sensibility that led to intellectuals in the first half of the century becoming increasingly skeptical of former wonders: "Nature abandoned loose customs for inviolable laws; the naturalist abandoned open-mouthed wonder for skeptical sangfroid."[188] The great value that had been placed on *copia* and *varietas* by late Renaissance naturalists in particular—their impetus for collecting factoids about monsters as comprehensively as possible initially—was a thing of the past. It now appeared more pressing to sift carefully through the available knowledge—especially that which concerned rare natural phenomena such as monsters—and to eliminate implausible factoids. In doing so, the burden of proof was all but reversed. The inclusion of false accounts was now regarded as more harmful than the exclusion of true ones.[189]

This change did not take place suddenly. It had been building over a period of decades. In the late seventeenth century, scholars had begun to draw on a new criterion in assessing the credibility of an observation. Previously, testimony had always been the central criterion. But now the evaluation of the intrinsic plausibility of an account appeared as a new criterion,[190] with major consequences for evaluating observations concerning monsters and other rare natural phenomena.

One significant precondition for this development was "a genuinely novel addition to the early modern repertoire of proof and persuasion: mathematical probability."[191] The origins of mathematical theories of probability are generally located in learned studies of gambling. But in the seventeenth century, writers who engaged with logic, the soul, or theoretical questions of historiographical knowledge also increasingly regarded it essential to assess the probable truth of a statement or historical event. The new tool of mathematical probability lent itself to this purpose.[192]

In *Logique de Port-Royal*, published in 1662, Antoine Arnauld and Pierre Nicole applied the new statistical methods to historiography. In assessing the credibility of an account, they advised, the historian should consider not only the credibility of the witness but also the intrinsic plausibility of the event itself.[193] John Locke transferred this distinction, which was already established in early modern rhetoric,[194] to natural philosophy. According to Locke, the probable truth of an account varied first according to "the number and credibility of testimonies," but second also according to the degree of its "conformity to our knowledge," "the certainty of observations," and "the frequency and constancy of experience."[195]

As Daston and Park point out, though, Locke's argumentation was not, in fact, directed again credulity. On the contrary, "as might be expected from a Fellow of the Royal Society, Locke used his criteria of probability to warn against excessive incredulity, relating the story of the King of Siam who had rashly dismissed the Dutch ambassador for tall tales about how water became hard in winter."[196] Some of his contemporaries, however, were already employing the new tool to combat what they regarded as credulity among naturalists confronting accounts of rare phenomena. Accounts of (supposed) prodigies or wonders were a particular target here, as illustrated by the work of the skeptical philosopher Pierre Bayle (1647–1706):

> Rotterdam professor Pierre Bayle argued apropos of the comet of 1680 that the sheer bulk of testimony was insufficient to warrant belief, for the "fabulous opinions" recently discredited in natural history had been supported by the testimony of innumerable persons. "One may rest assured," he asserted, "that an intelligent

man who pronounces only upon that which he has long pondered, and which he has found proof against all his doubts, gives greater weight to his belief, than one hundred thousand vulgar minds who only follow like sheep." Bayle also insisted that the content of the testimony should be inspected before assenting; reports of marvels and miracles were particularly suspect.[197]

In the spirit of this development, Albrecht von Haller surveyed the observations of others critically with regard to contentious questions of scholarship—particularly when the subject of the observation fell within the traditional categories of the preter- or supernatural. Haller's Renaissance predecessors typically reproduced the monstrous factoids available to them indiscriminately, and thus contributed to their wider circulation and continued existence in scholarly literature. They tended to do this even if they considered a degree of doubt as to the dependability and epistemic status of the source to be justified. Ulisse Aldrovandi, for example, had evident reservations about some of the cases of monstrosity reported and illustrated by Conrad Lycosthenes. Nevertheless, he very frequently made reference to Lycosthenes's *Chronicon* and even reproduced factoids from it that seemed questionable to him.[198] Aldrovandi, then, gave the monstrous factoid the benefit of the doubt. But Haller did not.

For Haller, the credibility of the witness, and therefore his membership in the *respublica literaria medica*, was an important criterion for the credibility of the observation, but not the only one. Like Pierre Bayle, he took into consideration the plausibility of the observation as a second criterion. This fact is well illustrated by his approach to one of the (supposed) effects of the maternal imagination, the cleft lip. In the *Lectures on Forensic Medicine*, Albrecht von Haller observes that the "ancients" had ascribed to the imagination a powerful effect on the figure of the baby. They believed, he claims, that "the sight of a hare was the reason that a woman might give birth to a baby with a harelip [*labium leporinum*], and that for this same reason it must be visible on the bottom of the baby if the mother was struck on hers during the pregnancy."[199]

Haller is referring here not only to the scholars of antiquity but also to early modern authors, as shown by the example that he then cites: the Flemish naturalist and polymath Jan Baptist van Helmont (1579–1644) recounted, Haller states, that "a baby was born without a head because the mother had taken fright at the sight of the beheading of the Swedish Count Horn, who is known in history."[200] Haller cannot accept this account. First, because he fundamentally rejected the theory of the influence of the maternal imagination on the fetus. The substance of the observation was therefore implausible. In addition to this the witness lacked credibility. The tale bore "the character of fable all the more because Helmont was not himself a witness to the incident, and furthermore wrote it down in good faith as related by his great-grandmother."[201]

Helmont reproduced an observation, then, that stemmed from his great-grandmother, which in Haller's view further diminished its credibility. Haller's use of the collective subject "the ancients" moreover suggests that, if in doubt, he placed more trust in the observation of a contemporary, "enlightened" naturalist than in that of a scholar from bygone and—in his view—unenlightened times.

Enlightened Reading

For all its emphasis on the central importance of observation, the empiricism of the eighteenth century still displayed clear traits of the "learned empiricism" of the early modern period. In this respect, too, Albrecht von Haller was typical of his time. In addition to observation, reading was an essential component of his working practice. He surveyed the earlier and more recent literature on each subject systematically and published bibliographies—his *Bibliothecae*—for several fields of nature study, too.[202]

The contemporary term for such systematic surveys of existing literature was *historia literaria*. They flourished from the second half of the seventeenth century through the eighteenth century, particularly in the Holy Roman Empire.[203] Programmatically, the *historia literaria* modeled itself on Francis Bacon's program for an as-yet unrealized *Historia Literarum*, which he set out in the second book of *The Proficience and Advancement of Learning, Divine and Human* (published in 1605, and as *De Dignitate et Augmentis Scientiarum* in 1623).[204] But its roots were deep in the sixteenth-century humanist world of learning, as has been shown by Wilhelm Schmidt-Biggemann in *Topica Universalis*.[205]

As conceived by Bacon, the *historia literaria* was not a documentation of inherited knowledge for its own sake. Rather, it was aimed at facilitating and guiding future research, and thus enabling progress, by critically reviewing existing knowledge. As Helmut Zedelmaier has observed, "The notion of a critical review and orientation of this kind—albeit within a different conceptional framework and with a different methodological approach—is constitutive, too, for the texts that call themselves 'Historia Literaria,' or make reference to this term. Unlike Bacon, however, they deal largely only with scholarly knowledge."[206] For Bacon and for most authors who, taking their lead from him, wrote *historiae literariae* (or textbooks that gave an introduction to this field), the primary meaning of the term *historia* in this context was "experience," rather than "history" in the narrower sense. It describes "the descriptive-atemporal, pre-scientific status of the knowledge [*notitia*] conveyed by it."[207]

Beginning in the early eighteenth century, Bacon's style of *historia literaria* experienced a boom, not least because it was well suited to the early Enlightenment project of "critique."[208] In *historia literaria* textbooks such as August Ludwig Heumann's (1681–1764) *Conspectus Reipublicae Literariae* (1718), the standard work of *historia literaria* in the eighteenth century,[209] a "historical model of a progressive development of the sciences and humanities through history"[210] was now often to be found. Over the course of the rest of the century, *historia literaria* became increasingly more differentiated. *Historia literaria specialis*, which unlike *historia literaria universalis* surveyed only one part of the history of scholarly knowledge—for example, that of one specific field or discipline—was particularly successful. Albrecht von Haller's *Bibliothecae* fall into this category.[211]

His *Bibliothecae* clearly demonstrate the historicization of knowledge that was typical of contributions to *historia literaria* at the time. Haller understood the history of academic knowledge to be one of continual progress and therefore considered it "normal that many things be revised in the course of twenty to twenty-five years."[212] For example, the purpose of his two-volume *Bibliotheca Anatomica*,[213] published between 1774 and 1777, was "as a bibliographical counterpart to the Elementa [*physiologiae*], so to speak, and one that was even more focused on the documenting of historical development—to provide the scholar with all of the literature on the questions that were of interest to him. Despite its documentary character, the *Bibliotheca Anatomica* is, first and foremost, a work oriented to the scholarship of the future."[214]

In addition, if one compares Haller's reading practice to that of many sixteenth- and seventeenth-century naturalists, it strikes one that the *collecting* aspect had lost some of its importance while the use of the *iudicium* had become more prominent. Haller's reading practice was "critical" and, closely connected to that, *selective*. As Helmut Zedelmaier summarizes with regard to scholarly reading practice in the age of Enlightenment, "knowledge merely recalled from memory is increasingly devalued and replaced by the model of 'independent thought.' Corresponding to this is a form of reading that, as it states in *The Art of Reading Books* by the Kantian Adam Bergk, the 'self-acting mind' must learn to 'master' if it is not to be 'smothered' by the 'collection of knowledge.' "[215] The *Bibliotheca Anatomica* illustrates this trait in Haller's reading practice, too. On 1,680 quarto pages, it brings together the anatomical and physiological knowledge published until this point and gives a critical commentary on it.[216] A further material correlate of this altered reading practice was a shift toward non-book-based techniques for ordering material gathered by reading, such as freely arrangeable slips of paper, as Carl von Linné used. Albrecht von Haller also sometimes used this method.[217]

The flourishing of review writing in the eighteenth century is also an expression of the altered reading practice of scholars, which was no longer predominantly aimed at collecting but frequently also at critical reflection. Although some reviews in the eighteenth century provided little more than a summary of the work under discussion, the bulk of these texts offered a mixture of summary and critical reflection. The great importance of the genre is palpable in Albrecht von Haller's scholarly output: from 1747 to 1753, Haller was the managing editor of the *Göttingischer Gelehrter Anzeiger*, a review journal to which he himself contributed some nine thousand articles.[218]

And, finally, the call for observations to be repeated meant that naturalists' reading was linked to their observational practice in a new way. It was, after all, primarily through literature that scholars were acquainted with the observations of their colleagues. So, when Haller reenacted Malpighi's observations of hens' eggs, it was, in a sense, a part of his reading practice.

In comparison to the working methods of earlier generations of naturalists, then, Haller's reading practice was disproportionately more critical and more strongly selective. Martin Stuber, Stefan Hächler, and Luc Lienhard characterize his working methods aptly: "The hallmark of his science is direct and preferably systematic experience by means of observation and experiment, which, in combination with a comprehensive and critical survey of the literature, is intended to provide the foundations of a field of study."[219] What concrete form the critical survey and selection of existing knowledge took, and which sources of knowledge Haller considered acceptable in relation to monsters, is demonstrated well *ex negativo* by the subchapter of the *Lectures on Forensic Medicine* in which the centaur rumor is discussed.

Haller's Sources of Knowledge

Haller mentions the rumor in the third subchapter, which is concerned with the "classification of deformed creatures."[220] In contrast to the preceding subchapter, which clarifies the term *Misgeburt* (deformed creature, or monster), Weber omitted Teichmeyer's remarks entirely in this case. He offers the reader no explanation for this decision, but it seems Weber proceeded in this way because Teichmeyer's and Haller's classification of monsters were fundamentally contradictory: the corresponding subchapter in Teichmeyer's text, entitled "Quomodo Dividuntur Monstra?" adheres significantly more closely than Haller to the literature of the late Renaissance, to Gaspard Bauhin, Fortunio Liceti, and Martin Weinrich, for instance. A further significant difference from Haller's remarks is that Teichmeyer discusses at length precisely monsters whose monstrosity consists in a transgression of the boundaries between species, particularly between humans and animals. Haller, by contrast, simply did not believe these monsters to exist.[221]

Albrecht von Haller distinguishes three classes of monsters. According to him, the first consists of "deformed creatures in whom one part of the body is in excess."[222] Such "deformed creatures" "appear to come into being by means of fusion (coalitus), so that one deems them to be two babies who are joined at the chest or the abdomen."[223] Drawing among other things on his own observations, Haller divides the monsters in this class into several kinds, depending on what part of their body is duplicated and where the bodies are connected. Haller then goes on to discuss two borderline cases of individuals with surplus limbs, who—despite these deviations from the usual anatomy—did not qualify as monsters.[224]

Albrecht von Haller's second class consists of those "deformed creatures" who are distinguished by a "lack of essential parts."[225] In contrast to the case of the first class, Haller evidently considered this definition to be self-explanatory. Rather than describing it or bringing it to life with case histories, he merely remarks briefly, "We have already given examples of this kind on occasion."[226]

In the case of the third class, the situation is very different. *If* they existed, the monsters whose existence Haller doubted would almost all have to be placed in this class. This fact prompts Haller to a lengthy excursus that casts light on his sources of knowledge. Almost before he has even defined the class in the briefest of terms—it consisted "of creatures deformed by unshapeliness," he claims[227]—he addresses the monsters of questionable existence and their supporters: "When talk is of these [creatures deformed by unshapeliness], there is easy opportunity to convince oneself of the knowledge of the distaff philosophers [*Rockenphilosophen*]: for they then reveal stories of deformed creatures who possess the body of a lion, an elephant's trunk, eagles' feet, or cloven hoofs, and so on."[228] *Rockenphilosophie* was an established term in the eighteenth century. It was used derisively and referred primarily to "the wisdom of old wives at the distaff."[229]

The term is a translation of *philosophia colus*, a phrase coined by the learned writer (and naturalist) Johannes Praetorius (originally Hans Schultze, 1630–1680) in his eponymous work of 1662. The title of the book itself indicates that the "old wives" stand here as a pars pro toto for the entirety of the uneducated and—in the view of the author—superstitious populace: *Philosophia colus or pfy/ lose vieh of the women: In which as many as one hundred multifarious and habitual superstitions of the common man are shown to be ridiculous.*[230] Haller's use of the term *Rockenphilosophen* points in this direction, too. He was less interested in disparaging the tales told by women than in critiquing those authors whose accounts of monsters, in his view, no longer deserved any credence.

For Albrecht von Haller, tales of human-animal hybrids fall more than any within the category of "distaff philosophy." It is telling that he addresses this group of (supposed) monsters in detail but makes no mention whatsoever of any of the other kinds of monster that belong to the third class. He cannot help himself, so burning an issue are the monsters of the "distaff philosophers" for him: "I must take this occasion to disclose my opinion on the births that are said to come about through human and animal congress. It is my view that no such births ever occur, and this is for the following reasons."[231]

The remarks that follow show that he continued to use two of the three sources of knowledge, at least, that had characterized scholarly texts on monsters since the Renaissance: he, too, draws on the firsthand accounts of trustworthy authors and on experiences. However, the third source of knowledge that was frequently consulted by Renaissance naturalists—historiographical texts[232]—is very seldom used by Haller.

In general, Haller's approach to literature differs from that of a Renaissance naturalist in the sense that he no longer grants any authority about nature to authors who were not themselves naturalists. It is also conspicuous that ancient authors no longer possessed any particular authority in his eyes. In fact, he tended to suspect the "ancients"—that is, authors of earlier times—of succumbing to their greater credulity when it came to monsters.

These last two points are evident in his first argument against human-animal hybrids: "First, we do not have a single example of such a birth that has absolute historical certainty, for the fact that writers of antiquity claim that a live centaur existed in Egypt during the reign of Emperor Tiberius, and that, furthermore, somewhat more recent ones give us news of a human who was completely human despite having been born of a donkey, far from settles the matter."[233] Haller does not tell the reader where the news of the human born of a donkey comes from any more than he tells the origin of the rumor of the London centaur, which is mentioned immediately afterward. It is very probable, however, that the expression "ancient writers" refers to Pliny.[234] This demonstrates that Albrecht von Haller was willing to call into question the credibility of the accounts of the (former) ancient authorities of nature study. If one reads such statements alongside his numerous testimonies of his own observations of monsters, it becomes apparent how significantly direct observation—at least that of contemporary naturalists—had gained in epistemic weight over the testimony of the "ancients" over the course of the seventeenth and early eighteenth centuries.

Haller's second argument relates to his insight that the mixing of animals was subject to stricter limitations than Renaissance natural history had believed: "A second reason, deriving from the nature of the matter, for which I reject the old opinion of such births is rooted in the certain perception that animals can never reproduce except with animals that are of the same family as them and differ from them only in such ways as varieties may differ from the main form."[235] Haller did concede that animals of differing species within the same family, such as horses and donkeys, could produce "mongrels"—and that this possibility applied to humans, too. But the animal in question would have to be very similar to a human, and only the orangutan could be considered an option.[236] Haller describes the source upon which he bases his second argument as "certain perception." We can assume that he regarded it as reliable because it was experienced repeatedly and by different observers. As such, this is an appeal to supraindividual experience, to *experientia* in the old Aristotelian sense.

Haller makes no mention of the numerous accounts and illustrations of human-animal hybrids in the texts of early modern naturalists—such as the woodcut of the famous Monk Calf in the work of Jacob Rueff, to which Weber refers in his notes.[237] Clearly, he considered these "ancients" unreliable informants because of their credulity. He does, however, go on to briefly mention a contemporary observer, Thomas Shaw, who in 1738 published a highly regarded description of his travels through Asia:[238] "It is true that the travel writer Shaw speaks of a human who was said to have given birth to an animal, as well as of the jumarts, which are said to be bred by the covering of a mare by a bull. But when one examines the entire tale in its context, it is clear that Shaw must have been uncommonly credulous towards the account of the Arabian herdsmen who made him believe the story of the jumarts."[239] This reasoning is reminiscent of James Parsons's criticism of naturalists who relied on information given by traveling showmen. As a contemporary, Shaw himself may have been a more credible witness in Haller's view than the "ancients." But he made the mistake of trusting all too "credulously" in the words of unenlightened third parties, instead of reporting only what he himself or reliable witnesses had observed.

Inherent in this argumentation is a cognitive approach to time that was prevalent in the middle decades of the eighteenth century and played a significant role in distinguishing trustworthy from untrustworthy ("credulous") witnesses. This distinction, which was central to the reading practice of enlightened naturalists, can be found even before Haller—in the work of his Swiss countryman, the natural historian Johann Jacob Scheuchzer, for example. Franz Mauelshagen characterizes Scheuchzer's approach to historical accounts of wonders—and thus to the *Wickiana*—in his *Natural History of Switzerland* (1717–1718):

> The book of nature was read backwards ... The intention was clear: the comparison of the "histories" with observations from the present—not only concerning the sightings, but also the reactions of the "common people" and their interpretations of what they had seen—was intended to make it possible, at least in individual cases, to reveal—even through the language of superstition—the phenomenon obscured behind it, and to position it within natural history in its intended place. By means of this hermeneutical groundwork, the book of natural history could be rewritten and the texts and images from the sources of the past could be processed accordingly. The most important methodological prerequisite for this was the distinguishing of two classes of observers that had not existed in this form two centuries previously: the superstitious general populace on the one hand and the enlightened scholars on the other.[240]

In Scheuchzer's view, Wick was "credulous" and had therefore accumulated "fraudulences" in his *Wickiana*. Scheuchzer's reading of these volumes was accordingly critical.[241]

For both Haller and Scheuchzer, only enlightened contemporaries were now truly trustworthy when it came to rare natural phenomena. Both "unenlightened" contemporaries such as Shaw's Arabian herdsmen and the "ancients," who had likewise not had the benefit of the Enlightenment, were seen as prone to enthusiasm, superstition, and the pathological forms of imagination. Cognitively speaking, "the ancients," together with unenlightened contemporaries, lived in pre-Enlightenment times.

It was nothing new at this point, admittedly, for members of earlier generations of scholars to be accused of having contributed to widespread and (for this very reason) "vulgar" errors. As early as 1578, the influential *Erreurs populaires*, by the French physician Laurent Joubert, included "vulgar" opinions from the works of a number of ancient philosophers and church fathers. Joubert even excuses certain popular notions held by his contemporaries in relation to beavers and pelicans on the basis that equivalent stories could be found in the texts of several great philosophers and ancient physicians. But Joubert did not target a specific section of the population. The terminological restriction of "vulgarity" to the part of the population most prone to error was not yet evident in his work; in his portrayal, *erreurs populaires* could be found in all social classes.[242]

By contrast, English physician Thomas Browne's *Pseudodoxia Epidemica*, published seventy-five years later in 1646, already shows a narrower concept of "vulgarity." Browne uses "vulgar" to describe "the most deceptible part of mankind," which was willing "with open armes to receive the encroachments of Error."[243] Ancient authors play a similar role for him as they do for Joubert. He, too, traces individual errors that were widespread among the common people back to ancient authors such as Pliny, Ctesias, and Strabo. The primary target of his criticism, however, was the "vulgar people" of his own times. He accuses them of being too willing to accept the tales of the venerable authors at face value.[244]

Into the eighteenth century, criticism of the errors of the "ancients" was, for the most part, only selective: it was typically aimed at individual errors or authors. This was true, too, of the seventeenth-century attempts prompted by Francis Bacon and others to purge natural history of untrustworthy accounts of wonders. Pliny, for instance, became a target of such efforts as early as the late sixteenth century. The ideas of the common people, by contrast, did not play any significant role in this context.[245] For the above-described cognitive approach to time to become a possibility, the idea that the uneducated, common people were particularly vulnerable to error—which is perceptible at the latest in the writings of Browne—had first to be joined by the notion that a well-trained intellect and will would offer protection from the dangers of enthusiasm, superstition, and pathological imagination.

Observation, then, was considered by Albrecht von Haller to be the most reliable source of knowledge—provided that the observer was not credulous or otherwise unsuited to undertake it successfully. But, as demonstrated not least by Haller's remarks on the deceptive power of the imagination, the outer appearance of a (supposed) monster could itself be misleading. In such cases, mere external examination was not enough. Neither did an observation of the living object offer ultimate certainty when it came to monsters such as those that, at first glance, could be taken for lions, toads, or other animals. Only after the death of the individual in question could ultimate clarity be obtained by means of dissection: "If, then, accordingly, something is found on a baby already that departs from the rule of the usual structure, one must not be too quick to bestow the title of monstrosity [*Misgeburt*], but rather await the time when one can decide by making dissections whether a birth deserves this name."[246] Haller was not alone in this view. As we have seen, James Parsons, in his monograph on hermaphrodites, also refers repeatedly to dissection as the ultimate authority on truth for the same reasons.

To conclude, let us consider the question of whether Haller put forward similar arguments to his fellow scholars as he did in the summer semester of 1751 to his Göttingen students. Did the learned audience, and the epistemic genres by means of which Haller sought to reach it, require a different treatment of the subject?

Haller's *De Monstris Libri II.*

The *Lectures on Forensic Medicine* are a particularly informative source in part because the format of the lecture gave Albrecht von Haller scope for extensive comments that do not feature in the same way in his Latin texts, which were intended primarily for fellow scholars. This is particularly true of the rumor of the London centaur. When, in 1768, Haller collated his numerous individual articles on the physiological subject of monsters and published them in his three-volume *Opera Minora* as *De Monstris Libri II.*,[247] he refrained from mentioning the London centaur at all. But does a reference to the differing epistemic genres sufficiently explain this absence? The target audience also seems to have played a part.

Haller, we can assume, viewed his students as not yet fully enlightened. Unlike the readers of his scholarly treatises on monsters, they still required training in enlightened reading and observation. This may have been a significant factor in mentioning the rumor—which from his point of view was ludicrous—in his lecture at all: the example was particularly well suited to demonstrating to his audience the importance of an enlightened approach to reports of monsters.

Apart from the absence of the centaur from *De Monstris Libri II.*, the commonalities between his lecture and his scholarly discourse are astonishing. This is true particularly of the sources from which the knowledge about monsters was said to be drawn in each case. Let me first clarify what kind of text *De Monstris Libri II.* is.

As Haller explains to the reader at the start, the subject matter of these two books had previously been scattered across numerous shorter publications on the topic of monsters that he had published from the 1730s on.[248] It is presented here in the form of two books on monsters, with the addition of things that he himself had seen (*vidi*) or reflected on (*meditatus sum*) since its publication. The two books, Haller continues, were thus being published for the first time, with the advantage— as compared to the earlier publications—of a better order (*ordo melior*) and several additional sections and reflections.[249] We can therefore assume that the work quite closely approximated the ideal of how a comprehensive and well-structured naturalist text on this subject was, in the view of the Swiss polymath, to be written.

The improved structure that Haller extols so confidently is, ultimately, a classical division into empirical knowledge, *historia*, on the one hand, and philosophical knowledge (in the broad sense of *scientia*), building upon it, on the other. In place of the classical term *philosophia naturalis*, however, Haller uses the word "physiology." The first, larger book is thus entitled *De Monstris Liber Primus Historicus* and the second *De Monstris Liber II: Physiologicus*.[250]

The adjective *physiologicus* gives a first indication of the debates to which Albrecht von Haller wished to contribute. Since the seventeenth century, the term "physiology" had described "the theory of the functioning of the healthy organism . . . , i.e. the theory of blood circulation, breathing, digestion, etc."[251] As it was understood at the time, physiology had its foundation in anatomical knowledge: "The functioning of the living body could only be explained on the basis of a precise understanding of its structure."[252] Accordingly, Haller carried out a great number of dissections of human bodies over his long career. By 1757, he had already dissected approximately 350 bodies, he claimed.[253] And he went to great effort to ensure that the students at the University of Göttingen—which, because it was newly founded, was hoping to attract students by offering ideal teaching conditions—had as many corpses as possible available to them for dissection.[254]

At the beginning of the first book, Haller initially defines his subject matter in a commonly accepted way reminiscent of Fortunio Liceti's definition in his *De Monstrorum Natura, Caussis, et Differentiis*. The Latin word *monstrum* appears, he writes, to have its origin in the description of a deviation (*aberratio*) in an animal from the usual structure (*fabrica*) of its species that is so pronounced as to be evident even to the inexperienced.[255] Haller goes on to introduce his own definition. For him, the word refers in the same way (*perinde*) to a structure of—again—fully grown and visible body parts that is deviated from the usual anatomy. As in his lectures, he is quick to point out that the dividing line between the monstrous on the one hand and the variability that exists in the nerves, the vessels, the muscles, and the bones *within* the boundaries of usual and natural anatomy on the other hand is not easy to draw, because, he argues, anatomists discovered that all smaller parts of the body, and in particular the vessels, differed from one body to the next.[256]

In Haller's opinion, then, it cannot possibly be obvious to every observer what a *monstrum* is. Only an observer trained and experienced in anatomy can distinguish monstrosity from variety. Here, as in his lectures, anatomical knowledge constitutes the foundation and ultimate truth on which basis the physiology of monsters is to be investigated. It should be mentioned in this context, however, that—although this definition suggests otherwise—the text also includes some discussion of plants in addition to monstrous humans and animals.[257]

The section that immediately follows also conforms to Haller's approach as we know it from his *Lectures on Forensic Medicine*. Once again, Albrecht von Haller is much concerned to separate the imaginary monsters from the real ones. Under the heading "Fictitious Monsters" ("Monstra Fictitia"), he excludes several kinds of (supposed) monsters from his discussion from the outset. He stresses that these monsters are quite numerous (*numerosa satis*) and gives several possible reasons for their exclusion. He rejects implausible monsters (a) that have nothing monstrous about them aside from the effects of decay or (b) a single deformation (*difformitas*), as well as (c) when what is monstrous has been exaggerated in wondrous anecdotes (*in miraculum historiolas*) or (d) when they have been entirely fabricated with fraudulent intent.[258]

Of each of these four forms of fictitious monsters, Albrecht von Haller gives at least one example. It is noteworthy here that he refers exclusively to naturalist writers and texts from the late seventeenth and eighteenth centuries. As an example of the first case, for instance, he cites the above-mentioned article by Philipp Jakob Hartmann, in which he attempts to demonstrate that a suspected satyr was not in fact a *monstrum*.[259]

The position of the *monstra fictitia* section in Haller's work is telling. Naturalists of the late Renaissance were certainly also familiar with "fabulous" monsters. Ulisse Aldrovandi's natural history, for example, includes such a section.[260] But whereas Aldrovandi mentions the objects classed as "fabulous" in the *middle* of his natural history, and thus presents them as a legitimate part of what was known about nature, Haller positions the fictitious monsters section as a prefix to his study, so as to exclude implausible accounts in advance.

Haller's chapter on fictitious monsters has the same thrust as the monograph on hermaphrodites published by Parsons. It is an expression of the tendency of enlightened intellectuals to banish elements of what had once constituted the subject of monsters to the realm of fables. The fact that accounts of such fictitious monsters even existed was attributed by them to the pathologies of perception.

In this sense, the category "fable" was used to mark a supposed phenomenon as unreal, and over the course of the late seventeenth and eighteenth centuries, parts of the classical canon of monsters found themselves there. This was true not least of those monsters with a counterpart among the exotics at the edge of the world—

which included the centaurs. As early as 1724, Bernard de Fontenelle reasoned that one should no longer search for knowledge hidden in ancient fables. In his view, the only question still worth asking of accounts of human-animal hybrids and other such *extravagances* was what had led the ancient Greeks and Phoenicians to invent them in the first place.[261]

Even if there were already doubts as to the existence of centaurs during Aldrovandi's lifetime, stories of them still claimed a place in his natural history. But for Fontenelle, Haller, Parsons, and many of their learned contemporaries, it was not only certain that such beings did not exist and never had: earlier affirmative statements about them now only formed the point of departure for their inquiry into the genesis of such ideas and the conditions under which they came into being. There was no longer any knowledge to be found in such statements that might benefit the study of nature—and they certainly provided no guarantee of the existence of such beings.

Haller's one-sided reference to (near) contemporary authors is a recurring feature throughout the first book. In the sections on the different kinds of real monsters, which follow the passage on fictitious monsters, he does occasionally refer to scholars—and particularly naturalists—of the late Renaissance for individual cases of monstrosity. He makes use, for instance, of the work of Ulisse Aldrovandi, Amato Lusitano, Johann Georg Schenck von Grafenberg, Fortunio Liceti, and occasionally Conrad Lycosthenes. But such recourse to early modern scholars occur significantly more rarely than those to more recent authors. And Haller does not mention ancient authors in his first book at all.

By contrast, a great deal of space is given to Haller's own observations of individual monsters. He describes them in detail, and they are depicted with lavish etchings. The numerous etchings typically include letters indicating specific details of the respective monster's anatomy, which are named in a legend. This method of illustrating and highlighting the relevant parts of the body was already widespread in the naturalist journals of the late seventeenth century. However, Haller's images are often larger and of better quality than the images in the *Ephemerides*, for example.

In cases where Albrecht von Haller refers to a study of his own on a specific kind of monster, he often contextualizes his findings: in an initial chapter he presents what he himself has observed, and in the following chapter he compares it with similar cases found in literature.[262] One is reminded, here, of the spatial separation of observation and scholium in the medical *observationes* genre, which was also the inspiration for the format of the *Miscellanea Curiosa*. In another respect, however, the eighteenth-century observation differs from the "strange fact" of the academies: it was often conducted with a specific hypothesis already in mind.

The second book corresponds with the first in terms of the knowledge sources cited. What was the underlying objective of this part of the work? In book 2, Albrecht von Haller takes the empirical findings presented in the first book as a basis to discuss the question of the genesis of monsters. At the very start, in the first chapter of the second book, he explains why this question was key. It touched on the most intimate mysteries of *generatio*, he argues, and had therefore to be weighed with the greatest of care. The way it was answered would confirm either the theory of epigenesis or that of preformation.[263] In deciding to follow the first, "historical" book with a *liber physiologicus*, then, Haller exemplified the embryological and anatomical or physiological appropriation of interest in monsters that was widespread in nature study at the time.

Book 2 discusses various kinds of monsters and monstrosities one after another and clarifies their respective causes. In the individual chapters, Haller refers predominantly to contemporary or almost contemporary authors and observations, as in the first book. In comparison, for example, to Fortunio Liceti's *De Monstrorum Natura, Caussis, et Differentiis*, ancient writers are very much underrepresented in this part of the work, too.

It is only in the second chapter that they play any appreciable part at all. The chapter is entitled "The Hypothesis of the Ancients" ("Veterum Hypothesis"), and it summarizes the positions held by the "ancients" on *generatio*, devoting space to Empedocles, Aristotle, and Plutarch, as well as to early modern authors such as Fortunio Liceti and William Harvey.[264] As in the lectures, then, the naturalists of the Renaissance are viewed as "ancients" and along with the scholars of antiquity now constitute only the historical background of the question discussed. It was necessary, though, to be familiar with the historical background of a subject—with this particular section of *historia literaria*—to generate new knowledge about it. Haller and his contemporaries no longer operated *within* the theories of the former ancient authorities of nature study but rather made use of them in the advancement of their own theses. They referred to ancient texts when they saw precursors to their own positions in them but not because they wished, like Liceti, to prove or perfect the supremacy of Aristotelian theories, for example.

The third chapter in the second book also deals with the scholarly prehistory of the contemporary discussion. But it is written with Haller's own position on the etiology of monsters—and hence also on the debate surrounding epigenesis and preformation—already in mind. It is entitled "Authors Who Allow Originally Formed Monsters" ("Qui Monstra Primigenia Admiserint"), and it brings together authors who considered the preformation of monsters to be a possibility—in addition, at least, to the genesis of monsters as a result of accidental causes. Ancient authors no longer play a role here: only those of the seventeenth and eighteenth centuries are mentioned. As the subsequent chapters show, Haller himself traced most monsters back to an originally existing form. Only in comparatively few cases does he support accidental causes.[265] The third chapter thus serves to strengthen his position by enumerating his predecessors.

The same applies to the short chapter that concludes the volume. It addresses objections to Haller's argumentation and seeks to refute them. These objections are very much typical of the early modern debate on preformation. Haller is responding here to religiously motivated concerns that originally existing monstrous forms were irreconcilable with the Christian Creator.[266]

In conclusion, Haller's approach to sources of knowledge in his *Lectures on Forensic Medicine* corresponds with his approach in the printed work intended for his fellow scholars. But in *De Monstris Libri II.*, a key feature of his use of observations—typical of his time—emerges even more clearly. Enlightenment observation was not only intentionally one-sided in terms of who, in the view of the intellectuals, was qualified to observe *correctly*—which is to say ultimately only the enlightened naturalists themselves, who were particularly intellectually gifted and appropriately trained. In addition, the observation was no longer carefully separated from the reflections that followed concerning, for instance, the causes of the observed phenomenon. As Lorraine Daston summarizes, observation in the eighteenth century itself became a "tool of conjecture."[267] When Haller carried out his observations of monstrous babies, it was always with the pressing questions of *generatio* and his position in the debate on epigenesis and preformation in mind. This becomes particularly conspicuous in the annex to his *De Monstris Libri II.*

෴

In contrast to their predecessors in the seventeenth century, eighteenth-century naturalists like Albrecht von Haller and James Parsons no longer only questioned the credibility of *individual* monstrous factoids or authors. They viewed many earlier monstrous factoids and current observations with skepticism and surveyed them critically on the basis of exemplary individual cases. This had a limiting effect on the cases that were accepted, and indirectly also on the kinds of monsters that were deemed possible. Much of the ancient knowledge concerning monsters that for Gessner and Aldrovandi was still integral to the study of nature was now banished to the category of "fable."

The relationship between contemporary scholarship and the authors of antiquity had undergone a fundamental change. Unlike their Renaissance predecessors, the naturalists of the eighteenth century no longer operated *within* the theories of the (former) ancient authorities of nature study when discussing monsters. At best, they might search their works for *precursors* to their own hypotheses.

The conviction that a critical survey of existing knowledge about monsters (and other former wonders) was necessary had a close connection with the devaluation of the emotions of curiosity and wonder in the study of nature. Reservations of this kind were less pronounced among common folk. What is more, they seemed particularly prone to credulity and pathologized forms of perception—as the scholars of earlier generations had been. The perceptions and views of large sections of the population, as well as those of the scholars of times gone by, was therefore more than ever subject to question.

Even at the close of the seventeenth century, the physicians publishing observations on monsters in naturalist journals vehemently rejected the prevalent view that monsters constituted divine omens. But a fundamental difference between them and naturalists of the eighteenth century is that the latter, animated by the Enlightenment imperative of (self-)advancement, attempted to explain the genesis of perceptions that they considered false and thereby to challenge them. Sometimes they even intervened directly in the consumption of wonders in their own time. It was not so much the belief in prodigies that was their focus: their criticism was increasingly directed at the belief that specific monsters existed at all.

The changes that the subject of monsters underwent in the context of this development were not primarily the result of the rise of observation as a scholarly practice and epistemic category. Nor did this rise occur at the expense of reading. Albrecht von Haller was undoubtedly a disciplined observer and as advanced in the practice of this art as anyone of his time. But this did not mean that he read any less than his sixteenth- and seventeenth-century predecessors. He read *differently* than Liceti, Aldrovandi, or Johann Georg Schenck von Grafenberg. From the late seventeenth century onward, freer forms for the ordering of reading notes and (as compared to the Renaissance) more critical and selective ways of reading could be found among scholars for whom the *iudicium* of the reader played a more important role. Reading for the purpose of collecting was superseded by reading for the purpose of critique and selection.

Furthermore, the idea (originating in astronomy) had become firmly established in nature study that an observation—whether one's own or that of another observer—must be conducted repeatedly to minimize error. Ideally, then, to read an observation meant to repeat it oneself. Observation thus continued to be intimately connected to reading, though in a new sense.

Conclusion

Let us return to the question with which we began. How did a centaur end up in early modern London? From the perspective of the eighteenth century, and of intellectuals of that time such as Albrecht von Haller and Richard Bentley, the answer, as we have seen, is clear: *it didn't*. In the eyes of "enlightened" scholars, this figure belonged in the category of purely fabulous knowledge. Challenging such knowledge, and what they saw as the dangerous pathologies of perception that underlay it, was among their highest aims. Fabulous knowledge was now interesting only in relation to the historical causes, and those pertaining to the theory of perception, that gave rise to it.

To make it to London and from there into the naturalists' writings—not as a rumor but as a possibility at least worthy of discussion—a centaur would have had to embark on its journey no later than the late seventeenth century. In immediate proximity to other "strange facts" in the *Miscellanea Curiosa, Acta Medica & Philosophica Hafniensia, Philosophical Transactions*, and similar periodicals, the past or present existence of a human-horse hybrid could perhaps have been discussed. After all, these journals often described curiosities that naturalists in the early eighteenth century would later relegate to the realm of fables.

Observatio CXXVIII in the third volume of the 1673 *Ephemerides*[1] is a telling example of this. In this article, Georg Wolfgang Wedel (1645–1721), a member of the Academia Naturae Curiosorum experienced in medical practice who was appointed professor of theoretical medicine in Jena in 1672, discusses the existence of basilisks. Traditional accounts of this extremely dangerous creature, depicted as a combination of a chicken and serpent, had long been questioned.

Wedel's introductory remarks reflect this. Among all animals, Wedel writes, none were of more uncertain origin than the unicorn and the basilisk, an unparalleled evil upon the earth. People often invoked its name, but very few believed that this animal could annihilate a person with a single glance or a single hiss or breath, as Galen attested. Many, he continues, including the Italian Renaissance scholar Gerolamo Cardano, rejected these claims outright as frivolities and idle chatter. Nevertheless, it was by no means settled in Wedel's mind that this being did not exist. And in this he was not alone.

Wedel mentions two recent sightings of live basilisks about which he had read. A number of people were said to have died in a cellar in Warsaw in 1587: according to a report by the physician Johannes Pincier, this was attributable to the presence of a basilisk. And more recently, in an unspecified location in France, a pigeon fell into a well, upon which several people died after climbing down into it. For these deaths, too, a basilisk was allegedly responsible.

Wedel was not credulous in the sense that he would have believed any and every such report to be true. He dismisses the classic explanation for the origin of the basilisk—according to which it hatched from the egg of an old rooster that had been incubated by a toad—as an old wives' tale, something simple folk believed. This explanation, he argues, ran counter to the Harveyan principles, according to which a rooster and a hen were required to fertilize an egg. But the basilisk did actually exist. In this respect, he writes, the authority of the ancients was consistent with experience. Wedel then seeks to prove this with a detailed review of corresponding statements by ancient, medieval, and early modern authors.

Wedel also presents his readers with an image of a basilisk said to have been born, killed, and preserved in Africa before being brought to Europe (fig. 27). He does mention the possibility that the owner may have been deceived. However, having been assured by friends that they had seen similar basilisk skins in the curiosity cabinets of Italy, he considers this unlikely.

With regard to the powers attributed to the basilisk, Wedel takes a position of compromise. He considers it out of the question that a person could be endangered by contact with the poison of even a dead basilisk. And it was the venomous

Fig. 27. Etching accompanying Georg Wolfgang Wedel, "Observatio CCXXVIII: De Basilico," *Miscellanea Curiosa Medico-Physica Academiae Naturae Curiosorum, sive Ephemeridum Medico-Physicarum Germanicarum* [. . .]. Annus Tertius (1673), inserted between p. 202 and p. 203, Forschungsbibliothek Gotha der Universität Erfurt: Med 4° 00146 (1).

bite of the creature, rather than its gaze, that was to be feared. By way of evidence for these theses, he cites an Italian proverb and passages in the works of Avicenna and Aristotle. Georg Wolfgang Wedel, then, was convinced that the authority of the ancients, borne out by the experience of later writers, proved the existence of basilisks.

The fact that this *observatio* was included in an academy journal even though the existence of basilisks had been disputed for quite some time is an expression of the verification standards typical of these periodicals in the second half of the seventeenth century. Wedel's *observatio* did receive criticism from Johannes Jänisch in a scholium appended to it—on the basis of his own firsthand inspection of the basilisk skin mentioned by Wedel and of Konrad Gessner's assertions that basilisk specimens were forged using ray fish, Jänisch holds basilisks to be nonexistent[2]— but at the same time he concedes the existence of winged dragons.[3] What is more, the editors of the *Ephemerides* also included another statement in the volume here. It appears in the form of a passage dated 1672, with no heading or indication as to the author, perhaps again Georg Wolfgang Wedel, who strongly rejects the sugges- tion that the basilisk skin was a forgery. The assertion that it had been put together artificially and made to look like a basilisk at the instigation of the (unnamed) Ham- burg owner is, he says, false.[4]

The inclusion of *observationes* about natural phenomena whose existence was disputed was possible because, for various reasons, the academies lowered their standards of verification in the second half of the seventeenth century. The *obser- vationes* about preternatural objects, in particular, seemed too valuable to be pre- maturely excluded. From this point of view, the risk of false observations creeping into natural history had to be accepted.

What was criticized, however, as we saw in chapter 3, was the interpretation of monsters as divine omens. This expression of the *credulitas* of the *vulgus* was, in their view, to be vehemently opposed. In doing so, parts of a monster's alleged appearance were often explained away. These interventions resulted indirectly in changes to the definition of the object of the *monstrum*. The great consensus among scholars on this point stemmed from the fact that the events of the Thirty Years' War had taught the secular and spiritual elites of the Holy Roman Empire just how dangerous divination could be to public order.

The elites of European countries acquired this experience at different times. Ac- cordingly, the criticism of divination did not take place everywhere simultaneously. Scholars' rejection of the belief in prodigies did not lead to a naturalization of monsters in the sense that the ability of God (or—as we saw in the case of Liceti— the devil) to create monsters deliberately was fundamentally called into question. Nor did academies like the Academia Naturae Curiosorum and the Royal Society conduct any systematic investigation into the natural causes of monsters or other (former) prodigies.[5]

In the Renaissance, statements about centaurs and other beings subsequently deemed fabulous were an integral part of natural history. This was due not least to how broadly the likes of Aldrovandi and Gessner defined the forms of knowledge that merited a place in the study of nature. If in doubt, Aldrovandi opted in favor of including a monstrous factoid in his "Pandechion Epistemonicon" or his extensive collection of drawings. His printed natural history is also notable for the fact that it does not suppress factoids from literature about which—as in the case of the centaur—opinion among authors differed or whose sources were considered unreliable. He did not carefully qualify statements that he judged to be fabulous or give them short shrift. They stand side by side with other statements at the heart of his natural history and are given equal weight.

The centaur is just one example of many formerly canonical monsters that disappeared from the ranks of "the true" in nature study over the course of the European early modern period. As we have seen with Albrecht von Haller, this applied in particular to the hybrids said to have both human and animal traits or body parts, of which early modern authors reported cases dating from both the distant past and their own era. By the eighteenth century, the (perfect) hermaphrodite, likewise a hybrid, in this case of a man and a woman, also had few supporters. Aristotelians, in particular, had in fact raised doubts as to the existence of perfect hermaphrodites since antiquity. Around 1800, the balance between the advocates and the doubters shifted clearly in favor of the latter group.

Do these developments fit into the classic explanatory framework of a Scientific Revolution whose actors, in a heroic act of liberation, cast book knowledge unceremoniously aside in favor of direct observation, of empiricism? Were the devaluing of book knowledge and the consequent diminished epistemic weight of the former ancient authorities of nature study responsible for the changes outlined here?

My research suggests otherwise. It has brought to light a number of factors that were responsible for these developments that have not received sufficient attention to date. There are several reasons why creatures that were subsequently deemed fabulous can be encountered in the scholarly journals of the late seventeenth century. What these reasons have in common is that they do not easily fit into the picture suggested by the contemporary rhetoric of the rise of empiricism at the cost of book-based scholarship. Nor can the presence in late Renaissance natural history of what would later be viewed as unbelievable factoids be explained by the time scholars spent in the study (rather than in the "field"). Let us now recapitulate the central findings of this study with regard to the Scientific Revolution and the naturalization thesis.

In the period in question, three forms of describing and depicting monsters can be distinguished. They do not follow one another in tidy sequence, however, but overlap. They are closely linked to the changes undergone by "scientific" observation on the one hand and "scientific" reading on the other. The way reading and observation related to each other changes from one stage to the next.

Factoids

In the late Renaissance, naturalists—particularly those with a university medical background—drew their factoids about monsters from a great variety of sources. Alongside naturalist writings, they referred particularly frequently to historiographical genres. Their representation of monsters was heavily influenced by the iconography of the exotics from the edge of the world and the authority of ancient authors. Once they found their way into the discourse, textual and pictorial factoids were reused time and again by subsequent generations of authors.

These circulation patterns may be explained to some extent by the use of humanist practices of reading and of ordering material. At this time, as the example of the Italian scholar Ulisse Aldrovandi has shown, physicians and natural historians employed scholarly practices and tools that were in use in other fields of scholarship, too. They extended the use of humanist techniques and their corresponding tools for reading and ordering material, such as commonplace books, to the study of nature.

In the late sixteenth and early seventeenth centuries, the epistemic value ascribed by naturalists to their own firsthand experience varied depending, among other things, on their background. One might think here, for instance, of the self-conception of early modern anatomists, which was characterized by "disciplined seeing," to use Peter Dear's happy term.

All in all, the rhetoric of the actors should be taken with a grain of salt. The apparent contradiction between this rhetoric and the practices demonstrably utilized reveals that for many authors at this time, reading and experiential knowledge were *not* at odds. Ulisse Aldrovandi was therefore able to stress repeatedly that he had included nothing in his natural history that was not based on his experience, despite the fact that most of his monstrous factoids were taken from literature.

The images used in Renaissance natural history were not illustrated observation. The circulation patterns of images of monsters in this period are remarkably similar to those of the textual factoids. And in many respects the techniques and tools chosen for the gathering and ordering of the images also resemble their counterparts in reading and the ordering of written material.

These findings can, up to a certain point, be generalized beyond the subject of monsters. With the rise of the rhetorical and epistemic categories of *copia* and *varietas*, exhaustively gathering factoids about the potentially infinite variety of nature came to appear generally desirable and important. But monsters were an extreme case inasmuch as they were by definition rare. When it came to this subject matter, natural historians depended even more than usual on colleagues sharing their observations. An amalgamation of *one's own* observation with that *of others*, mostly documented in writing, was inevitable.

Given these findings, the criticism by early modern scholars of an "inundation of books," often discussed in more recent research in terms of "information overload," must—at least in relation to the study of nature—be viewed in a different light. The experience underlying these complaints cannot be attributed solely to the invention of the printing press and the expeditions of the early modern age. This abundance of knowledge was produced by the actors themselves, not least as a result of the rhetorical and epistemic ideals described above. Their gathering of factoids about monsters can be understood in many cases as a sort of initial survey of the subject matter, with a view to the subsequent generation of verified knowledge (in the sense of *scientia*). But for many late Renaissance authors, the accumulation of factoids about monsters and other rare natural phenomena also represented a value in its own right.

The fact that European naturalists from the late sixteenth century on were so concerned with monsters at all was due in part to the comparative novelty of the category of the preternatural. Academically educated physicians like Johann Georg Schenck, and philosophically trained colleagues like Fortunio Liceti even more so, seized the opportunity to make their mark in this new field and to present themselves as experts in demarcating the preternatural from the supernatural. So it was not so much a rejection of the supernatural—the divine part potentially present in rare phenomena—as the promise of the new field of the preternatural. The significance ascribed by late Renaissance scholars to exploring the preternatural reached its initial culmination in Francis Bacon's program for the reform of natural philosophy, in which the natural history of the preternatural played a central role.

Of all the objects classed as *praeter naturam*, it was the monsters that were to advance to become an object sui generis. This was because—and this is the second reason for the striking increase in the number of naturalist publications on the subject—they were already much discussed, and it seemed particularly important in their case, given the widespread belief in prodigies, to draw a clear line between the preternatural and the supernatural. Furthermore, a systematic study of monsters promised important findings for anatomy.

Observatio

The self-presentation of the scientific academies in the second half of the seventeenth century is characterized by a rhetoric of experience that might easily lead one to believe that the members of these societies had abandoned reading for good. However, recent research has shown that reading remained central to the new experimental philosophy of the Royal Society and other institutions, which in fact often treated objects and technologies remarkably similarly to texts.[6] With regard in particular to the observations of monsters that emerged from these institutional settings, it is evident on closer examination that they are not so much the expression of a sudden increase in the importance of "empiricism" as a new *form* of experience and its associated epistemic genre.

It was not least the result of the rise of the epistemic genre of the *observatio* and similar formats that the circulation of monstrous factoids decreased beginning in the late seventeenth century. Daston and Park have conceptualized these formats as "strange facts." The *observatio* required that its authors clearly distinguish their own firsthand observation from received accounts. In the *Miscellanea Curiosa*, this practice was reflected visually and spatially in the typographical separation of the scholium from the observation itself.

Skill in this genre and its concomitant observational and writing practices had first to be acquired by the authors. My review of the correspondence and statements of the *curiosi* responsible for the academy journal has also shown that even in the late seventeenth century, despite the journal being understood as a kind of curiosity cabinet, the *curiosi* were already beginning to have qualms about the inclusion of an overly large number of reports on (similar) monsters. This was only partly because monsters must inevitably have come to seem increasingly less rare in the light of their intensively pursued documentation over many centuries. Monsters also lost their appeal for the *curiosi* because the interpretation of prodigies in the context of the Thirty Years' War had lent them a dangerous aspect. They were therefore unanimous in rejecting the interpretation of monsters as portents—though they did not undertake any systematic investigation of their natural causes.

Repeated Observation

By 1751, when Albrecht von Haller mentioned the rumor of a London centaur to his Göttingen students, belief in prodigies had largely been quashed. Intellectuals now directed equally vehement criticism at other perceptions of monsters that were classed as *false*. Attention focused particularly on the "pathologies of perception" of unenlightened contemporaries and earlier generations of scholars.

While the scholars of the Renaissance aspired to a comprehensive *collection* of the factoids about monsters available to them, and their successors in the seventeenth century increasingly documented *new* cases, naturalists in the middle decades of the eighteenth century undertook a critical *review* and *selection* of the existing knowledge. Both new and old accounts of monsters were critically examined, in the process of which the boundaries of the object of the *monstrum* were redrawn once again. What was left for Albrecht von Haller and others was a sorry remnant of severely deformed babies with no will or vitality of their own, who had nothing of the animal about them but also nothing divine.

The skepticism toward the perceptions of others that was at the heart of this critical review was concerned primarily with the phenomena and events previously classed as preternatural or supernatural, and it was decades in the making. It was closely linked to a general rise in the verification standards used to assess the observations of rare phenomena in the study of nature, which in turn was related both to a shift in the concept of nature and to a desire on the part of the intellectuals to distance themselves from the curiosity and wonder of the common folk.

Still rarely encountered in the naturalist monographs on monsters published around 1600, the concept of *credulitas*—credulity—played a central role from the second half of the seventeenth century. At first, it appeared mainly in the naturalist approach to the notion—widespread among the ranks of the *vulgus*—that monsters were divine portents or warnings. In the eighteenth century, intellectuals no longer only wielded this concept against uneducated contemporaries. They often accused scholars from earlier times of credulity, too. The shifting usage of the term went hand in hand with a change in its content: education or membership in the Republic of Letters no longer afforded protection from *credulitas*. Educated individuals and even scholars could now also be credulous. One of the aims of this study was to show that we should not adopt Haller's approach to the concept unchallenged, but rather we should view it within the context of its genesis.

The transformations that the naturalist discourse on monsters underwent in this period are not attributable solely to the rise of observation as a scientific practice. The changing observational practices of the scholars undoubtedly contributed as well. Particularly noteworthy here is the idea, generally accepted by naturalists in the middle decades of the century, that observations should be repeated to minimize errors in one's own observation or those of others. In parallel, however, the prevailing reading practices of scholars also changed. "Free" reading increased. The ordering of reading notes in bound form was replaced by more flexible organizing systems and tools. This corresponded to a greater emphasis on *iudicium* in cognitive practices: readers were to exercise their judgment. Their reading was to be defined by the principle of criticism and the related principle of selection.

What is more, the nonexistence of a centaur or basilisk could scarcely be proven by observation. How indeed could one have used direct observations to verify the literary accounts of such beings? Particularly for monsters now classed as fabulous, the intellectuals' new approach to literature, in particular the more rigorous evaluation of the credibility of reports about unusual phenomena or events, was at least as far reaching in its implications as their new observational practices.

How should we view these findings in light of the current discussion of the relationship between "scientific" and other forms of knowledge, the relationship between the history of science and the history of knowledge?[7] The grand narrative of the Scientific Revolution was built on the premise that the natural sciences emerged in the European Renaissance by emancipating themselves from other forms of scholarship. The rise of scientific observation and of the experiment seemed to come at the cost of the significance of scholarly practices like reading. The latter, this narrative suggests, continued to exist in other fields of scholarship that we now know as the humanities. One might say that the explanatory model of the Scientific Revolution relocated the proverbial separation of the "two cultures" to early modern Europe.

In the light of the present study, this argument proves anachronistic. The emphasis placed on experience by both early modern actors and later historians has long obscured the fact that reading continued to be of great importance in the study of nature. And it has given undue weight to the differences between the practices and sources of knowledge that were employed in nature study in the sixteenth to the eighteenth centuries and those used in other fields of scholarship. This can be the starting point for an integrated history of the humanities *and* the natural sciences[8]—whether we wish to call it the history of "science," in the broad sense of the German term *Wissenschaft*, or the history of knowledge.

Acknowledgments

This work began with a fascination with certain texts whose texture had me puzzled: there was something strange about many of the early modern natural history and medical books on rare things of nature that I had read. They all seemed to be woven from the same textual building blocks. I could often scarcely make out the voice of the author in the tapestry of these elements. For me, these texts raised questions about their authors' reading and writing practices, and about their understanding of authorship.

Lorraine Daston and Helmut Zedelmaier helped me to take the step from this curiosity to the formulation of the question that would guide my research over the following years. And they accompanied the entire process through to the finished book with their unique expertise and generosity. I examined the sources from which the authors of these texts drew their knowledge about monsters. The result of my study is a contribution to the historiography both of scholarly reading and of scientific observation.

Numerous colleagues and institutions supported me during my work on this book. Most of the research on which the study is based was conducted during my time as a PhD student on the research project The History of Scientific Observation at the Max Planck Institute for the History of Science (MPIWG). I benefited enormously from the discussions with my colleagues at the MPIWG—particularly Gianna Pomata, who was so generous as to read and comment on every chapter. Parts of the study were critically proofread by Daniel Andersson, Francesca Bordogna, Oliver Gaycken, Daryn Lehoux, Jeffrey Schwegman, Skúli Sigurdson, and Thomas Sturm, and—outside the MPIWG—Ulrike Klöppel, Mark Kammerbauer, Andreas Koch, Alexandra Petrović, and Aleksandar Zivanović. I am also indebted to many others who were then at the MPIWG, among them Elisa Andretta, Domenico Bertoloni Meli, Estelle Blaschke, Stefan Borchers, Christina Brandt, Mirjam

Brusius, Mechthild Fend, Nils Güttler, Stefanie Klamm, Bernhard Kleeberg, Richard Kremer, Nicolas Langlitz, Wolfgang Lefèvre, Susanne Lehmann-Brauns, Rhodri Lewis, Hannah Lotte Lund, José Ramón Marcaida, Erika Lorraine Milam, Tania Munz, Claudia Stein, Fernando Vidal, Annette Vogt, Christine von Oertzen, Kelly Whitmer, and Yossi Ziegler.

Periods spent researching at the Herzog August Library (HAB) in Wolfenbüttel, the International Centre for the History of Universities and Science (CIS) at the University of Bologna, and the Gotha Research Centre were profitable and productive, likewise due in no small part to my colleagues there. After submitting my doctoral thesis, a two-semester research position at the Italian Academy for Advanced Studies in America at New York's Columbia University allowed me to delve more deeply into Ulisse Aldrovandi's reading practices. For this I am grateful to the director of the Italian Academy, David Freedberg, and his deputy Barbara Faedda. My "fellow fellows" and New York colleagues—including Pamela H. Smith and Matthew L. Jones, Alix Cooper, James Delbourgo, and Daniel Margocsy—provided valuable input. I would especially like to thank Monica Calabritto and Nancy G. Siraisi. A term as Frances A. Yates Fellow at the Warburg Institute in London also contributed to a deeper understanding of Aldrovandi's working methods. The doors of Guido Giglioni's, Jill Kraye's, and Peter Mack's offices were always open to me; discussions with Sietske Fransen, Richard Oosterhoff, and Denis Robichaud were equally important.

My research was funded by many organizations. In addition to the above institutions, I would like to thank the Dr. Günther Findel Foundation for funding a research visit to the HAB, the Fritz Thyssen Foundation for a Herzog Ernst scholarship at the Gotha Research Library at the University of Erfurt, the FAZIT Foundation, and the Gerda Henkel Foundation for their writing-up fellowships.

I am also grateful to the many colleagues who have lent me their ears and their judgment at colloquiums, at conferences, and in private conversations, and helped me to answer the countless questions that arise in the course of a major research project. They include Andrea Bernardoni, Ann Blair, Isabelle Charmantier, Rosa Costa, Palmira Fontes da Costa, Caroline Duroselle-Melish, Paula Findlen, Volker Hess, André Holenstein, William John Kennedy, Sachiko Kusukawa, Kathleen Perry Long, Hanspeter Marti, Christoph Meinel, Staffan Müller-Wille, Urs Leu, Herbert Mehrtens, Katharine Park, Lise Camilla Ruud, Wolfgang Schivelbusch, Claus Spenninger, Rüdiger vom Bruch, and Bettina Wahrig. The same is true of the librarians, archivists, and nonacademic colleagues at the institutions where I had the opportunity to conduct my research, especially Josephine Fenger, Regina Held, and Carola Kuntze.

When Ann Blair, Tony Grafton, and Earle Havens offered to include my book in their Information Cultures series with Johns Hopkins University Press, I did not hesitate to accept. The present text is an abridged and revised version of the monograph originally entitled *Ein Zentaur in London: Lektüre und Beobachtung in der frühneuzeitlichen Naturforschung* and published by Didymos-Verlag. The translation of this work was provided by Shivaun Heath in collaboration with Joy Titheridge and Fiona Mizani, and funded by Geisteswissenschaften International— Translation Funding for Work in the Humanities and Social Sciences in Germany, a joint initiative of the Fritz Thyssen Foundation, the German Federal Foreign Office, the collecting society VG WORT, and the Börsenverein des Deutschen Buchhandels (German Publishers & Booksellers Association). My wonderful PhD student Dominik Knaupp provided invaluable assistance in preparing the manuscripts of both editions for printing. Special thanks also to Thomas Richter of Didymos-Verlag, who played a key part in the German edition of the book turning out so well that the Geisteswissenschaften International jury awarded it the 2018 special prize in first place.

Notes

Introduction

1. Blair, "Humanist Methods in Natural Philosophy," 545. See my full explanation of the term in chapter 1.

2. See Zedelmaier and Mulsow, "Einführung."

3. Nick Popper positions himself similarly in his notable study on the related subject of late Renaissance historiography: Popper, *Walter Ralegh's History of the World*, 9–10.

4. On Haller and the rumor about the centaur in London, see also chapter 4 of this book.

5. The lectures are preserved in the German translation of a Latin transcript made by his eldest son. See the "Vorrede des Uebersezers," in Haller, *Vorlesungen über die gerichtliche Arzneiwissenschaft*, vol. 1, 1, fol. *2r.

6. Haller, *Vorlesungen über die gerichtliche Arzneiwissenschaft*, vol. 1, 205–209.

7. Cf. Haller, *Vorlesungen über die gerichtliche Arzneiwissenschaft*, vol. 1, 190–193.

8. On the distinction between *être vrai* and *être dans le vrai* in Canguilhem's work, see Grond-Ginsbach, "Georges Canguilhem als Medizinhistoriker," 240.

9. Haller, *Vorlesungen über die gerichtliche Arzneiwissenschaft*, vol. 1, 191. Unless otherwise indicated, all translations are by Shivaun Heath, Joy Titheridge, and Fiona Mizani. For the original citations, see Kraemer, *Ein Zentaur in London*.

10. My use of the term "naturalist" in this book follows that of Lorraine Daston and Katharine Park. The description refers to authors from a variety of fields (particularly medicine, natural history, and natural philosophy) who attempted a systematic examination of nature. This choice of terminology allows me to avoid the anachronistic term "scientist." In the period under examination, the boundaries between academic subjects were markedly more permeable than the disciplinary boundaries of the modern academic system; they were frequently crossed by scholars. As we shall see, university-educated physicians, natural historians, and natural philosophers shared many of their practices and sources with scholars who were concerned primarily with subject areas that do not—from a modern perspective—have any immediate connection with the study of nature. On Daston and Park's use of the term "naturalist," see *Wonders*, 373n4.

11. See Steinke, "Anatomie und Physiologie," 237–241. Cf. also Jahn, *Geschichte der Biologie*, 256 and passim.

12. Haller, *Vorlesungen über die gerichtliche Arzneiwissenschaft*, vol. 1, 191. Wherever possible in the following, all quotations are reproduced as they are found in the sources. Thus, for example, abbreviations are left unexpanded and now unfamiliar diacritical marks are retained, since the exact form of the text is often crucial to their analysis.

13. The most comprehensive study to date on the reception of Pliny's *Natural History* in the early modern period documents convincingly how great its authority remained even into the seventeenth century: Doody, *Pliny's Encyclopedia*, particularly 11–39. In the context of the humanist publication efforts of the Renaissance, isolated criticisms of ambiguities and weaknesses in the content of Pliny's *Natural History* were voiced from the late fifteenth century onward. A particularly well-known debate was ignited by the humanist and physician Niccolò Leoniceno in 1492, which continued intensively for several years in humanist circles, concerning errors in the medical section of the *Natural History*. Leoniceno pointed out weaknesses in the botanical terminology of the *Natural History* for which he also held Pliny personally responsible. See Nauert, "Humanists, Scientists, and Pliny," 81–83. Nauert has rightly made the point, however, that criticism of this kind from the humanist publishers and commentators of the late fifteenth and sixteenth centuries did not lead to an explicit but at most subconscious questioning of the authority of the text (85). In the seventeenth century, criticism of individual assertions in Pliny's *Natural History* grew; see chapter 4.

14. Aldrovandi, *Monstrorum Historia*, 30; original italics. Translation of the Pliny quote within the quotation after Rackham; see Plinius Secundus, *Natural History*, vol. 2, 528 (Latin) and 529 (English). On Aldrovandi's writings on monsters, see chapter 2.

15. Cf. the conceptual history overview in Shapin, *The Scientific Revolution*, 1–4.

16. Shapin, *The Scientific Revolution*, 68–69.

17. See, for example, Moscoso, "Vollkommene Monstren und unheilvolle Gestalten," 58. Strictly speaking, Canguilhem himself refers to the "naturalizing" of monsters, not to the "naturalization of the monstrous." He argues that the same era that—according to Michel Foucault—naturalized insanity also applied itself "à naturaliser les monstres." Canguilhem, "La monstruosité et le monstrueux," 228. Cf. also 227, where he writes among other things that the *monstrum* was "naturalized" from the moment at which *monstruosité* became a biological concept.

18. For a summary, see Park and Daston, "Introduction," 12–13. A detailed discussion of the current state of research on the Scientific Revolution can be found in chapter 3.

19. See, in particular, Céard, *La nature et les prodiges*, chapter 18; Canguilhem, "La monstruosité et le monstrueux"; and/or Canguilhem, "Monstrosity and the Monstrous"; and Daston and Park, "Unnatural Conceptions."

20. Daston and Park, "Unnatural Conceptions," 23. In parts, the summary above follows that in Daston and Park, *Wonders*, 175–176.

21. With this in mind, Alan W. Bates extends Céard's coinage of the "golden age" of prodigies—Bates speaks of "monsters" rather than of prodigies—to include the whole of the sixteenth and seventeenth centuries. See Bates, *Emblematic Monsters*, 11. The state of research on this point is addressed more fully in chapter 1. Unlike other text genres, the *alte Dissertationen*, "university dissertations," were long neglected. From the late sixteenth century on, monsters were frequently the subject of disputations—or academic debates—particularly at the Protestant universities of the Holy Roman Empire. For a documentation of the many dissertations on monsters that were written for these occasions from the sixteenth century on, see Kraemer, *Ein Zentaur in London*, appendix.

22. Céard, *La nature et les prodiges*, 159.

23. This point is discussed in more detail in chapter 1.

24. See Daston and Park, *Wonders*, particularly 175–176.

25. See Bates, *Emblematic Monsters*, particularly 8 and 11–12. The state of research on the naturalization thesis is addressed more fully in chapter 3.

26. See Daston and Park, *Wonders*, chapter 6. This point is discussed in more detail in chapter 4.

27. See Daston and Park, *Wonders*, chapter 9. This is addressed in greater depth in chapter 4.

28. Gianna Pomata coined this term to highlight the fact that certain literary forms, in particular the *observatio*, influenced the cognitive practices of the scholars by shaping and focusing their attention. See Pomata, "Sharing Cases," 197. Cf. also Park, "Observation in the Margins," particularly 48.

CHAPTER 1: Three Monstrous Factoids

1. Katharine Park and Lorraine Daston give a concise summary of the controversy surrounding this term since the 1980s—and the resulting cautiousness in its use in the history of science—in the context of explaining why their volume of the *Cambridge History of Science* on the early modern period dispenses with the term entirely: "But the omission that is likely to arouse the most surprise is in the title itself: Where is the Scientific Revolution? Our avoidance of the phrase is intentional. The cumulative force of the scholarship since the 1980s has been to insert skeptical question marks after every word of this ringing three-word phrase, including the definite article. It is no longer clear that there was any coherent enterprise in the early modern period that can be identified with modern science, or that the transformations in question were as explosive and discontinuous as the analogy with political revolution implies, or that those transformations were unique in intellectual magnitude and cultural significance." Park and Daston, "Introduction," 12–13. For a recent attempt to revive the Scientific Revolution, see Wooton, *The Invention of Science*.

2. On the growing interest in hermaphrodites among naturalists in France, see Daston and Park, "The Hermaphrodite." On the Holy Roman Empire, see Kraemer, "'Under so viel wunderbarlichen und seltsamen Sachen.'"

3. Varchi, *La prima parte delle lezzioni*. Varchi's text has only recently received the attention it deserves. See Ciseri, "A lezione con i mostri."

4. On the history of this book, see the excellent critical edition Paré, *Des monstres et prodiges*.

5. Weinrich, *De Ortu Monstrorum Commentarius*.

6. Schenck von Grafenberg, *Monstrorum Historia*; Schenck von Grafenberg, *Wunder-Buch*.

7. Bauhin, *De Hermaphroditorum Monstrosorumque Partuum Natura*.

8. Liceti, *De Monstrorum Natura*. Liceti's treatise was published in three further Latin editions (1634, 1665, and 1668) and one French edition (1708).

9. See the documentation of *alte Dissertationen* on monsters in Kraemer, *Ein Zentaur in London*.

10. For a detailed discussion of the naturalization thesis and the arguments against it, see Kraemer, *Ein Zentaur in London*, chapter 3.

11. Céard, *La nature et les prodiges*, 159.

12. Cf. Daston and Park, *Wonders*, 177 and 180. The growing number of reports of monstrous births was interpreted by some readers and authors as a sign that the end times were near. A good example of this is Lycosthenes, *Prodigiorum ac Ostentorum Chronicon*; on this, see later in this chapter.

13. Cf. Daston and Park, *Wonders*, 180. According to Alan W. Bates, too, the technical availability of printing encouraged the growth in interest in monsters that began around 1500. Bates, *Emblematic Monsters*, 15.

14. On these examples, see Daston and Park, *Wonders*, 187–188; and Spinks, *Monstrous Births and Visual Culture*, 59–79. Cf. Daston and Park, *Wonders*, 180.

15. Lycosthenes, *Prodigiorum ac Ostentorum Chronicon*; Lycosthenes, *Wunderwerck oder Gottes unergründtliches vorbilden*; Batman, *The Doome*. On the popularity and commercial success of vernacular translations of this book, see Bates, *Emblematic Monsters*, 67. Cf. Daston and Park, *Wonders*, 182–183.

16. "From the very beginning of the world, until this our time." Lycosthenes, *Wunderwerck oder Gottes unergründtliches vorbilden*, front page.

17. See Bates, *Emblematic Monsters*, 67.

18. See Daston and Park, *Wonders*, 149; and Bates, *Emblematic Monsters*, 75.

19. A comprehensive analysis of the early modern understanding of nature is offered in Daston, "The Nature of Nature."

20. On these aspects, see Daston and Park, *Wonders*, chapters 4 and 5. On the lack of conviction in the late Renaissance that nature was confined by fixed rules, see also Daston, "The Nature of Nature"; and Roger, *The Life Sciences*, 26. Cf. also Findlen, "Jokes of Nature."

21. See particularly Blair, "Humanist Methods in Natural Philosophy," 545.

22. The parallels can only be mentioned briefly in this chapter; they are considered more closely in chapter 2.

23. The alternate meaning of "factiod," "false fact," is of no significance in this study. The veracity of the morsels of knowledge examined is not relevant to the question addressed here. Cf. "Factoid," *Oxford English Dictionary* (n.d.), accessed December 16, 2010, http://www.oed .com/view/Entry/67511?redirectedFrom=factoid#.

24. Cf. Blair, "*Historia* in Zwinger's *Theatrum*," 285.

25. The circulation of knowledge concerning monsters in the early modern period has hitherto only been analyzed from the following three perspectives: First, individual accounts have been examined with the aim of establishing the extent to which they were in circulation in other works. One of the better known examples is the monster of Ravenna. See particularly Bates, *Emblematic Monsters*, 22–24; Daston and Park, *Wonders*, 177–178; Niccoli, *Prophecy and People*, 35–51; and Schenda, "Das Monstrum von Ravenna." Second, the sources for what is today perhaps the best-known early modern naturalist monograph on monsters, Ambroise Paré's *Des monstres et prodiges*, have been studied in detail. See Céard, introduction to *Des monstres et prodiges*, ix–xlvi; and, on the sources of Paré's woodcuts, Daston and Park, *Wonders*, 149. Third, Wolfgang Harms and Michael Schilling, or rather the authors of the individual chapters in their volumes, have investigated substantially which authors took up and reused the events reported in the illustrated broadsheets they edited. Many of these broadsides discuss the birth or observation of a monster. See particularly Harms and Schilling, *Die Sammlung der Zentralbibliothek Zürich*, parts 1 and 2. A larger-scale study of the way in which factoids about monsters circulated in the texts of early modern naturalists has not hitherto been conducted.

26. For a recent overview of research literature on the topic, see Kraemer, "Die Individualisierung des Hermaphroditen."

27. There are good reasons for avoiding the anachronistic term "race" in this context. The term *genus*, which—following Augustine, *De civitate dei* XVI, VIII—was commonly used in the Middle Ages and the early modern period to refer to the "monstrous human races" (Augustine speaks of "genera hominum monstrosa"), was not clearly defined before Carl von Linné. *Genus* had the approximate meaning of "kind" or "type." For the most part it meant an overarching species, which in turn encompassed several kinds. Similarly, it was not until the eighteenth century that the English term "race" received a narrower definition in the modern biological sense—i.e., that of a unit within a species that is distinguished by hereditary characteristics. Before the eighteenth century, the English term "race" was used analogously to the Latin *genus* in a broad and comparatively vague sense. Since John Friedman's influential monograph on *The Monstrous Races in Medieval Art and Thought*, however, the term "monstrous races" has become the standard term for this subject or group of phenomena in medieval and early modern literature, particularly in Anglophone research. For this reason I use it here despite the reservations mentioned. On the history of the exotics from the edge of the world, see also particularly Wittkower, "Marvels of the East"; and Austin, "Marvelous Peoples or Marvelous

Races?" I am grateful to Staffan Müller-Wille for his assessment of this terminological question, which confirmed my own thoughts and on which some of the remarks above are based.

28. Jennifer Spinks stresses that individual monstrous births and the monstrous human races were different phenomena, which is true inasmuch as the exotics from the edge of the world could not be interpreted as prodigies. But Spinks, too, acknowledges that the visual depictions of individual monstrous births and those of the monstrous races resembled one another even from the fifteenth century onward. In the following, I make these similarities a central focus and examine their causes. Cf. Spinks, *Monstrous Births and Visual Culture*, 13–36, particularly 18.

29. Schenck von Grafenberg, *Monstrorum Historia*, 1–3.

30. Schenck von Grafenberg, *Observationes Medicæ de Capite*, fol.):(2r. Cf. Pomata, "*Praxis Historialis*," 133. The term *doctrina* was used by physicians with a great variety of meanings during the period under examination. But its core meaning was clearly tied to teaching and learning. See Maclean, "Doctrines médicales à la Renaissance." The paper has since been published as Maclean, "La doctrine selon les médecins." As in the above case, I reproduce the passages quoted from contemporary sources with all their typographical particularities, because it is essential here that we consider not only their content but also their physical composition.

31. Schenck von Grafenberg, *Monstrorum Historia*, 3.

32. Bauhin and Liceti were among those who referred to Fincel's work in their monographs on monsters. Liceti, for example, also with reference to Fincel, mentions the headless baby from Meissen. Liceti, *De Monstrorum Natura*, 58.

33. Cf. Hammerl, "Prodigienliteratur."

34. Fincel, *Wunderzeichen*, vols. 1–3.

35. Fincel, *Wunderzeichen*, vol. 1, front page.

36. Fincel, *Wunderzeichen*, vol. 1, fol. bvv.

37. Johannes Schenck von Grafenberg's *Observationes* collection does not contain a woodcut; the textual description of the case is identical to that of Johann Georg Schenck. See Schenck von Grafenberg, *Observationes Medicæ de Capite*, 25.

38. Cf. Daston and Park, *Wonders*, 187.

39. See Daston and Park, *Wonders*, 187. At least one early modern reader felt called—even without an explicit prompt in the form of several blank pages bound into the book—to continue the contemporary German edition of the *Prodigiorum ac Ostentorum Chronicon*: in loosely chronological order and over almost an entire blank and unpaginated page at the end of one (still existing) copy, he recorded additional prodigies up to the year 1562. I am referring here to Lycosthenes, *Wunderwerck oder Gottes unergründtliches vorbilden* (Berlin State Library; 4° Lg35536).

40. Lycosthenes, *Prodigiorum ac Ostentorum Chronicon*, 645.

41. Cf. also chapter 2.

42. Lycosthenes, *Prodigiorum ac Ostentorum Chronicon*, 4.

43. Lycosthenes, *Prodigiorum ac Ostentorum Chronicon*, 9.

44. Lycosthenes, *Prodigiorum ac Ostentorum Chronicon*, 670. The woodcut can also be found there.

45. Cf. Bäumer-Schleinkofer, "Ulisse Aldrovandi," 195. For a detailed discussion of Ulisse Aldrovandi's *Monstrorum Historia*, see chapter 2.

46. On the posthumous publication of Ulisse Aldrovandi's *Monstrorum Historia* by Bartolomeo Ambrosini, see chapter 2.

47. On Aldrovandi's concept of *differentiæ*, see Aldrovandi, *Monstrorum Historia*, particularly 4 and 5.

48. Aldrovandi, *Monstrorum Historia*, section "Differentiæ," 4–43. On Aldrovandi's discussion of the monstrous races, cf. Céard, *La nature et les prodiges*, 456–457.

49. Cf. Céard, *La nature et les prodiges*, 457.

50. Aldrovandi, *Monstrorum Historia*, 8. Cf. Céard, *La nature et les prodiges*, 457.

51. Aldrovandi, *Monstrorum Historia*, 402.

52. Aldrovandi is referring here to the travel account published in 1599 of a German mercenary who spent many years in the New World in the service of the Spanish: Schmidel, *Vera Historia*.

53. Schenck von Grafenberg, *Observationes Medicæ de Capite*, 24–25.

54. For instance, two of Johannes Schenck von Grafenberg's *observationes* address hermaphrodites. Both are structured very similarly to the section on headless humans described above: they, too, begin with quotations on the race in question—in this case, a race of hermaphrodite humans—and only then are individual cases of hermaphroditism presented. See Schenck von Grafenberg, *Observationum Medicarum*, 5–16. On the description and visual representation of hermaphrodites around 1600, see Kraemer, "Die Individualisierung des Hermaphroditen"; and Kraemer, "Hermaphrodites Closely Observed."

55. Zedelmaier, "Wissensordnungen der Frühen Neuzeit," 836.

56. Park, "Nature in Person," 50.

57. Zedelmaier, "Wissensordnungen der Frühen Neuzeit," 838–839.

58. Zedelmaier, "Wissensordnungen der Frühen Neuzeit," 836. Cf. Schmidt-Biggemann, *Topica Universalis*.

59. As a qualification, it should be said that experienced scholars were often not content to restrict themselves to the *loci* technique but experimented with a variety of techniques and classification principles for the ordering of the material they gathered. See Kraemer, "Ein papiernes Archiv"; and Kraemer, "Ulisse Aldrovandi's *Pandechion Epistemonicon*." On the concept of paper technology, see te Heesen, "The Notebook"; and Hess and Mendelsohn, "*Paper Technology* und Wissenschaftsgeschichte." On the use of paper technology by scholars such as Ulisse Aldrovandi, see chapter 2.

60. Blair, *The Theater of Nature*, 5. On early modern reading guides, see, for example, Zedelmaier, "Wissensordnungen der Frühen Neuzeit," 840–841; and Zedelmaier, "Johann Jakob Moser."

61. On the coexistence of factoids obtained from reading and from observations in early modern commonplace books, see chapter 2.

62. On the relationship of the "New Science" to scholarly knowledge, cf. Zedelmaier, "Wissensordnungen der Frühen Neuzeit," 836 and 841–842.

63. See Findlen, *Possessing Nature*, 58–59.

64. Aldrovandi, *Monstrorum Historia*, 8. On this whole section, cf. Céard, *La nature et les prodiges*, 457.

65. Cf. Serjeantson, "Proof and Persuasion," 135.

66. The statement in question is one in Cicero's *De partitione oratoria* 2.1 that describes rhetoric as "probabile inventum ad faciendam fidem." See Serjeantson, "Proof and Persuasion," 147.

67. See Serjeantson, "Proof and Persuasion," 147–149.

68. Aldrovandi, *Monstrorum Historia*, 8.

69. Schenck von Grafenberg, *Observationes Medicæ de Capite*, fol.):(2r.

70. Aldrovandi, *Monstrorum Historia*, 41. On Colombo's own contribution to the discourse on monsters, see below.

71. See, for example, the list of sources entitled "AUTHORES Qui Citantur" in Bauhin, *De Hermaphroditorum Monstrosorumque Partuum Natura* (1614), 12–36.

72. The concept of authorial function stems from an influential essay by Michel Foucault. See Foucault, "Was ist ein Autor?" In recent years, the history of authorship in the sciences has at times been the subject of intensive debate—though central questions such as those discussed above are yet to be answered satisfactorily. A good overview of the current state of research is offered by Biagioli and Galison, *Scientific Authorship*.

73. Chartier, "Foucault's Chiasmus," 16.

74. See Höfele and Laqué, introduction to *Humankinds*, 15.

75. The literature on Caliban is almost too extensive to survey. On Caliban and numerous other "public animals" in Shakespeare's work, see, for example, Yachnin, "Shakespeare's Public Animals." A large-scale study of the relationship between humans and animals in the work of Shakespeare is offered by Höfele, *Stage, Stake, and Scaffold*.

76. Lycosthenes, *Prodigiorum ac Ostentorum Chronicon*, 437–438.

77. Schenck von Grafenberg, *Monstrorum Historia*, 100.

78. The self-containedness of this part of the *Monstrorum Historia* is reflected, too, in the mention it receives in the title of the volume:

MONSTRORUM
HISTORIA
MEMORABILIS . . . *Accessis Analogicum Argumentum*
DE *MONSTRIS BRUTIS.*
SUPPLEMENTI LOCO AD OBSERVATIONES MEDI-
cas SCHENCKIANS edita
à
IOANNE GEORGIO SCHENCKIO A GRAFENBERG FILIO . . .

While part 1 extends from p. 1 to p. 96, this second part occupies only pp. 97 to 135. All in all, Johann Georg Schenck intended his volume to be understood as an addition to his father's *Observationes*; see the dedication (fol. (:)2v).

79. Pomata, *"Praxis Historialis,"* 133.

80. Schenck von Grafenberg, *Monstrorum Historia*, 99.

81. Pictorius, *Sermonum Convivalium Libri X.*; and Pictorius, *Sermonum Convivalium Apprimè Utilium, Libri X.*

82. Neither of the two editions of Pictorius's work contains a corresponding illustration. It seems likely, therefore, that the etching was produced on the basis of Pictorius's textual description of the monster.

83. Pictorius, *Sermonum Convivalium Libri X.*, 84–85.

84. See Kraemer, *Ein Zentaur in London*, chapter 3.

85. The chapter title itself explains the hypothesis underlying the list; namely, that of the existence of such monsters and consequently of the validity of the corresponding category: "Monstra multiforma, diversas animalium species in eodem genere proximo referentia, non esse figmenta, sed in rerum natura reperiri." Liceti, *De Monstrorum Natura* (1634), 180.

86. Liceti, *De Monstrorum Natura*, 183–184. The printed marginal note with the year 1254 is found on p. 184. In the inner margin of the preceding page there is a short reference "l. 24. c. I." In all probability, this refers to the second edition of Ambroise Paré's *Œuvres*, of which (as was the case in the first edition) the twenty-fourth book is his *Des monstres et prodiges*, which contains the case in question. On the various editions of *Des monstres et prodiges*, see Paré, *Des monstres et prodiges*, xlix–xlx. Incidentally, Liceti was also familiar with Lycosthenes's chronicle and makes reference to it on several occasions. See, for example, Liceti, *De Monstrorum Natura*, 25.

87. Paré, *Des monstres et prodiges*, critical note on line 49, p. 66.

88. See Paré, *Des monstres et prodiges*, 7 and 153n17.

89. This statement refers to all editions of his treatise that include woodcuts from the second edition onward. The first edition does not have any illustrations.

90. Little is known about de Tesserant. See Céard, *La nature et les prodiges*, particularly 319n7.

91. Boaistuau, *Histoires Prodigieuses*.

92. See Paré, *Des monstres et prodiges*, 153n17. On Boaistuau's *Histoires prodigieuses*, cf. Bates, *Emblematic Monsters*, particularly 65–66 and 72–73. On the central importance of this text in Paré's monograph, see also Céard, introduction to *Des monstres et prodiges*, xi–xiv.

93. See Paré, *Des monstres et prodiges*, 153n17.

94. Mussatus, *Historia Augusta Henrici VII. Caesaris*.

95. Liceti, *De Monstrorum Natura*, 190–191.

96. Plutarch, *Septem sapientium convivium*, 149. See Plutarch, *Moralia* II, 364–367.

97. In chapter LXVIII of book II, where the passage from Plutarch is quoted in its entirety, Liceti makes more of this point: after the quotation he notes that the case originates from Plutarch, who expresses it under the guise (*persona*) of Thales. Liceti, *De Monstrorum Natura*, 216. But this constellation is not discussed any further here either; rather, Liceti proceeds to other cases of monsters who combine human and animal traits, and whose origin he sees in a mixing of human and animal seeds. Even though he is aware that Plutarch presents the case via his character Thales—one of the seven wise men of his *convivium*—Liceti nevertheless uses it as evidence for this tenth of the causes of monsters with human and animal body parts in his monograph.

98. The title begins as follows:

SERMONUM
CONVIVALIUM LIBRI X.
Historicæ, poëticæ ac medicæ rei studiosis
valde utiles: autore D. GEORGIO
Pictorio Villingano medico

Pictorius, *Sermonum Convivalium Libri X.*, front page.

99. See Pictorius, *Sermonum Convivalium Libri X.*, fol. a2v.

100. See Pictorius, *Sermonum Convivalium Libri X.*, no pagination.

101. See the paratext "CANDIDO LECTORI PICTORIUS S. D." in Pictorius, *Sermonum Convivalium Libri X.*, no pagination.

102. See Föcking, "Mißverständnisse der Rezeption als Innovationsfaktoren," 198–199. On dialogic text genres in the Renaissance, see Cox, *The Renaissance Dialogue*; and Guthmüller and Müller, *Dialog und Gesprächskultur*; on the genre of table talks, see B. Müller, "Die Tradition der Tischgespräche."

103. Thucydides, *History of the Peloponnesian War*, 85–113.

104. Similar texts exist by both Plato and Xenophon. The wise men, however, vary from one text to the next. Cf. the introduction in Plutarch, *Moralia* II, 346–347.

105. Pictorius, *Sermonum Convivalium Libri X.*, 84–85. Plutarch's version of this episode can, for its part, be described as an "original" only to a limited extent. One of the fables (III,3 Aesopus et rusticus) of the Roman author Phaedrus (first century BC), a contemporary of Plutarch, has a strikingly similar story line: lambs with human heads born in a herd of sheep frighten their owner because he takes them for a divine omen. So he asks the oracles for their opinion. But they offer various different explanations for the phenomenon that only frighten him all the more. Then, however, wise old Aesop gives him the following advice: if he wishes to

avoid such an *ostentum* in the future, he should ensure that his shepherds are provided with wives. See Schenda, "Wunder-Zeichen," 17n10.

106. See Aldrovandi, *Monstrorum Historia*, 433.

107. Lycosthenes's account is summarized on p. 433; the woodcut that appears on the following page has the caption "I. Pullus equinus humana facie" and occupies almost an entire folio page. The head and tail of the foal are strongly reminiscent of Paré's woodcut. One of the front legs is raised, however, and the *monstrum* is depicted as a mirror image. See also below on the artist's obvious use of Paré's woodcut as a basis for his own.

108. Ulisse Aldrovandi died before the first edition of Johann Georg Schenck's *Monstrorum Historia* was published. This is therefore obviously one of Ambrosini's numerous alterations to Aldrovandi's manuscript. Cf. chapter 2 on his role in the publication of the work.

109. Aldrovandi, *Monstrorum Historia*, 433.

110. Aldrovandi, *Monstrorum Historia*, 433.

111. This supposition is based—in addition to the author's critical approach to the two all too similar versions of the factoid—on the fact that Johann Georg Schenck von Grafenberg's *Monstrorum Historia*, which was not published until 1609 and therefore after Aldrovandi's death, is used as a source here, too. Aldrovandi also mentions the monstrous horse born with a human head in 1254 in the section "OMINA. SOMNIA. OSTENTA. Prodigia." (127–134) of the extensive chapter on the horse ("Cap. I.," 1–294) in his *De Quadrupedibus Solidipedibus*, which was not published by Ambrosini. The only source that Aldrovandi mentions here is Paré. No reference is made, either, to the numerous other versions of this factoid in the literature. The woodcut tracing back to Paré that is used in the *Monstrorum Historia* is employed here, too. Aldrovandi informs the reader that it is taken from Paré's "libro de monstris et prodigijs." Aldrovandi, *De Quadrupedibus Solidipedibus*, 129.

112. Medical texts on other topics also frequently contain "fictional" and historiographical sources. This is demonstrated in Calabritto, "Examples, References and Quotations."

113. Bauhin, *De Hermaphroditorum Monstrosorumque Partuum Natura* (1614); and Bauhin, *De Hermaphroditorum Monstrosorumque Partuum Natura* (1629). I have not been able to verify the existence of a third edition (Frankfurt 1600) mentioned in the literature. Unless otherwise stated, I am using the 1614 edition.

114. Bauhin, *De Hermaphroditorum Monstrosorumque Partuum Natura*, 357.

115. See Pomata, *"Praxis Historialis,"* 112, 115, 124, and 130.

116. See, for example, Blair, *"Historia* in Zwinger's *Theatrum."*

117. For this reason, an identification of the concept of the medical case history with the *exempla* in Renaissance medical texts seems problematic. Gaspard Bauhin, for example, gives us no cause to believe that he used the term in a narrow, purely medical sense that excluded other forms of historical knowledge. Indeed, Theodor Zwinger's *Theatrum Vitae Humanae*, used by Bauhin as a source for the *exempla* that he compiled, is a particularly prominent example of exemplary historiography; see below. Michael Stolberg's recent and helpful writing on early modern case histories sadly ignores the question of the characteristics of the *medical* case history as distinct from other short narratives such as Zwinger's *exempla* and does entirely without a definition of his subject. Cf. Stolberg, "Formen und Funktionen Medizinischer Fallberichte." The connections between the medical case report and those in other discourses and fields have been the subject of intensive debate in recent years—mostly, however, with a later focus. See, for example, Süßmann, Scholz, and Engel, *Fallstudien*, Behrens and Zelle, *Der Ärztliche Fallbericht*; and Wübben and Zelle, *Krankheit Schreiben.*

118. The first printed edition was published as early as 1508, however. On Julius Obsequens, cf., for example, Bates, *Emblematic Monsters*, 66.

119. According to the index of authors, Bauhin consulted a 1604 edition of the *Theatrum*. On this list of sources, see Bauhin, *De Hermaphroditorum Monstrosorumque Partuum Natura*, 12–36.

120. Indeed, according to Mauelshagen, in the two centuries following its publication, Lycosthenes's chronicle was cited more frequently by naturalists than by theologians or parish clergy; Mauelshagen, *Wunderkammer auf Papier*, 75.

121. Siraisi, *History, Medicine*, 2. Cf. also Siraisi, "Anatomizing the Past."

122. Siraisi, *History, Medicine*, 3. Elsewhere, Siraisi sums up parts of this line of thought as follows: "Learned physicians who interested themselves in history and antiquities brought to these topics an intellectual formation that emphasized such significant methodological features as the use of narrative, description, and a measure of attention to material evidence" (19).

123. Large parts of this section are based on Siraisi, *History, Medicine*, 1–24.

124. See Bacon, *The Advancement of Learning*, 69.

125. Wilson, *The Making of the Nuremberg Chronicle*, 207–224. Cf. Siraisi, *History, Medicine*, 19.

126. See Ogilvie, "Natural History, Ethics, and Physico-Theology," particularly 81–82.

127. Ogilvie, "Natural History, Ethics, and Physico-Theology," 91.

128. Ogilvie, "Natural History, Ethics, and Physico-Theology," 82.

129. See Schenda, "Wunder-Zeichen." On the appropriation of elements of scholarly texts by the authors of broadsides, see also Daston and Park, "Unnatural Conceptions," 30.

130. On this view, see Harms and Schilling, "Einleitung," x; and Bates, *Emblematic Monsters*, 45, where Bates gives a critical examination of Dudley Wilson's more traditional position on this question. Cf. Wilson, *Signs and Portents*, 32.

131. On Wick's description of his collection, see Harms and Schilling, "Einleitung," vii.

132. Mauelshagen, *Wunderkammer auf Papier*, 13.

133. Mauelshagen, *Wunderkammer auf Papier*, 13. Further evidence of the mixing of (seemingly) popular and elite culture is offered by the mix of scholarly and vernacular sources that Ambroise Paré employed in the production of his monograph on monsters and prodigies. Céard, introduction to *Des monstres et prodiges*, particularly xix–xxv.

134. Cf. the example in the following chapter and Kraemer, "Die Individualisierung des Hermaphroditen."

135. This can be attributed in part to particularly widespread interest in prodigies and their interpretation in the regions in question. It is also significant that Zurich and Basel were flourishing centers of printing in the sixteenth century, which implies an availability of competent printers and woodcutting artists able to realize such publication ventures. Leu, "Book and Reading Culture in Basle and Zurich, 1520 to 1600." See Beyer, "Tagungsbericht 'The Book Triumphant.'"

136. This is true, for example, of the overwhelming majority of illustrated broadsides chronicling preternatural phenomena in the *Wickiana* with a known place of printing. They originate conspicuously often from Basel, Zurich, Strasbourg, Augsburg, and Nuremberg, in particular. By contrast, little of this material comes from Catholic centers of printing such as Vienna and Venice. Cf. Harms and Schilling, *Die Sammlung der Zentralbibliothek Zürich*, parts 1 and 2.

137. See Harms and Schilling, *Die Sammlung der Zentralbibliothek Zürich*, part 2; and Mauelshagen, *Wunderkammer auf Papier*.

138. Harms and Schilling, *Die Sammlung der Zentralbibliothek Zürich*, part 2.

139. See Harms and Schilling, "Einleitung," viii. One of the few counterexamples of the sixteenth century, a broadsheet by the Catholic Johann Nas published in Ingolstadt in 1569, is discussed by Jennifer Spinks. Her finding—that Nas used the imagery developed by Protestant writers or artists to further the purposes of the Counter-Reformation—is further, indirect

evidence, however, of the dominance of Protestant authors in this field. See Spinks, *Monstrous Births and Visual Culture*, 105–129. On monstrous births in Protestant printing in England, see Crawford, *Marvellous Protestantism*.

140. On the latter point, see Mauelshagen, *Wunderkammer auf Papier*, 13.

141. On a more recent thesis in Reformation history put forward by Robin Bruce Barnes and Volker Leppin, according to which eschatological *expectation* (understood as the assumption of an imminent, datable end to the world) was, around the middle of the sixteenth century, only widespread in a number of less dominant Lutheran movements, see Mauelshagen, *Wunderkammer auf Papier*, 88.

142. Cf. Cunningham and Grell, *The Four Horsemen*, particularly 19.

143. Seifert, *Der Rückzug der biblischen Prophetie*, 7.

144. Seifert, *Der Rückzug der biblischen Prophetie*, 8.

145. See Schenda, "Die deutschen Prodigiensammlungen," 672–674.

146. Schenda, "Die deutschen Prodigiensammlungen," 668. Still seminal on French prodigy literature: Schenda, *Die französische Prodigienliteratur*.

147. Rueff, *De Conceptu et Generatione Hominis*; Rueff, *Ein schön lustig Trostbüchle*.

148. Chapter III in book V, entitled "De Imperfectis Infantibus, nec non Monstrosis & Prodigiosis Partubus."

149. Among other things, Ulisse Aldrovandi had a woodcut of a hermaphrodite from Rueff's chapter on monsters reproduced for his *Monstrorum Historia*. See Kraemer, "Die Individualisierung des Hermaphroditen," 52. Alongside many other writers, Fortunio Liceti and Giovanni Benedetto Sinibaldi refer to Rueff. See Liceti, *De Monstrorum Natura*, where Rueff is mentioned several times; and Sinibaldi, *Geneanthropeiae sive de Hominis Generatione Decateuchon*, col. 878.

150. See Rueff, *De Conceptu et Generatione Hominis*, fol. 47r and 47v.

151. Rueff, *De Conceptu et Generatione Hominis*, fol. 42r–42v.

152. The same conclusion is reached by Costa, "Fremde Wunder oder vertraute Fehler?"

153. Bates, *Emblematic Monsters*, 14.

154. See Céard, introduction to *Des monstres et prodiges*, ix–x; Bates, *Emblematic Monsters*, 76; and Daston and Park, "Unnatural Conceptions," 41. Rueff's contemporary, the Huguenot Ambroise Paré, makes a similar argument. In his view, too, "natural" *and* "divine" causes could be responsible simultaneously for a specific *monstrum*.

155. Cf. Niccoli, *Prophecy and People*. Niccoli's thesis, according to which belief in prodigies came to an abrupt end with the Sack of Rome in 1527, has been convincingly refuted by Paula Findlen. According to Findlen, the "secularization" of prodigies in Italy took place over a longer period of time in the late sixteenth and seventeenth centuries. Findlen, *Possessing Nature*, 18n6.

156. A form of Aristotelianism developed in Padua between the fourteenth and sixteenth centuries that attached a comparatively high epistemic value to experience. Still seminal in this regard: Poppi, *Introduzione All'aristotelismo Padovano*. On Liceti's position within this Aristotelian school, see also Kraemer, *Ein Zentaur in London*, chapter 3.

157. Reske, *Die Buchdrucker des 16. und 17. Jahrhunderts*, 1043–1044.

158. See Kraemer, "Die Individualisierung des Hermaphroditen," 51; and Kraemer, "Hermaphrodites Closely Observed," 45–48. Among the illustrated broadsides in the *Wickiana* that discuss rare natural phenomena and interpret them as divine omens, three can be attributed to the Froschauer *Offizin*: the broadside concerning the hermaphrodite born in Zurich in 1519, the above-mentioned broadside about conjoined twins that was printed in 1543, and one about an occurrence of grain rain allegedly witnessed in Carinthia in 1550. See Harms and Schilling, *Die Sammlung der Zentralbibliothek Zürich*, part 1, 12–13, 58–59, and 86–87.

159. Rosa Costa points out the coexistence of supernatural and natural explanations in this broadsheet, too, and on this basis takes a position against the appropriation of Rueff to support the theory of the naturalization of monsters during this period. See Costa, "Fremde Wunder oder vertraute Fehler?"

160. Harms and Schilling, *Die Sammlung der Zentralbibliothek Zürich*, part 1, 58. All above quotations refer to PAS II 15/27, Zurich Central Library—Prints and Drawings Collection; reproduced on p. 59. As Kuechen has additionally pointed out, this broadside is among the texts listed under Rueff's name in Konrad Gessner's *Bibliotheca universalis*. Rueff and Gessner were friends (58). In general, Gessner included printed and unprinted texts from a wide variety of genres in his bibliography, some of which he did not have the opportunity to examine himself. As such, no particular regard for this specific example or for the genre can be inferred from the broadside's inclusion. Conversely, the low number of broadsides that found their way into the *Bibliotheca Universalis* does not suggest disregard for the genre either, because Gessner generally included only texts written in the languages of scholarship—Latin, Greek, and Hebrew—while most broadsides were written in the vernacular. On Gessner's *Bibliotheca Universalis*, see Zedelmaier, *Bibliotheca Universalis*.

161. PAS II, 2/25, Zurich Central Library—Prints and Drawings Collection, reproduced in Harms and Schilling, *Die Sammlung der Zentralbibliothek Zürich*, part 1, 63. This pamphlet does not give any printing details.

162. Harms and Schilling, *Die Sammlung der Zentralbibliothek Zürich*, part 1, 58.

163. See Pomata, "Sharing Cases," 221.

164. See Harms and Schilling, "Einleitung," ix.

165. See Harms and Schilling, "Einleitung," ix. On Lycosthenes's Zurich connections, see Mauelshagen, *Wunderkammer auf Papier*, 75; on the Zurich ruling elite's intensive discussion of wonders, see pp. 72–74.

166. The following three pairs of monographs are particularly noteworthy here: Rueff, *De Conceptu et Generatione Hominis*, and Rueff, *Ein schön lustig Trostbüchle*; Lycosthenes, *Prodigiorum ac Ostentorum Chronicon*, and Lycosthenes, *Wunderwerck oder Gottes unergründtliches Vorbilden*; Schenck von Grafenberg, *Monstrorum Historia*, and Schenck von Grafenberg, *Wunder-Buch*.

167. Cf. Chartier, "Foucault's Chiasmus," 20.

168. Chartier, "Foucault's Chiasmus," 20. A similar argument is made in Johns, "The Ambivalence of Authorship."

169. See Johns, "The Ambivalence of Authorship," particularly 73–82.

170. For the most part, broadsides such as the illustrated broadsheets in the *Wickiana* that are documented by Wolfgang Harms and Michael Schilling do not contain details of their author. Some of the authors known to us are naturalists such as Jacob Rueff, while others are scholars from other backgrounds such as the Reformer Calvin. See, for example, PAS II 12/22, in Harms and Schilling, *Die Sammlung der Zentralbibliothek Zürich*, part 1, 34–35. In still other cases, the printer of the broadside was also its author. See, for example, PAS II 9/14, on pp. 16–17. And in at least one recorded case, a farmer who was eyewitness to the phenomenon addressed in an illustrated broadside is named as the broadside's author—which is not necessarily to say, of course, that he did indeed write it himself. See PAS II 12/5, on pp. 76–77.

171. See Shapin, "The House of Experiment"; and Shapin, *A Social History of Truth*. Cf. Chartier, "Foucault's Chiasmus," 21–22.

172. On this linguistic exclusion criterion in Gessner's work, see n160 above; on his engagement with pamphlet printing, see Mauelshagen, *Wunderkammer auf Papier*, 186–187 and 213.

173. See Daston and Park, "Hermaphrodites in Renaissance France"; Daston and Park, "The

Hermaphrodite"; Kraemer, "'Under So Viel Wunderbarlichen und Seltsamen Sachen,'"; and Kraemer, "Die Individualisierung des Hermaphroditen."

174. See, for example, Schenck von Grafenberg, *Observationum Medicarum*, 16; Schenck von Grafenberg, *Monstrorum Historia*, 49; Bauhin, *De Hermaphroditorum Monstrosorumque Partuum Natura*, 363–364; Aldrovandi, *Monstrorum Historia*, 41. Noteworthy examples of later publications include Schott, *Physica Curiosa* (1697 [1662]), 396; and Mollerus, *Discursus de Cornutis et Hermaphroditis Eorumque Jure* [. . .]., 150–151 and 154.

175. See Park, "Observation in the Margins," 61 and 77n85. On the frequent omission of sources from medieval *florilegia*, see Wallis, "The Experience of the Book."

176. Park, "Observation in the Margins," 17. On the medieval concept of *experimentum/experientia*, see pp. 16–18.

177. Blair, *The Theater of Nature*, particularly 65–77; and Blair, "Humanist Methods in Natural Philosophy." Cf. also Pomata, "*Praxis Historialis*," 133.

178. See Pomata, "*Praxis Historialis*"; Pomata, "Sharing Cases"; and Stolberg, "Formen und Funktionen Medizinischer Fallberichte."

179. Schenck von Grafenberg, *Observationes Medicæ de Capite*, frontispiece.

180. *Catalogus Authorum*, no pagination. Michael Stolberg also emphasizes that Schenck's *Observationes* are a significant example of case histories gathered from more recent literature. Stolberg, "Formen und Funktionen Medizinischer Fallberichte," 81n3.

181. Schenck von Grafenberg, *Observationes Medicæ de Capite*, 253.

182. *Observatio CCXVII*, 251–253.

183. *Catalogus Authorum*, no pagination.

184. I am referring here to the copy in the Berlin State Library with the catalog number Jc3410.

185. Translated with the masculine personal pronoun because of the use of *qui* in the Latin original.

186. See Bacon, "Historia Ventorum," 73–76 and 92; Bacon, "Historia Vitae et Mortis," 123–124, 131–134, and 142–145; and Bacon, "Parasceve," 423, where he speaks of *observationes generales* or *catholicae*, however. On the *observationes majores* in the work of Francis Bacon, cf. Daston, "The Empire of Observation," 89.

187. Montuus, *De Medica Theoresi Liber Primus*, 35–36.

188. Monteux's dates of birth and death are unknown. However, all the writings published by him date from the middle of the sixteenth century.

189. Along with a further text by Jérôme Monteux, it was also published as Montuus, *Opuscula Juvenilia, D. Hieronymi Montui*.

190. See Pomata, "Observation Rising"; and Pomata, "Sharing Cases."

191. Pomata and Siraisi, *Historia*. Still seminal in this regard: Seifert, *Cognitio Historica*.

192. See Seifert, *Cognitio Historica*; and Pomata and Siraisi, introduction to *Historia*, 17. In medical literature at this point, the terms *historia* and *observatio* were converging, so to speak. In the early seventeenth century, then, they were used for the most part synonymously. Pomata, "*Praxis Historialis*," 134.

193. Even significantly later writers thus learned—via Schenck—of Monteux's firsthand account. See, for example, Mollerus, *Discursus de Cornutis et Hermaphroditis Eorumque Jure* [. . .]. , 150–151 and 154.

194. Blair, *The Theater of Nature*, 4–5.

195. Though printed *loci communes* and *florilegia* met with some criticism, it is clear that they were nevertheless used frequently, since they spared scholars from—or at least supplemented— time-consuming note taking. See Blair, "Reading Strategies," esp. 21–23. On printed common-

place books, their place in the pedagogy of the early modern period, and their influence on scholarly ways of writing, see Moss, *Printed Commonplace-Books*. On the part played by paper technology such as *loci communes* collections in the circulation of monstrous factoids, see chapter 2.

196. As a qualification, it should be said here that etymology was not without an impact on the ontology of the subject. Authors such as Bauhin, Aldrovandi, and Liceti did not view the connection of a specific signifier to a signified as arbitrary, and a terminological discussion could therefore lead to insights into the ontological status of the thing itself. For this reason it was, for example, of some significance whether the term *monstrum* was derived from the Latin verb *monere*—"warn"—or *monstrare*—"show." While the first root might suggest that monsters were divine omens, the second implies they are shown in public because they are of interest as curiosities. Fortunio Liceti, for example, took advantage of this second possibility. On this, and on literature on this topic, see Kraemer, *Ein Zentaur in London*, chapter 3.

197. Park, "The Rediscovery of the Clitoris," 179.

198. I use "autopsy" here in the general sense of seeing for oneself, not in the narrower sense of postmortem examination. The Greek term *autopsia* (αὐτοψία, from αὐτός, "self," and ὄψις, "seeing") is used very little, however, in the sixteenth- and early seventeenth-century texts examined here.

199. Duval, *Traité des hermaphrodits, parties génitales, accouchements des femmes* [. . .].

200. Riolan, *Discours sur les hermaphrodites*; Duval, *Responce au discours fait par le sieur Riolan*.

201. Daston and Park suggest that this agreement is explained by the prestige of post-Vesalian anatomy. Daston and Park, "The Hermaphrodite," 429. On the case of Marie/Marin le Marcis and the Duval-Riolan debate, cf. Daston and Park, "Hermaphrodites in Renaissance France," 1–3; Daston and Park, "The Hermaphrodite," 426–429; and Park, "The Rediscovery of the Clitoris," 179–184.

202. The journal of the Academia Naturae Curiosorum, for example, the *Miscellanea Curiosa*, contains numerous accounts of dissections performed on conjoined twins. On the *Miscellanea Curiosa*, see chapter 3.

203. With this in mind, it is no coincidence that the autopsy, the postmortem examination of the human body, which has occupied so central a place in the self-conception of anatomists since this period, bears a name derived from the Greek *autopsia* in the major European languages (English: "autopsy"; French: *autopsie*; Italian and Spanish: *autopsia*). See, for example, the lemmata "autopsy," "Autopsie," and "autopsie," in Simpson and Weiner, *The Oxford English Dictionary*; Wahrig-Burfeind, *Gerhard Wahrig, Deutsches Wörterbuch*; and Rey, *Le grand Robert*.

204. This biographical sketch draws on Carlino, *Books of the Body*, 60–61.

205. See Carlino, *Books of the Body*, 61–63.

206. On the concept of the "scientific persona," see Daston and Sibum, "Introduction."

207. See Dear, "The Meanings of Experience," 111–112.

208. See Park, "The Criminal and the Saintly Body."

209. Carlino, *Books of the Body*, 202.

210. Heseler, *Andreas Vesalius' First Public Anatomy*. On these conflicts, cf. Carlino, *Books of the Body*, 204–205.

211. Dear, "The Meanings of Experience," 112. Dear draws here on Baroncini, *Forme di esperienza*, chapter 5: "Harvey e l'esperienza autoptica."

212. On this whole section, see Wear, "William Harvey." Even within anatomy, the epistemic value that individual writers ascribed to *autopsia*, seeing for oneself, varied to a certain degree. On the differences among Vesalius, Hieronymus Fabricius (also called Girolamo Fabrici or Fabricius ab Aquapendente, ca. 1533–1619), and Harvey in this regard, see French, *William Har-*

vey's Natural Philosophy, 256. See also Pomata, "*Praxis Historialis*," 117–118; and Wear, "William Harvey," particularly 227–228 and 224n23. In his explicit and outright rejection of the authority of earlier authors, Harvey was more radical than all of his predecessors. The new method he proposed was to be based exclusively on observation. In his actual research practice, he nevertheless often referred back to Aristotle and Galen, and among his contemporaries particularly to Fabricius and Colombo (235).

213. On the concept of *lusus naturae*, which was widespread among Renaissance scholars, see Findlen, "Jokes of Nature."

214. On this whole section, cf. Siraisi, "Vesalius and Human Diversity in *De Humani Corporis Fabrica*." Siraisi stresses that the anatomists of the sixteenth century agreed neither on the prevalence of anatomical "anomalies" nor on how they were to be dealt with intellectually. The extreme caution and the understatement with which Andreas Vesalius included passages on anatomical variations in his famous *De Humani Corporis Fabrica* demonstrate the enormous influence exerted on this discussion by the widely felt urgency to establish the "normal" body as the uniform subject of the field. See particularly pp. 61 and 62.

215. On the whole paragraph, see Siraisi, "Vesalius and Human Diversity in *De Humani Corporis Fabrica*," 61. Andrea Carlino comes to a similar conclusion to Siraisi with regard to Colombo's position on anatomical variations. See Carlino, "L'exception et la règle," particularly 170. In this chapter, Colombo makes particular use of the terms *varietas* and *diversitas* to describe the variety or diversity he observes in the human anatomy. See, for example, Columbo, *De Re Anatomica*, 264.

216. Columbo, *De Re Anatomica*, 268. On Colombo's discussion of hermaphroditism, cf. also Daston and Park, "Hermaphrodites in Renaissance France," 5.

217. Columbo, *De Re Anatomica*, 268.

218. Columbo, *De Re Anatomica*, 268. *Considerare* was used even by twelfth- and thirteenth-century astronomers for "observe." See Park, "Observation in the Margins," 27.

219. Columbo, *De Re Anatomica*, 169 (misprint of 269).

220. The diversity of usage expands even further if additional authors are considered, too. Thus, *observare* could also be used at the time in a less specific sense to describe individual acts of seeing. See, for example, Bauhin, *De Hermaphroditorum Monstrosorumque Partuum Natura*, 346, where Bauhin uses the verb to describe his observations in relation to one specific living hermaphrodite. Like Colombo, however, he too—in the passage that introduces his lengthy quotation from Colombo's examination report—uses the verbs *notare* and *observare* in different ways. See below. It is difficult, then, to arrive at generalizations with regard to the use of the terms *observare* and *observatio* around 1600. As observation during this period was on the point of rising to become a key practice in nature study, it stands to reason that this rise was reflected in a proliferation of different usages, and consequently an engagement with the concept of observation itself, as it were.

221. Park, "Observation in the Margins," 16.

222. Park, "Observation in the Margins," 16–17.

223. As Katharine Park has shown, for much of the European Middle Ages, a prescriptive meaning of *observatio* dominated that was already established in classical Latin and that referred less to observing than to the following of rules derived from observation. It was expressed in the term "observance" in the English and French vernaculars, and *Observanz* in German. By contrast, the meaning of *observatio* discussed here—in the sense of the observation of natural phenomena—was, by comparison, marginal in the Middle Ages and is ultimately only demonstrable—and even here only to a limited extent—in relation to the occupation of naturalists with the celestial bodies. See Park, "Observation in the Margins." On Augustine's part in the redefinition of the *observatio* in the Middle Ages, see p. 20.

224. Park, "Observation in the Margins," 19. On this whole line of thought, cf. p. 19.

225. See Pomata, "Observation Rising," particularly 67; and, on the first appearance of the authorized observation in late fifteenth-century astronomy, Park, "Observation in the Margins," 32–35.

226. See Daston, "The Empire of Observation," 87.

227. Daston, "The Empire of Observation," 91. On this whole section, see pp. 87–91.

228. Cf. Céard, *La nature et les prodiges*, 440.

229. The first edition contains 572 paginated pages, along with plates and unpaginated paratexts.

230. Kathleen P. Long mistakenly reads the passage as if Bauhin himself were speaking here. Cf. Long, "The Cultural and Medical Construction," 65–66.

231. Bauhin, *De Hermaphroditorum Monstrosorumque Partuum Natura*, 338.

232. Bauhin, *Institutiones Anatomicæ, ad Lectorem*, fol. a3v, quoted in Wear, "William Harvey," 247n55.

233. Bauhin, *Institutiones Anatomicæ, ad Lectorem*, fol. a4v; see Wear, "William Harvey," 234 and 235.

234. Schenck von Grafenberg, *Observationum Medicarum*, 5.

235. "Gignuntur & utriusque sexus quos, Hermaphroditos vocamus, olim Androgynos vocatos, & in prodigijs habitos, nunc vero in delicijs. Plinius lib. 7. cap. 3. Nat. Hist." Schenck von Grafenberg, *Observationum Medicarum*, 5.

236. "Supra Nasomonas confinesque illis Machlyas, Androgynos esse utriusque naturæ, inter se vicibus coeuntes, Calliphanes tradit. Aristoteles addijcit, dextram mammam ijs virilé, lævam muliebrem esse. Idem lib. 7. cap. 2." Schenck von Grafenberg, *Observationum Medicarum*, 5.

237. *Observatio* III covers pp. 5–15. *Observatio* IV on the perfect hermaphrodite follows immediately afterward on p. 16; see above.

238. Pomata and Siraisi, "Introduction," 17. It is striking that Pomata and Siraisi refer precisely to the empiricism of anatomy as "learned" here, since post-Vesalian anatomy is, after all, considered to be the epitome of experience-based medical research.

239. Schenck von Grafenberg, *Monstrorum Historia*, 44–48.

240. Lorraine Daston has rightly emphasized that it was inconceivable for Gessner, despite his empiricism, to give space to individual observations that could not be categorized within a system. Daston, "Warum sind Tatsachen kurz?," 138.

241. Aldrovandi, *Monstrorum Historia*, 39–43; Monteux and Colombo are mentioned on p. 41.

242. Overall, the section covers pp. 513–519; Colombo's account is paraphrased on pp. 517–518.

243. See Liceti, *De Monstrorum Natura*, 48.

244. Liceti, *De Monstrorum Natura*, 169.

245. Liceti, *De Monstrorum Natura*, 168–169.

246. Liceti, *De Monstrorum Natura*, 169–170.

247. In Liceti's view, hermaphrodites are only monstrous if they can be classed as sitting "perfectly" between the sexes because of the minimal expression of each, or if one of the two external sex organs is far removed from its natural position. One of Colombo's two living hermaphrodites is placed by Liceti in the first of these two categories. Liceti, *De Monstrorum Natura*, 170.

CHAPTER 2: Ulisse Aldrovandi's Twofold "Pandechion"

1. In the history of science, the "practical turn," suggesting that knowledge production be examined in the practices of the actors themselves, took place some years ago already. In applying this perspective to the early modern period, we must necessarily pay particular attention

to "scholarly practices." Helmut Zedelmaier and Martin Mulsow first prompted this approach. Zedelmaier and Mulsow, *Die Praktiken der Gelehrsamkeit*. A contribution to the debate that takes a similar position, though in relation to the history of "science" in the narrower, English sense, is offered by Blair, "Scientific Readers."

2. Blair, "Note Taking," 85.

3. Alessandro Tosi has examined the picture's elements with a view to their function in the way scholars presented themselves. See Tosi, *Portraits of Men and Ideas*, 73.

4. The core meaning of this term (Latin: *simplicia*) referred to medicine ingredients, particularly herbs. Sixteenth-century naturalists also used it more broadly for the natural history specimens they collected. See Findlen, *Possessing Nature*, 7n13.

5. On Aldrovandi's *museum*, see Findlen, *Possessing Nature*, 7n13.

6. Still seminal in this regard: Olmi, *L'inventario del mondo*. Cf. also Céard, "La *Monstrorum Historia* d'Ulisse Aldrovandi ou La nature, l'art et le monstre," vii.

7. See particularly Fischel, *Natur im Bild*; and Fischel, "Zeichnung und Naturbeobachtung."

8. Olmi, "Ulisse Aldrovandi," 60; there are details here, too, of the size of the collection. An initial impression of the collection of drawings—also in terms of the major role that monstrous phenomena play in it—is conveyed by Antonino, *Animali e creature monstruose*. On Giuseppe Olmi's work on Aldrovandi's collection of drawings, see below.

9. See Fischel, *Natur im Bild*, 81.

10. See particularly Bacchi, "Ulisse Aldrovandi"; and Duroselle-Melish and Lines, "The Library of Ulisse Aldrovandi."

11. See Fischel, *Natur im Bild*, 118. For a fuller discussion of Aldrovandi's woodcuts and their relation to the drawings, see the recent publication Olmi and Somoni, *Ulisse Aldrovandi*.

12. Findlen, *Possessing Nature*, 17.

13. Findlen, *Possessing Nature*, 17.

14. See Findlen, *Possessing Nature*, 17.

15. Findlen, *Possessing Nature*, 21. An association of the appearance of the dragon with the papacy of Gregory XIII also suggested itself because the coat of arms of his family, the Boncampagni, depicted a winged dragon without a tail. On this connection, see p. 21. Even aside from the sighting of the dragon near Bologna, the dragon on the family crest presented a problem for Gregory in the light of the widespread association of this creature with the devil. Shortly after Gregory's enthronement, Giacomo Boncompagni therefore asked the scholars of the papal court to design an emblem that would invest the dragon on the family crest with a positive meaning. See Ruffini, "A Dragon for the Pope."

16. Aldrovandi, *Discorso naturale di Ulisse Aldrovandi*, BUB, fol. 516r. The *Discorso naturale* was first published in Tugnoli Pàttaro, *Metodo e sistema*, 173–232. On the entire passage, cf. also Findlen, *Possessing Nature*, 21.

17. See Findlen, *Possessing Nature*, 18. On the entire episode, cf. pp. 17–23. Unlike Findlen, Marco Ruffini considers the dragon of Bologna to be a "hoax"—though he does not supply any direct evidence of this. Aldrovandi, Ruffini maintains, created it himself in his *museum* from the air-filled abdomen of a large snake, the feet of a bird, and the head of a fish. See Ruffini, "A Dragon for the Pope," 86. And indeed, the practice of fabricating dragon or basilisk specimens was not uncommon in the sixteenth century, nor was it unknown to Aldrovandi. See Findlen, "Inventing Nature"; and Ruffini, "A Dragon for the Pope," 94–95n40.

18. Aldrovandi, *Serpentum, et Draconum Historiae Libri Duo*. See Ruffini, "A Dragon for the Pope," 94.

19. Cf. Findlen, *Possessing Nature*, 20.

20. Aldrovandi, *De Quadrupedibus Solidipedibus*; Aldrovandi, *Quadrupedum Omnium Bisulcorum Historia*.

21. Aldrovandi, *Quadrupedum Omnium Bisulcorum Historia*, 123–138.

22. Ashworth, "Emblematic Natural History."

23. Harms, "Bedeutung als Teil der Sache," 364.

24. Harms, "Bedeutung als Teil der Sache," 364; on the entire passage, see p. 364.

25. A comparison of Gessner's eight-part grid with that of Aldrovandi, along with references to the current literature, can be found in Fischel, *Natur im Bild*, 79–80.

26. Cf. Harms, "Bedeutung als Teil der Sache," 364–365. On Gessner's *philologica* category, see p. 353 and passim.

27. Harms, "Bedeutung als Teil der Sache," 365.

28. See Fischel, *Natur im Bild*, 80.

29. Foucault, *Order of Things*, 44.

30. The title of the corresponding section varies depending on the animal in question. In the chapter on horses in *De Quadrupedibus Solidipedibus Volumen Integrum*, it is headed "Omina. Somnia. Ostenta. Prodigia," whereas in the chapter on cattle in *Quadrupedum Omnium Bisulcorum Historia*, it bears the heading "Auspicia. Prodigia."

31. The chapter on horses covers pages 1–294.

32. See Aldrovandi, *Quadrupedum Omnium Bisulcorum Historia*, 130–138. In all of the previous sections in the chapter on cattle, which begins on p. 13, there are just eight, and in the following sections a total of two woodcuts, before the chapter ends on p. 346.

33. Cf. below.

34. On Hartmann Schedel's use of the same woodcuts for multiple towns, see Lefèvre, *Picturing the World of Mining*, 24–25.

35. We can retrace this double appearance most clearly by looking at the use of woodcuts of monstrous animals. Many of them were used twice, for example, two woodcuts of bovine animals with five and six hooves, respectively, on pages 540 and 541 of the *Monstrorum Historia*. They had already been used in the *Quadrupedum Omnium Bisulcorum Historia*. The first is found on p. 133 and the second on p. 138. But the context in which the woodcuts are placed varies from one text to the next. Unlike the other volumes of Ulisse Aldrovandi's zoological encyclopedia, the *Monstrorum Historia* is not organized by species. The two woodcuts appear in the second half of the volume, which is structured in morphological categories and where they were placed in the chapter on quadrupeds with more than four feet ("Multiplicatio Pedum in Bellvis Quadrupedibus").

36. On Ambrosini, see Antonelli, "Ulisse Aldrovandi e la metamorfosi," esp. 196–197n2 and 198. On the Studio Aldrovandi, see, for example, Tugnoli Pàttaro, *Lo studio Aldrovandi*; and Findlen, *Possessing Nature*, 24–31, particularly 25n25.

37. On the question of Ambrosini's faithfulness to Aldrovandi's plans and of which parts of the work reveal Ambrosini's hand, cf. Céard, "La *Monstrorum Historia* d'Ulisse Aldrovandi ou La nature, l'art et le monstre"; and Olmi, *L'inventario del mondo*, 45–46n81. Ezio Antonelli has aptly characterized the book itself as "a monster, a deformed hybrid," because two voices are perceptible in it. Antonelli, "Ulisse Aldrovandi e la metamorfosi," 196. Paula Findlen has pointed out that many of the administrators of the *studio Aldrovandi* "enhanced" the texts they published posthumously with their own scholarship. See Findlen, *Possessing Nature*, 26.

38. Aldrovandi, *Monstrorum Historia* (1658). On Ashworth's theory of the collapse of "emblematic natural history" around 1650, see Ashworth, "Emblematic Natural History," 35–36. On the fact that the works of naturalists such as Aldrovandi and Della Porta sometimes became outdated even during their lifetimes because of the "changing tenor of the scientific community," see Findlen, *Possessing Nature*, 7.

39. See chapter 3.

40. Aldrovandi, *Monstrorum Historia* (2002 [1642]), fol. 2v. Unless otherwise specified, citations to this text refer to this edition.

41. Olmi, *L'inventario del mondo*, 45 and 45–46n81, argues that all of them were produced during Aldrovandi's lifetime. Rebecca Carnevali shows in an unpublished master's thesis, however, that the two woodcuts on pp. 603 and 604 of the *Monstrorum Historia* were made after Aldrovandi's death. See Carnevali, "The Publication of the Monstrorum Historia," 12n45.

42. Aldrovandi, *Monstrorum Historia*, "Ordninis Ratio," 1.

43. For ease of reading, the author of this work is referred to in the following as Aldrovandi. On the complex matter of the authorship of the work, see above.

44. Aldrovandi, *Monstrorum Historia*, 1.

45. Aldrovandi, *Monstrorum Historia*, 516.

46. Aldrovandi, *Monstrorum Historia*, 513.

47. Aldrovandi, *Monstrorum Historia*, 518–519. The whole section covers pp. 513–519.

48. Bacon, *Novum Organum* II, X; original italics. On the making of lists of rare observations in the study of nature in the early modern period, cf. also chapter 3.

49. Blair, "Humanist Methods in Natural Philosophy," 542; italics added.

50. Ogilvie, *The Science of Describing*.

51. Giuseppe Olmi has thus described Aldrovandi's research activities as follows: "Aldrovandi's aim in his long and tireless activity as a scholar was accurately to identify and describe the largest possible number of animals, plants and minerals." Olmi, "Ulisse Aldrovandi," 59.

52. Aldrovandi, *Monstrorum Historia*, 1.

53. Aldrovandi, *Monstrorum Historia*, 721.

54. Aldrovandi, *Monstrorum Historia*, 733.

55. See Cave, *The Cornucopian Text*; and Cave, "Copia and Cornucopia." Erasmus's textbook was particularly popular and influential within Reformed Protestantism. See Leu, "Die Loci-Methode."

56. Fekadu, "Variation," col. 1009.

57. See Fekadu, "Variation," col. 1010.

58. See de Courcelles, *La varietas à la Renaissance*. On the dialectic between *varietas* and *ordo* in the Renaissance and baroque periods from a literary perspective, see Föcking and Huss, *Varietas und Ordo*.

59. See below.

60. Couzinet, "La variété," 105.

61. Pomata, "Observation Rising," 58.

62. With regard to the example of hermaphrodites, see Kraemer, "Die Individualisierung des Hermaphroditen," 52; and Kraemer, "Hermaphrodites Closely Observed."

63. On the close connection between historiography and medicine/nature study in the Renaissance, cf. chapter 1.

64. See Blair, *Too Much to Know*, 198. Cf. also p. 202.

65. Cf. Blair, *Too Much to Know*, 249.

66. It is thanks to the censorship efforts of the church that we know Aldrovandi had a copy of the *Chronicle* (which does not appear in his library catalogs). It is included in a list of books that were removed from his library after his death because they had been placed on the index of prohibited books. See ASB, Assunteria di Studio, 100. I am grateful to Caroline Duroselle-Melish, who gave me access to a reproduction of the source and checked whether the *Chronicon* was entered in the library catalogs.

67. This index takes the form of a two-volume book whose pages are filled with handwritten notes pasted into them in alphabetical order. Each slip of paper contains one subject and

the reference to a particular page. See BUB MS Aldrovandi 33. On this index and Aldrovandi's reading of the *Theatrum* in general, see Blair, *Too Much to Know*, 202 and 232, where she also discusses the fact that Aldrovandi specifically obtained permission to read this reference work, whose numerous positive references to Protestants had seen it placed on the index.

68. On Spross, see Harms and Schilling, *Die Sammlung der Zentralbibliothek Zürich*, part 1, 12.

69. Harms and Schilling, *Die Sammlung der Zentralbibliothek Zürich*, part 1, 12. On the history of the reception of this case, see Kraemer, "Die Individualisierung des Hermaphroditen"; and Kraemer, "Hermaphrodites Closely Observed."

70. Lycosthenes, *Prodigiorum ac Ostentorum Chronicon*, 524. The fact that Lycosthenes did not adopt the interpretation of the hermaphrodite as a prodigy should not be misinterpreted to mean that he was not similarly concerned with the theological significance of the case. Indeed, the case is unambiguously vested with a theological significance in his work, too, by means of paratextual framing and its embedding in the chronology of prodigies and wonders.

71. Cf. also Harms and Schilling, *Die Sammlung der Zentralbibliothek Zürich*, part 1, 12, where it is postulated that Lycosthenes had a reproduction made of the woodcut used in the broadsheet.

72. Quoted from Harms and Schilling, *Die Sammlung der Zentralbibliothek Zürich*, part 1, 12; on the whole section, cf. p. 12.

73. See Harms and Schilling, *Die Sammlung der Zentralbibliothek Zürich*, part 1, 12.

74. Johann Laurentius Bausch, *Collectaneorum Chronologicorum Suinfurtensium*, Stadtarchiv Schweinfurt, Handschriften, 102–104. See, for example, the broadsides that Bausch appended to the text *Cometa caudatus vel barbatus* about a comet sighted in various European towns in November and December 1664. See Müller, " 'Cometa caudatus vel barbatus.' " In addition to his own direct observations of the comet, Bausch also refers in the text to material from broadsides and pamphlets as well as other sources (64).

75. Wick left the university without graduating. On Wick's career, see Harms and Schilling, "Einleitung," x–xi.

76. It was not a given that less distinguished or canonical texts and genres were not indicated as sources in the late sixteenth and early seventeenth centuries. On the contrary, scholars often expected readers to be familiar with the canonical texts and thus did not mention them specifically, whereas they named their less canonical reading matter. On Jean Bodin in this context, see Blair, "Humanist Methods in Natural Philosophy," 545. On the history of the critical apparatus and particularly of footnotes in historiographical discourse from the Renaissance to the nineteenth century, see Grafton, *The Footnote*.

77. See Harms and Schilling, "Einleitung," ix.

78. Aldrovandi names these "three intelligent scribes" in his will. Aldrovandi, "Testamento di Ulisse Aldrovandi," 76. Cf. also Fischel, *Natur im Bild*, 76–77.

79. The note is, evidently with the exception of the correction of the term *muliebre*, not written in Aldrovandi's own hand. We regularly encounter the same handwriting in his commonplaces, so it is very likely to be that of an amanuensis, of his second wife, or of a regular correspondent. Alongside Aldrovandi's amanuenses, his second wife, Francesca Fontana, wrote many of the notes in the "Pandechion." She helped her husband with the compilation of the "Pandechion," as Paula Findlen has demonstrated. See Findlen, *Possessing Nature*, 65. On Francesca Fontana's part in her husband's work, see also Findlen, "Masculine Prerogatives."

80. On the "Pandechion," see Kraemer, "Ein papiernes Archiv"; Kraemer and Zedelmaier, "Instruments of Invention"; and Kraemer, "Ulisse Aldrovandi's *Pandechion Epistemonicon*." See also the mostly brief sections on the "Pandechion" in Tugnoli Pàttaro, *Metodo e sistema*, 19; Findlen, *Possessing Nature*, 65–66 and passim; Findlen, "Masculine Prerogatives," 44; Blair,

"Reading Strategies," 26–27; Bacchi, "Ulisse Aldrovandi," 305–307, and Blair, *Too Much to Know*, 105.

81. An overview of reading strategies in the early modern period, and of the techniques and media involved, is provided in Blair, "Reading Strategies"; and Blair, *Too Much to Know*.

82. Te Heesen, "The Notebook." On the concept of paper technology, see also chapter 1.

83. Michel Foucault thus read notes one-sidedly as egodocuments, and for a long time reading notes were analyzed primarily from this perspective. Cf. Blair, "Note Taking," 88–89. But studies on early modern reading guides by Helmut Zedelmaier in particular have demonstrated the supraindividual nature of at least the prescriptive discourse of reading in the early modern period. See particularly Zedelmaier, "Wissen erwerben"; Zedelmaier, "Wissensordnungen der Frühen Neuzeit," 840–841; Zedelmaier, "Johann Jakob Moser"; Zedelmaier, "Lesetechniken"; Zedelmaier, "De Ratione Excerpendi"; and Zedelmaier, *Bibliotheca Universalis*, particularly 75–99.

84. *Breve nota delle opere fatte dà Ulisse Aldrovandi*, appendix to Ulisse Aldrovandi to Ferdinando I dei Medici, April 1588 [day unknown]; reproduced in Mattirolo, "Le lettere di Ulisse Aldrovandi," 380–384, 381. Cf. Findlen, *Possessing Nature*, 64.

85. See Kraemer and Zedelmaier, "Instruments of Invention."

86. Ulisse Aldrovandi, "Pandechion Epistemonicon," BUB, vol. COE–COMP, fol. 265v; original italics.

87. Tugnoli Pàttaro, *Metodo e sistema*, 19.

88. See the entry on *silva* in Georges, *Ausführliches lateinisch-deutsches und deutsch-lateinisches Handwörterbuch*, vol. 2, cols. 2669–2670. On the significance of the term *selva universale* used by Aldrovandi, cf. also Tugnoli Pàttaro, *Metodo e sistema*, 19.

89. Adam, *Poetische und kritische Wälder*, 58; cf. Findlen, *Possessing Nature*, 64n55, where she adds that Aldrovandi's contemporary Tommaso Garzoni often employed the metaphor *selva universale*.

90. On the sources of knowledge on which Bacon's *Sylva Sylvarum* drew, see Colclough, "'The Materials for the Building,'" 192–193.

91. Bacon, *Sylva Sylvarum*, unpaginated preface. On the place of the *Sylva Sylvarum* within Bacon's oeuvre, see Colclough, "'The Materials for the Building.'" On Bacon's notion of *sylva*, see also Giglioni, "From the Woods of Experience."

92. Aldrovandi, "Pandechion Epistemonicon," BUB, vol. MIN–MU, fol. 234r.–297r.

93. Findlen, *Possessing Nature*, 64.

94. Gessner explains in the preface that his choice of the term is due to the veritably breathtaking sweep of his collection, stating that he called it *Pandectae* because it "ex omne genus authoribus confectae sunt, et quod studiorum omnium classes complectantur." Gessner, *Pandectarum sive Partitionum Universalium*, fol. *4r; quoted in Zedelmaier, *Bibliotheca Universalis*, 52n142. On Gessner's reasons for choosing this term, see p. 52n142.

95. Zedelmaier, *Bibliotheca Universalis*, 52.

96. Foxe, *Pandectae Locorum Communium*; a third edition with the same title was published in 1585. On this publication, see, for example, Vine, *Miscellaneous Order*, 44–46; and Yeo, *Notebooks*, 47. On printed commonplace books, see Moss, *Printed Commonplace-Books*.

97. Findlen, *Possessing Nature*, 64. He also used the term *pandechion* to describe the Grand Duke of Tuscany's natural history collection. See Findlen, "The Museum," 32.

98. On extremely rare occasions, a note is written directly on the page between the slips of paper.

99. Tugnoli Pàttaro, *Metodo e sistema*, 19.

100. See Tugnoli Pàttaro, *Metodo e sistema*, 20. Many thanks to Ann Blair for drawing my attention to this passage in Tugnoli Pàttaro's study.

101. See Tugnoli Pàttaro, *Metodo e sistema*, 20.

102. On the fact that notes were usually organized in bound form in the early modern period and into the eighteenth century, see Zedelmaier, "Buch, Exzerpt, Zettelschrank, Zettelkasten."

103. On this technique, though focusing on a later period, see te Heesen, "Cut and Paste um 1900."

104. See, for example, Aldrovandi, "Pandechion Epistemonicon," BUB, vol. A–AER, fol. 459r, where all the notes are missing, despite the fact that the left side of the page had clearly once been completely covered, or fol. 459v, where individual notes were removed and, in some instances, new ones pasted in their place.

105. Waller, *The Posthumous Works of Robert Hooke*, 64; quoted from te Heesen, "The Notebook," 587.

106. On Gessner's use of this and another method for temporarily ordering handwritten notes, see Wellisch, "How to Make an Index." On Gessner's habit of cutting up letters addressed to him and arranging the fragments thematically in his notes, see Ogilvie, "Observation and Experience in Early Modern Natural History," 276–277. On Cardano's use of the cut-and-paste method, see Blair, "Reading Strategies," 26. For a more general examination of the use of this technique in the ordering of information in the early modern period, see Blair, *Too Much to Know*, 213–229.

107. On Aldrovandi's relationship to Cardano, see Folli, "La natura scritta," 499; on his relationship to Gessner, which was also marked by envy of him as a rival, see, for example, Harms, "Bedeutung als Teil der Sache," 366–367.

108. Blair, "Reading Strategies," 26–27.

109. Ulisse Aldrovandi, guestbook, BUB; illustrations of the visitors' book are found in Findlen, *Possessing Nature*, 139; and in Vai, "Aldrovandi's Will," 79.

110. Zedelmaier, "Buch, Exzerpt, Zettelschrank, Zettelkasten," 39.

111. See Zedelmaier, "Wissensordnungen der Frühen Neuzeit," 840.

112. Zedelmaier, "Wissensordnungen der Frühen Neuzeit," 840–841.

113. Zedelmaier, "Buch, Exzerpt, Zettelschrank, Zettelkasten," 43. As Zedelmaier rightly notes, the repeated criticism of "secondary" memory systems can, however, be interpreted as an indication that the use of non-bound forms of organizing material was on the increase. Zedelmaier, "Wissensordnungen der Frühen Neuzeit," 841. On this point, see also Kraemer and Zedelmaier, "Instruments of Invention." On a set of building instructions for a filing cabinet that was published in 1689 in Vincent Placcius's *De Arte Excerpendi*, and that is considered the first record of a flexible filing system for excerpted material, see Cevolini, *Thomas Harrison*; Zedelmaier, "Buch, Exzerpt, Zettelschrank, Zettelkasten," 45–53; Zedelmaier, "Wissensordnungen der Frühen Neuzeit," 841; Zedelmaier, "Johann Jakob Moser"; Meinel, "Enzyklopädie der Welt"; and Malcolm, "Thomas Harrison and His 'Ark of Studies.'" There has been little research thus far on the use by naturalists of flexible filing systems for notes. On Carl von Linné's index cards, see Müller-Wille and Scharf, "Indexing Nature"; Müller-Wille, "Vom Sexualsystem zur Karteikarte"; and Müller-Wille and Charmantier, "Natural History and Information Overload."

114. See Zedelmaier, "Wissensordnungen der Frühen Neuzeit," 839–840; cf. also Zedelmaier, "Buch, Exzerpt, Zettelschrank, Zettelkasten."

115. Foxe, *Pandectae Locorum Communium*. On the research literature on this work, see above.

116. See Zedelmaier, "Wissensordnungen der Frühen Neuzeit," 839–840.

117. Yates, *The Art of Memory*. Cf. Blair, *Too Much to Know*, 75–76.

118. "Maior Est Apparatus Quam Emolumentum," BUB MS Aldrovandi 21, II, 166; quoted in Blair, *Too Much to Know*, 76 and 284n66. Blair, in turn, is quoting from the unpublished doctoral thesis Giudicelli-Falguières, "Invention et mémoire," 236.

119. See BUB MS Aldrovandi 21, II, 168–169; quoted from Blair, *Too Much to Know*, 76 and 282n30. Blair, in turn, is quoting from Giudicelli-Falguières, "Invention et mémoire," 247–248. On the use made by early modern scholars of the merchant practice of double-entry bookkeeping, see also Kraemer, "Ulisse Aldrovandi's *Pandechion Epistemonicon*"; and Blair, *Too Much to Know*, 69.

120. On the first two terms, see Brendecke, "Papierfluten," 21. On the concept of "information overload," see Rosenberg, "Early Modern Information Overload"; Werle, "Die Bücherflut"; with regard to the scholarly practice of reading, Blair, *Too Much to Know*; Blair, "Reading Strategies"; and regarding the study of nature, Ogilvie, "The Many Books of Nature."

121. See Neuber, "Topik und Intertextualität," 274n9.

122. Brendecke, "Papierfluten," 21.

123. This is not to dispute the virulence of the early modern discourse concerning the overabundance of printed knowledge. But this complaint was an old topos firmly rooted in scholarly tradition even at the time. And, furthermore, the rhetorical ideal of *copia* seems to have been partly responsible for the abundance of knowledge in circulation. On the topical character of the complaint that the times were afflicted by an inundation of books, cf. Brendecke, "Papierfluten." Ann Blair has also recently elaborated her thesis regarding the "information overload" of the early modern period, arguing that the "info-lust" of the scholars themselves played a key role in the information flood. See Blair, *Too Much to Know*, 11–12. Similarly, Blair elsewhere cites two reasons why scholars seem, from the fifteenth century on, increasingly to have taken notes: first, the availability of paper and, second, "habits of taking and saving notes among humanist scholars as well as scientific observers and travellers." See Blair, "The Rise of Note-Taking," 303.

124. Cf. Blair, "Restaging Jean Bodin," 4.

125. Cf. Findlen, *Possessing Nature*, 65. Findlen puts it as follows: "The real and the imaginary, the ordinary and the extraordinary, all found their place in his encyclopedic paradigm, undifferentiated by any criteria of truth." On Aldrovandi's relationship to Pliny, see Bäumer-Schleinkofer, "Ulisse Aldrovandi."

126. See Daston, "Warum sind Tatsachen kurz?," esp. 141–142; Daston's examples are Konrad Gessner and John Locke.

127. See Swan, "*Ad Vivum*," 358–360; Olmi, *L'inventario del mondo*, 21–117; and Fischel, "Zeichnung und Naturbeobachtung."

128. On the sources of the illustrations in Gessner's history of animals, which were often copies of copies, and the similarities in approach to text and images that are also evident in his work, see particularly Egmond and Kusukawa, "Circulation of Images"; and Kusukawa, "The Sources of Gessner's Pictures."

129. For a fuller discussion of the biography of this woodcut, see Kraemer, "The Persistent Image."

130. See Olmi, *L'inventario del mondo*, esp. 75. On the artist's workshop attached to Aldrovandi's *museum*, see chapters 1 and 2. Cf. also Fischel, "Zeichnung und Naturbeobachtung."

131. The woodcut based on Albrecht Dürer's famous rhinoceros is one of the better-known examples. In this case, the printing block was not wide enough, meaning that the woodcut version of the rhinoceros was squatter than the original. Both rhinoceros depictions were produced in 1515. See Cole, "The History of Albrecht Dürer's Rhinoceros," 340. No original drawing of the two-legged centaur is to be found among the 2,900 *tavole acquarellate* that were once in Aldrovandi's possession and are now preserved in the Biblioteca Universitaria di Bologna. They can be viewed via the database Il teatro della natura di Ulisse Aldrovandi (n.d.), accessed April 4, 2019, aldrovandi.dfc.unibo.it.

132. See Aldrovandi, *Monstrorum Historia*, esp. 4 and 5. On the function of the recording of

differentiae between different animals for comparative anatomy and physiology in Aristotle's work, see Jahn, *Geschichte der Biologie*, 66.

133. The fact that Aldrovandi names the advocates first and then the critics may reflect a preference for the critical point of view. Many thanks to Claus Spenninger for this observation.

134. Céard, "La *Monstrorum Historia* d'Ulisse Aldrovandi ou La nature, l'art et le monstre," xxiv. On the use of the term *historia* for early modern natural history encyclopedias, see Schneider, *Seine Welt Wissen*, 9. On the ways in which the term was used in the early modern period, cf. also chapter 1. Both the lengthy alphabetical index at the back of the volume and the extensive use of printed marginalia that briefly introduce the adjacent content—such as "Centauri qui." or "Centauri origo." (p. 30)—contribute to the *Monstrorum Historia*'s usability as a reference work.

135. Francis Bacon, *Novum Organum*, vol. 1, cxviii.

136. Aldrovandi, *Monstrorum Historia*, 30–34.

137. See Olmi, *L'inventario del mondo*, 46; see also Olmi, "'Figurare e descrivere,'" 106–109.

138. Parshall, "Imago Contrafacta," 563–564.

139. On the ways in which *contrafactum* and its vernacular equivalents were used, particularly in the art of the Northern European Renaissance, see Parshall, "Imago Contrafacta," 563–564.

140. See Swan, "*Ad Vivum*," 354. One of the many examples of the use of this phrase in natural history is found in the salamander chapter in the second volume of Konrad Gessner's *Historia Animalium*. Here, the caption for a depiction of the fire salamander declares that it was "figura ad vivum expressa" (an image made from the life). Gessner, *De Quadrupedibus Oviparis*, 74.

141. See Egmond, "The *ad Vivum* Conundrum."

142. Aldrovandi uses the term *icon* as was customary at the time. In Johannes Micraelius's 1662 edition of the *Lexicon Philosophicum*, for example, *icon* is defined simply as "rerum imago." Micraelius, *Lexicon Philosophicum Terminorum Philosophis Usitatorum*, col. 423.

143. The chapter on horses, the first in the volume, is also by far the longest. It begins on p. 1 and ends on p. 294. The section on centaurs covers pp. 195–198; the woodcut is on p. 198.

144. Aldrovandi, *De Quadrupedibus Solidipedibus*, 197.

145. See Aldrovandi, *De Quadrupedibus Solidipedibus*, 197–198.

146. See Folli, "La natura scritta," 495.

147. See Lefèvre, *Picturing the World of Mining*, 8–10. José Ramón Marcaida comes to a similar conclusion in his examination of early seventeenth-century gallery paintings: technical devices like the *perpetuum mobile* continued to be depicted in the same way regardless of the further developments they had undergone, and the *naturalia* in these paintings were also reproduced over and over again. See Marcaida, "Portraying Technology in Gallery Paintings."

148. See Wittkower, "Marco Polo and the Pictorial Tradition." On the persistence of elements of this medieval iconography in the depiction of monsters in the early modern period, cf. Bates, *Emblematic Monsters*, 46–47.

149. Schedel, *Register des buchs der Croniken*, fol. XIIv.

150. Note in particular the similarities in the depiction of the ground on which the centaur is standing.

151. See Wittkower, "Marvels of the East," 183. Elmar Locher has observed that the woodcuts illustrating the monstrous human races in the *Nuremberg Chronicle* are rooted in an older tradition of depiction and in turn inspired later representations, for instance, in the work of Sebastian Münster. See Locher, "Topos und Argument," 200–201.

152. Mandeville, *Das buch des ritters*, no pagination. The centaur is holding a human in his left hand, in reference to the species' unpleasant habit of catching people in order to eat them, as mentioned in the text.

153. Dante, "Divine Comedy," LLL, MS Holkham 514, fol. 18v–19r [*Inferno* XIII, 10–15] and BCC, MS 597, fol. 95r [*Inferno* XII, 11–12]. Though meant to depict Nessus, the centaurs in these two manuscripts are depicted as two legged. On these manuscripts, and on the early modern iconography of Nessus and how it relates to the two-legged centaur, see Kraemer, "The Persistent Image."

154. The ceramic tiles were made by Diego and Juan Pulido between 1543 and 1546 for the construction of the pavilion on the basis of the previous Moorish building. Many thanks to Lorraine Daston for drawing my attention to the tiles and to the information on the pavilion's history from an information board at the site.

155. Aldrovandi's use of the term "species" has little in common with a modern understanding, as Angela Fischel has rightly observed. Aldrovandi often also uses it to describe "individual creatures." Fischel, *Natur im Bild*, 83n340. In the case under discussion here, however, it does seem to mean species in the sense of a subdivision of a genus.

156. Lycosthenes uses the image of the *hippopos* twice. The multiple use of the same woodcut is typical of the illustrations in his *Chronicon* as well as of the *Nuremberg Chronicle*. The image is first used by Lycosthenes to illustrate a male human-horse hybrid that is said to have first lived for 120 years and then was brought back to life three times. Its second use matches that in Aldrovandi's work: A race by the name of Ipopodes is mentioned and illustrated as part of a list of monster races following Lycosthenes's report on the Tower of Babel. This race has a human body with horses' legs and is said to be native to Asia. Lycosthenes, *Prodigiorum ac Ostentorum Chronicon*, 3 and 8. The template for the image of Aldrovandi's "another species of centaur" is found in the appendix to Lycosthenes's work, where it illustrates a passage about a race of island people found by the Portuguese (664).

157. Only one further, identical edition of the *Monstrorum Historia* was published in the early modern period (1658). *De Quadrupedibus Solidipedibus Volumen Integrum* was published in three editions, the third also including the woodcut of the two-legged centaur: Aldrovandi, *De Quadrupedibus Solidipedibus* (1649), 198. Not only does the edition published in Frankfurt in 1623 lack said woodcut; the entire text was reset, deviating significantly from the Bolognese editions. See Aldrovandi, *De Quadrupedibus Solidipedibus* (1623).

158. Schott, *Physica Curiosa* (1662), no pagination. For a detailed discussion of Schott's etching, see Kraemer, "The Persistent Image," 334–337.

159. Schott, *Physica Curiosa* (1662); Schott, *Physica Curiosa* (1667), no pagination; Schott, *Physica Curiosa* (1697), no pagination.

160. Aldrovandi, *Discorso naturale di Ulisse Aldrovandi*, BUB, fol. 508r.

161. Bäumer-Schleinkofer, "Ulisse Aldrovandi," 184.

162. Illuminating surveys of the disparate earlier and more recent appraisals of Aldrovandi by historians—he is viewed by some as a mere compiler and by others as very independent in his judgment—can be found in Bäumer-Schleinkofer, "Ulisse Aldrovandi," 199n43; and in Fischel, *Natur im Bild*, 74–75.

163. Olmi, "Ulisse Aldrovandi," 59; original italics.

164. Olmi, "Ulisse Aldrovandi," 62.

165. Fischel has shown this in relation to Aldrovandi's drawings; see Fischel, "Zeichnung und Naturbeobachtung," 223.

166. Duroselle-Melish and Lines, "The Library of Ulisse Aldrovandi," 135–137; cf. Findlen, *Possessing Nature*, 122; and Bacchi, "Ulisse Aldrovandi," 262. On the design of the *studio*, see Tagliaferri and Tommasini, "Microcosmos Naturae."

167. Bacchi, "Ulisse Aldrovandi," 263.

168. See also Zedelmaier, "Wissen repräsentieren," particularly 95–98; and Clark, *Academic Charisma*, 297–335.

169. Aldrovandi to Cardinal Barberini, no date; quoted in Ogilvie, *The Science of Describing*, 174–175, who in turn found the quotation in Ley, *Dawn of Zoology*, 158. Ley does not provide any further information on his source.

170. The storing of art objects and *naturalia* together conformed to contemporary practice. In Aldrovandi's case, this is explained at least in part by the function of the *museum* as a place accessible to other scholars, where they could verify the veracity of his writings through first-hand inspection of the things described by him—whether *in pittura* or *al vivo*. See Aldrovandi, *Discorso naturale di Ulisse Aldrovandi*, BUB, fol. 508r–508v. He also used the collection in his teaching to "place before the eyes" (*ob oculos ponere*) of his students the things mentioned in his lectures. On both functions of the *museum*, see Olmi, "Die Sammlung," 175.

171. Findlen, "The Museum," 32; original italics. On this whole line of thought, cf. p. 32.

172. Aldrovandi, *Monstrorum Historia*, 30; Aldrovandi, *De Quadrupedibus Solidipedibus*, 198.

173. Nor is there, to my knowledge, any printing block preserved in the Biblioteca Universitaria di Bologna based on this watercolor, which would indicate that Aldrovandi intended to publish it.

174. The British Library has a copy of this etching that is neither hand colored nor cut out. BL, Prints & Drawings, 1841,0509.261. See Collection Online (n.d.), accessed May 21, 2013 www .britishmuseum.org/research/collection_online/collection_object_details.aspx?objectId=1570340 &partId=1&people=129768&peoA=129768-2-60&page=1.

175. For two of many other examples, see the drawing of a stick insect that was sent to him by a correspondent and that he pasted into his "Observationes" (BUB MS Aldrov. 136, vol. 27, fol. 236r), and a page in his drawings collection that has five images of insects glued onto it alongside sketched ones (BUB MS Aldrov., Tavole, vol. 4 unico, c. 57); reproduced in Olmi and Somoni, *Ulisse Aldrovandi*, 170.

176. See Lycosthenes, *Prodigiorum ac Ostentorum Chronicon*, 668.

177. See Perondino, *Magni Tamerlanis Scytharum Imperatoris Vita*; and Marlowe, *Tamburlaine the Great*, a play written sometime before 1587 about the Mongol conqueror's pursuit of power. On *Tamburlaine*, see Seeber, *Englische Literaturgeschichte*, 126 and 135–136.

178. Fischel, "Zeichnung und Naturbeobachtung," 222. Cf. also Fischel, *Natur im Bild*, 75–76.

179. See chapter 1.

180. Pinon, "Entre compilation et observation," 53. Änne Bäumer-Schleinkofer comes to a similar conclusion; see Bäumer-Schleinkofer, "Ulisse Aldrovandi," esp. 196–197.

181. See Aldrovandi, *Monstrorum Historia*, 558 and 559. Cf. also Fischel, *Natur im Bild*, 118n439.

182. Aldrovandi, *Monstrorum Historia*, 559; the presentation in two separate paragraphs corresponds to the setting of the text in the original.

183. Mauelshagen, *Wunderkammer auf Papier*, 190.

184. See chapter 4.

CHAPTER 3: Observing Correctly

1. The full name of the academy would change twice toward the end of the seventeenth century, once upon being officially recognized by the emperor in 1677 and, later, upon attaining the privileged status of a *Reichsakademie* in 1687. As the subsequent text deals predominantly with the period prior to these events and the relationship of the academy to the emperor is not the subject of this chapter, I will use its original name, Academia Naturae Curiosorum—that is, the Academy of Those Curious about Nature—throughout. A good overview of the academy's journey to the status of *Reichsakademie* is given by U. Müller, "Johann Laurentius Bausch."

2. Vollgnad, "Observatio CCXIX: De Monstroso Foetu," 519.

3. As one of two assistants, Vollgnad also took on various administrative tasks on behalf of the academy's president.

4. Frances Mason Barnett is alone to date in providing a detailed—if brief—examination of the academy's engagement with monsters. See Barnett, "Medical Authority and Princely Patronage," 182–195. Vollgnad's scholium escaped his notice.

5. See Daston and Park, *Wonders*, 1998, 176.

6. On these publications, cf. chapters 1 and 2.

7. Vollgnad, "Observatio CCXIX: De Monstroso Foetu," 519.

8. For a concise analysis of the theme of this section, see Kraemer, "*Richtig* Beobachten."

9. Vollgnad, "Observatio CCXIX: De Monstroso Foetu," 519.

10. On Perrault and the Paris academy's collection project, cf. Daston and Park, *Wonders*, 242–243.

11. On this whole section, see Perrault, *Memoires*, fol. ar.

12. Perrault, *Memoires*, fol. av. The distinction between encyclopedic and philosophical *histoire* can be seen in Perrault's work on the plans for the academy's never-completed *Histoire des plantes*. In the case of this project too, he supports the thesis that a *philosophical* (and not solely encyclopedic) *histoire* is needed. See Claude Perrault, "Projet pour la botanique," January 15, 1667, Archives de l'Académie de Sciences, Paris. I am obliged to Lorraine Daston for drawing my attention to this source.

13. M. Perrault, "Proiet pour les Experiences et observations Anatomiques," January 15, 1667, Registre, Procès-Verbaux, Registre de physique December 1666–April 1668, Archives de l'Académie des Sciences, Paris, 24. I am obliged to Lorraine Daston for drawing my attention to this source.

14. Vollgnad, "Observatio CCXIX: De Monstroso Foetu," 519. Similar emphases on the edifying Christian nature of observing are also sometimes found in the individual articles about monsters in the *Miscellanea Curiosa*. Their function, among other things, was to dispel suspicions of an idle curiosity concerned only with entertainment.

15. The first year of *Miscellanea Curiosa*, from 1670, contains a total of 160 articles, of which seven are on subjects classified as "monstrous" or, simply, monsters. The first volume of the journal that was most comparable in terms of its content, the *Acta medica & philosophica Hafniensia* from 1673—covering Bartholin's *observationes* of 1671 and 1672—features a total of 139 articles, of which 3 are, according to their titles, about monsters and two deal with "monstrous" things. In 1670, the *Philosophical Transactions* contained two articles that, according to their titles, deal with monsters—impressive, considering the wide range of subjects it covers and the total of forty-nine reports—but none that reported on "monstrous" phenomena. That the Parisian Académie Royale des Sciences was less interested in rare phenomena than its English counterpart is illustrated by a glance at the observations recorded in the *Histoires de l'Académie Royale des Sciences* for 1670: neither monsters nor other "monstrous" subjects appear at all; later years, however, do occasionally feature articles on monsters. The first year of the more thematically diverse academic monthly journal, the *Acta Eruditorum*, from 1682, contains only one article that reports on a *monstrum* and another that details a "monstrous" thing. The *Journal des sçavans*, which reported on scholarly matters in the broadest sense, also fits this pattern. The issue from February 10, 1670, contains four articles, with none about monsters or things categorized as "monstrous." The same is true for the issues from previous years. We have to go back to the first issue, from January 5, 1665, to find an article about a *monstre*. See the following for details: *Miscellanea Curiosa Medico-Physica Academiae Naturae Curiosorum sive Ephemeridum Medico-Physicarum Germanicarum Curiosarum Annus Primus*; Bartholin, *Acta Medica et Philosophica Hafniensia*; *Philosophical Transactions* 5 (1670); *Histoire de l'académie Royale*, 120–135; *Acta Eruditorum*; *Le journal des sçavans du lundy 10. fevrier M. DC. LXX*. On

the particular frequency of articles about monsters in *Miscellanea Curiosa*, see also Barnett, "Medical Authority and Princely Patronage," 183.

16. The former was the subject of eight articles in the first volume of 1670, and the latter of five. The level of interest in these subjects did not abate but remained more or less constant until the provisional end of the journal's publication in 1706. An overview of the illnesses described in the journal—but unfortunately not including subjects that fell outside this category—is offered by Hartmann, "Ärztliche Praxis und klinische Medizin," 384–385.

17. Hall and Hall, *The Correspondence of Henry Oldenburg*, vol. 6, no. 1267: Sluse to Oldenburg, August 6, 1669, 182 (Latin original) and 185 (English translation). For more on this letter, see also Scriba, "Auf der Suche nach neuen Wegen," 80.

18. Hall and Hall, *The Correspondence of Henry Oldenburg*, vol. 6, number 1329: Paisen to Oldenburg, November 27, 1669, 338 (Latin original) and 340 (English translation). For more on this letter and on this entire section, cf. Scriba, "Auf der Suche nach neuen Wegen," 80–81.

19. Cf. one of many writing on this subject, Scriba, "Auf der Suche nach neuen Wegen," 76–78.

20. Cf. Zedelmaier, "Wissensordnungen der Frühen Neuzeit," 842.

21. Leibniz, "Bedenken von Aufrichtung einer Akademie," 548–549. Cf. Berg and Parthier, "Die 'Kaiserliche' Leopoldina," 39; and on Leibniz's relationship to the Leopoldina generally, Streudel, "Leibniz und die Leopoldina."

22. On the contemporary criticism of the academy's original work program, cf. Berg, "Die frühen Schriften der Leopoldina," 70.

23. Roger, *The Life Sciences*, 133; on the new mentality cited by Roger, cf. pp. 133–204.

24. Cf. Roger, *The Life Sciences*, 138.

25. See Neigebaur, *Geschichte der Kaiserlichen Leopoldino-Carolinischen Deutschen Akademie der Naturforscher*, 1.

26. See Ule, *Geschichte der Kaiserlichen Leopoldinisch-Carolinischen Deutschen Akademie der Naturforscher*, 5 and 12.

27. Recent research has paid more attention to the academy; this monograph and other publications by the author bear witness to this. On the earlier neglect of the history of the Academia Naturae Curiosorum in the historiography of the German and European academy movement, cf. Toellner, "Im Hain Des Akademos," 21–24.

28. See Zedelmaier, "Wissensordnungen der Frühen Neuzeit," 842. On the early history of the *Royal Society*, see Johns, "Reading and Experiment," particularly 256, where Johns shows that "experimental philosophy" was by no means able to forgo reading or text either. On the contrary, he describes the combination of reading and experiment as characteristic of the "new philosophy." On the method of commonplaces in the Renaissance, see chapter 1.

29. Still seminal on this: Shapin, *The Scientific Revolution*, chapter 2.

30. See Park and Daston, "Introduction," 16.

31. Zedelmaier, "Wissensordnungen der Frühen Neuzeit," 842. Cf. also Blumenberg, *Der Prozeß der Theoretischen Neugierde*, 193; and, on the declared self-perception of many seventeenth-century naturalists that they were undertaking something new and thoroughly sweeping aside the old, Shapin, *The Scientific Revolution*, chapter 2.

32. Toellner, "Im Hain des Akademos," 19.

33. Cf. Toellner, "Im Hain des Akademos," particularly 19 and 25–28. Cf. Toellner, "Im Hain des Akademos," on this whole section.

34. Cf., among many other texts who make this point, Parthier, *Die Leopoldina*, 15.

35. Boehm, "Studium, Büchersammlung, Bildungsreise," 142.

36. Heunisch, *Panacea Apostolica*, 23; quoted by U. Müller, "Johann Laurentius Bausch," 19. On the biographical details, cf. pp. 18–19.

37. Boehm, "Studium, Büchersammlung, Bildungsreise," 135.

38. An exhibition catalog provides an excellent documentation of the collection. Its contents underwent a first review in the Leopoldina meeting of June 1998 and the resulting anthology. See Müller et al., *Die Bausch-Bibliothek*; and Folkerts, Jahn, and Müller, *Die Bausch-Bibliothek*.

39. Boehm, "Studium, Büchersammlung, Bildungsreise," 118.

40. U. Müller, "Johann Laurentius Bausch," 17. On the whole section, cf. p. 17.

41. Johann Laurentius Bausch's modest legacy of handwritten documents is not well suited to answering the question of his actual reading. See Boehm, "Studium, Büchersammlung, Bildungsreise," 123. To my knowledge, evidence of reading—such as underlining and marginal notes—in his surviving books have not yet been systematically examined.

42. See Müller et al., *Die Bausch-Bibliothek*, 9.

43. Toellner, "Der Arzt als Gelehrter," 58. One striking aspect is the interdenominational nature of the collection. For more on this and possible explanations for it, see Boehm, "Studium, Büchersammlung, Bildungsreise."

44. Richard Toellner interprets the absence of texts by Bacon, Descartes, and Galileo as proof of the intellectual conservatism of the two Bausches. In light of the following findings, this analysis must be relativized. Cf. Toellner, "Im Hain des Akademos," 29.

45. Cf. chapter 1.

46. Boehm, "Studium, Büchersammlung, Bildungsreise," 118–119. On the conspicuous parallels between the *peregrinationes academicae* undertaken by Leonhard and Johann Laurentius Bausch, cf. Müller et al., *Die Bausch-Bibliothek*, 13–14.

47. See Kuhn, *Venetischer Aristotelismus*.

48. Müller et al., *Die Bausch-Bibliothek*, number 5557, 620.

49. Müller et al., *Die Bausch-Bibliothek*, number 4322, 498.

50. Müller et al., *Die Bausch-Bibliothek*, number 4577, 526 and number 6163, 676–677.

51. See chapter 2.

52. Gianna Pomata has documented a total of sixty-three titles from this genre from the period 1551 to 1676 but notes that her list is incomplete. Pomata, "Sharing Cases." The number thirty-eight results from the following calculation: the printed catalog of the Bausch library as compared with the fifty-five publications listed by Pomata that appeared before Johann Laurentius Bausch's death. Only where more than one edition of a work appears in her list were the specific editions noted and each counted as an individual title.

53. For example, Lusitanus, *Curationum Medicinalium Centuriae Septem*; and de Heer, *Observationes Medicae*. In both cases, the markings indicate that they belong to Johann Laurentius Bausch date from 1636—that is, the year of his father Leonhard's death. See Müller et al., *Die Bausch-Bibliothek*, number 2411, 310 and number 4830, 548.

54. For instance, de Heer, *Observationes Medicae Oppido Rarae*. On this volume, which bears no owner's marking, see Müller et al., *Die Bausch-Bibliothek*, number 5729, 636.

55. On the history of the *leges* of the academy, see U. Müller, "Die Leges der Academia Naturae Curiosorum."

56. See Berg and Thamm, "Die systematische Erfassung der Naturgegenstände," 285.

57. See Keller, *Dr. Johann Laurentius Bausch (1605–1665)*, 52. Cf. also Berg and Thamm, "Die systematische Erfassung der Naturgegenstände," 286.

58. See Berg and Thamm, "Die systematische Erfassung der Naturgegenstände," 286.

59. Berg, "Die frühen Schriften der Leopoldina," 69.

60. Cf. the bibliography in Berg and Thamm, "Die systematische Erfassung der Naturgegenstände," 295–303. Of course, this does not mean that the individual authors did not address examples of the subjects covered by them that were classified as monstrous. For example, Philipp Jacob Sachs von Lewenhaimb took the opportunity, in section 7, "Vitis Variæ Species,"

of chapter 1 of *Ampelographia*, the first monograph to be published according to the stipulations of the Academia Naturae Curiosorum, to report on the sighting in 1599 of a monstrous, "bearded" vine (*Vitis monstrosa, barbata*) in the Palatinate region that he had read about in a text by Gaspard Bauhin. See Sachs von Lewenhaimb, *Ampelographia sive Vitis Viniferae*, 55.

61. The German translation and Latin original can be found in Uwe Müller's bilingual edition of the earliest known version of the *leges* of the academy from 1651/52. Müller, "Die Leges der Academia Naturae Curiosorum," 249. On the different versions of the *leges* in the second half of the seventeenth century, see p. 249.

62. U. Müller, "Die Leges der Academia Naturae Curiosorum," 249.

63. Sachs von Lewenhaimb to Johann Laurentius Bausch, May 15, 1660, LAH, MM 17.

64. Grafton, *Inky Fingers*, introduction, esp. 23–24.

65. For a current assessment of Sachs's central role in reforming the work program of the academy, see U. Müller, "Johann Laurentius Bausch," particularly 33–34. The following sources, which have now been edited, are still largely awaiting evaluation: Kefalas, "Philipp Jacob Sachs von Lewenhaimb."

66. On the importance of journals for the Republic of Letters and the dissemination of the new theories arising from the "new mentality," see Roger, *The Life Sciences*, 145–147.

67. See Laeven, *The "Acta Eruditorum,"* 19.

68. See Toellner, "Im Hain des Akademos," 37. For a good overview of the development of journal publishing, beginning with the first academic journal, the *Journal des sçavans*, see Jaumann, *Critica*, 253–263.

69. Hubertus Laeven blames this narrow focus for the fact that the *Miscellanea Curiosa* were known only to a relatively small circle of specialists compared to the *Acta Eruditorum*. See Laeven, *The "Acta Eruditorum,"* 19. On the *Philosophical Transactions* and the *Journal des sçavans* as models for the *Miscellanea Curiosa*, cf. p. 19.

70. Jaumann, *Critica*, 256.

71. See Laeven, *The "Acta Eruditorum,"* 19. Jaumann, too, criticizes the fact that the *Miscellanea Curiosa*, as the "first natural science journal in Germany," have mostly been overlooked in research on scholarly journals. Jaumann, *Critica*, 255. On the unprecedented subject range of the *Acta Eruditorum* for a periodical of its time, cf. p. 261.

72. Cf. Daston, "Die Akademien," 18.

73. See Daston, "Die Akademien," 18.

74. Cf. also Daston and Park, *Wonders*, chapter 6.

75. For more on this whole line of argument, see Daston, "Die Akademien," 221–222.

76. On this whole section, see Daston, "Die Akademien," 227.

77. See Daston, "Die Akademien," 228–229.

78. With regard to the Royal Society and the Académie Royale, see Daston, "Die Akademien," 230–231.

79. On this whole section, see Daston, "Die Akademien," 240–246.

80. Cf. Toellner, "Im Hain des Akademos," 19.

81. On the aims of the Academia Naturae Curiosorum, cf. among others U. Müller, "Die Leopoldina," 51.

82. Bausch, "Epistola invitatoria and draft of statutes from 1652," 343. For a German version, see the translation of the entire letter in U. Müller, "Die Leopoldina," 50.

83. German translation and Latin original to be found in Uwe Müller's bilingual edition of the earliest known version of the *leges* of the academy from 1651/52. Müller, "Die Leges Der Academia Naturae Curiosorum," 250.

84. An older view that Bacon's *New Atlantis* served as inspiration for the founding of the Academia Naturae Curiosorum has since proved untenable. See Minkowski, "Die Neu-Atlantis

des Francis Bacon." On the current state of research, see Gerber, "Die Neu-Atlantis"; Toellner, "Im Hain des Akademos," 27–28; and Parthier, *Die Leopoldina*, 15–16.

85. Daston, "Die Akademien," 20. Daston is referring to the *Historia Succincta*, a short history of the academy written by the academy's second president, Johann Michael Fehr and published in the *Miscellanea Curiosa* in 1671. It names Bacon's *The New Atlantis* (1627) as a model for the academy in Schweinfurt.

86. Cf. U. Müller, "Die Leopoldina," 57–59.

87. Daston, "Die Akademien," 22; on this whole section, cf. pp. 20–22.

88. Seifert, *Cognitio Historica*.

89. On this section, cf. Pomata, "*Praxis Historialis*."

90. On this section, see Pomata, "*Praxis Historialis*," 114–122. On the concept of anatomy espoused by Fabricius, cf. also Cunningham, "Fabricius and the 'Aristotle Project' "; and chapter 1.

91. See Pomata, "*Praxis Historialis*," 122–123.

92. On this whole section, see Pomata, "*Praxis Historialis*," 123–131.

93. Pomata, "*Praxis Historialis*," 131.

94. Lorraine Daston and Katharine Park also point out the similarities between the *observationes* found in medical compendia of rare phenomena—and the facts of early modern historiography—and the "strange facts" of the academies in the second half of the seventeenth century. See Daston and Park, *Wonders*, 422n84.

95. "Epistola Invitatoria," 4. Cf. Pomata, "Sharing Cases," 225. The letter of invitation had already appeared along with selected *observationes* from the first volume in a type of publicity text that was printed for the 1670 Frankfurt Book Fair. See Parthier, *Die Leopoldina*, 17.

96. See Pomata, "Observation Rising," 54–57; and, on the paradigmatic nature of *Centuriae Curationum*, 58.

97. Pomata, "*Praxis Historialis*," 136.

98. See Pomata, "Observation Rising," 61.

99. Welsch, *Sylloge Curationum et Observationum Medicinalium*.

100. Fehr and Welsch, *Epistolae Mutuae*. On this whole section, see Pomata, "Observation Rising," 62–64 and 78n99.

101. "Epistola Invitatoria," 5–6; original italics.

102. Mulsow, *Prekäres Wissen*. Mulsow calls these two forms of precarity "precarious status of the knowledge medium" and "precarious status of the speaker role and claims." See pp. 15 and 16–18.

103. Mulsow, *Prekäres Wissen*, 6; original italics.

104. Jaumann, *Critica*, 256. On the following comments, cf. Mulsow, *Prekäres Wissen*, 256–257.

105. Jaumann, *Critica*, 270. The whole section is based on pp. 270–272.

106. On *Acerra Philologica*, see also Bürger, "Die *Acerra Philologica*."

107. Jaumann, *Critica*, 272.

108. On the customary arrangement of *observationes* in numbered lists, cf. Pomata, "Sharing Cases," particularly 206.

109. See above. To give two more examples from many: Fabricius Hildanus, *Observationum & Curationum Chirurgicarum Centuriae*; Coberus, *Observationum Castrensium et Ungaricum*.

110. I am obliged to Richard Kremer for this information.

111. Cf. chapter 2, where it is also pointed out that Bacon's lists are rooted in an understanding of nature similar to that of Aldrovandi and therefore also to that of the *curiosi*.

112. See "Protocollum," LAH, P1–03, 12.

113. On this, see the corresponding passages in the revised *leges* of the academy from 1671 in

the bilingual edition of the 1671 *leges* in U. Müller, "Die Leges der Academia Naturae Curiosorum," 255–260.

114. Cf. Kraemer, "Faktoid und Fallgeschichte."

115. On Major, cf. also Scriba, "Auf der Suche nach neuen Wegen," particularly 69; and more recently, with a comparison of the working methods of Wedel and Sachs von Lewenhaimb, Keller, "Professionalizing Doubt."

116. Quoted in Barnett, "Medical Authority and Princely Patronage," 217n40.

117. Pomata, "Sharing Cases," 197. On the cognitive nature of epistemic genres, cf. also Park, "Observation in the Margins," particularly 48.

118. "Epistola Invitatoria," 4.

119. Rayger, "Observatio VII: Anatomia Monstri Bicipitis."

120. Greisel, "Observatio LV: Anatome Monstri Gemellorum"; an etching is bound between pp. 152 and 153; Greisel, "Observatio LVI: Ren Succenturiatus Monstrosus"; de Graaf, "Observatio CXXVIII: Monstrosus Uterus."

121. *Observatio* CXV also falls into this category: Jung, "Observatio CXV: Lapis Bezoar Monstrosus." The following report provides details of Leopold I's consent to have parts of the imperial collection described in the *Ephemerides*. Here, the reader also learns that the emperor offered to have etchings made of the objects. Sachs von Lewenhaimb, "Observatio CXVI: Radix Miranda," 268–269.

122. The legend on the etching clarifies when the beet—a wild radish—had been sighted: "Aō 1628. IS DESĒ RADŸS DER HEYDEN IN DEN GARDEN GEWASSEN."

123. The illustrations in the *Miscellanea Curiosa* have so far received little attention from historians. A statement by Sachs von Lewenhaimb is interesting in this context. On October 29, 1671, he wrote in a letter to the secretary of the Royal Society, Henry Oldenburg, that the publication of the second volume of the *Ephemerides* would be delayed. The reason he gave was the death of the Frankfurt engraver who had been commissioned to provide the volume's total of forty etchings. Now the printing plates had been lost, delaying the publication of the volume— which seemed otherwise to be printed already—by about a month. See Hall and Hall, *The Correspondence of Henry Oldenburg*, vol. 8, no. 1809, Sachs to Oldenburg, October 29, 1671, 321–323 (Latin original) and 323–325 (English translation). The fact that the academy was responsible for the etchings also explains why the problem of their expense was repeatedly brought up by the *curiosi*. For more on this, see the correspondence evaluated below between the president, Johann Michael Fehr, and Johann Georg Volckamer the Elder. In the presidents' invoicing preserved from the early years of the Leopoldina, etchings appear repeatedly as expenditure. Unfortunately, it is not clear from these documents, however, which specific etchings are meant. See Aeltere Rechnungsstellung der Akademie 1677–1798 Sektio XIII no. 1, LAH 26/21/1.

124. Sachs von Lewenhaimb, "Observatio XLVIII: Rapa Monstrosa Anthropomorpha."

125. On the role of this strategy in *Philosophical Transactions*, see da Costa, *The Singular*, 75.

126. See Jänisch, "Observatio CII: Buglossum Silvestre Monstrosum," 233.

127. Jänisch, "Observatio CII: Buglossum Silvestre Monstrosum," 234.

128. Jänisch, "Observatio CII: Buglossum Silvestre Monstrosum," 235.

129. Jänisch, "Observatio CII: Buglossum Silvestre Monstrosum," 235; original italics.

130. See Vollgnad, "Observatio CCXIX: De Monstroso Foetu," 518. A similar line of argument, which makes an etching of an object superior to a description of it, can be found in the *observationes* of other contributors. See, for example, Wurfbain, "Observatio CCVI: De Folio Lactucæ."

131. See Vollgnad, "Observatio CCXIX: De Monstroso Foetu," 518.

132. Fehr to Volckamer, June 18, 1674, UBEN.

133. Fehr is referring here to a letter from Oldenburg to Vollgnad and Jänisch. See Hall and Hall, *The Correspondence of Henry Oldenburg*, vol. 9, number 2031: Oldenburg to Vollgnad and Jänisch July 22, 1672, 169–171 (Latin original) and 171–173 (English translation); cf. Barnett, "Medical Authority and Princely Patronage," 187–188.

134. On the *collectores'* first contact with Oldenburg, see Hall and Hall, *The Correspondence of Henry Oldenburg*, vol. 9, number 2005: Vollgnad and Jänisch to Oldenburg, June 21, 1672, 125–127 (Latin original) and 128–129 (English translation).

135. Hall and Hall, *The Correspondence of Henry Oldenburg*, vol. 7, number 1620: Oldenburg to Sachs von Lewenhaimb, February 2, 1670/71, 432–434 (Latin original) and 434–435 (English translation), 434.

136. Cf. Daston, "Die Akademien," 30. On the tendency that can be seen in other seventeenth-century academies, too, to lower verification standards solely to ensure the continued existence of the fragile collective of naturalists, see p. 30. Both sides of this tension are expressed in a paratext inserted at the end of the volume covering the ninth and tenth year of the first decury of the *Miscellanea Curiosa*, in which the *collectores* react to criticism of the inclusion in the *Ephemerides* of what, in the eyes of the critics, were *observationes* that lacked credibility. They refer, among other things, to the impossibility of conducting checks on the ground as to the veracity of *observationes* from India, for instance. See "Postloquium," particularly 328–329.

137. Fehr to Volckamer, July 30, 1679, UBEN. Regarding this letter, cf. Barnett, "Medical Authority and Princely Patronage," 188–189.

138. Cf. Barnett, "Medical Authority and Princely Patronage," 190.

139. On Hofman, who was to become *director ephemeridum* in 1721, see Mücke and Schnalke, *Briefnetz Leopoldina*, 17.

140. Fehr to Volckamer, January 25, 1681, UBEN.

141. Fehr to Volckamer, January 25, 1681. Cf. Barnett, "Medical Authority and Princely Patronage," 190–191.

142. On Francis Bacon's criticism of those who engaged in the study of nature (particularly in relation to miracles) for pleasure's sake, see Daston and Park, *Wonders*, 228; and Daston, "Die Akademien," 26.

143. Twenty-two of the 160 *observationes* in the first volume are accompanied by illustrations—21 of which by one or more etchings and one by a woodcut inserted in the body of the text. In total, the volume contains nineteen unpaginated etchings. Of a total of 7 *observationes* in the first volume that deal with monsters or "monstrous" phenomena, 4 are accompanied by etchings.

144. See da Costa, *The Singular*, 71; da Costa is concerned with the period from the first volume to the middle of the eighteenth century.

145. Fehr to Volckamer, October 10, 1682, UBEN. On this letter, cf. Barnett, "Medical Authority and Princely Patronage," 191.

146. Mentzel to Volckamer, May 28, 1681, UBEN. Cf. Barnett, "Medical Authority and Princely Patronage," 190.

147. Valentini, "Observatio XC: De Monstris Hassiacis," 190–191. Despite an extensive search, I was unable to find the admonitions in question—neither in the *Miscellanea Curiosa* themselves, nor among the handwritten collections of the Leopoldina archive, nor in Wedel, "Progressus Academiae Naturae Curiosorum."

148. See "Vorbericht," fol. a3v–a3v.

149. "Vorbericht," fol. a3v–a4r.

150. Of the first eight volumes of the *Breslauische Sammlungen*, which were published between 1718 and 1720, only the second does not contain an article about a *monstrum*. All the other issues contain at least one such report. However, these reports were often summarized

together with other assorted observations in an article entitled "Miscellaneous Observations." This is a further indication that interest in such phenomena was waning.

151. "Classis IV. Von eintzelnen Natur-Geschichten des Julii 1717. Artic. X. Historia eines Monstri," 96.

152. "Classis IV. Von eintzelnen Natur-Geschichten des Julii 1717. Artic. X. Historia eines Monstri," 96. These general conclusions drawn from the individual case are then laid out in detail. See pp. 96–97.

153. See chapter 2.

154. See Park, "Observation in the Margins," 58.

155. Bausch, "Epistola invitatoria and draft statutes of 1652," 343.

156. Bausch, "Epistola invitatoria and draft statutes of 1652," 343.

157. Bausch, "Epistola invitatoria and draft statutes of 1652," 344.

158. "Dedicatory Epistle to Leopold I," fol. b2r; original italics.

159. The fact that the cabinet of curiosities constituted an important reference model for at least one other late seventeenth-century journal has been shown by Flemming Schock using the example of *Relationes Curiosae*, a weekly journal that appeared in Hamburg between 1681 and 1691 and was aimed at a wider readership. Schock, *Die Text-Wunderkammer*, particularly chapter 6. The *Miscellanea Curiosa* were an important source of contributions for *Relationes Curiosae*. See Schock, *Die Text-Wunderkammer*, 151–152 and 258, 263.

160. "Serenissimum, et Invictum Ferdinandum II. Magnum Hetruriæ Ducem in Dedicatione Operis Monstrorum Bartholomæus Ambrosinus hoc Tetrasticho Veneratur." Aldrovandi, *Monstrorum Historia*, fol. ✳3r.

161. See Park, "Nature in Person," 51. On Diana of Ephesus in art, see Goesch, *Diana Ephesia*. Allegories of nature similar to the one presented here can be found in the frontispieces of other contemporary works by naturalists, for example, in the second volume of Kircher's *Mundus Subterraneus*. See Kircher, *Mundi Subterranei*, vol. 2. A reproduction of this frontispiece can be found in Goesch, *Diana Ephesia*, 318.

162. See Park, "Nature in Person," 60–64.

163. See Bernitz, "Observatio LII: Muscus Terrestris"; original italics.

164. "Habet *Ulysses Aldrovandus l. de Monstr. historiâ* varias vitulorum bicipitium figuras & descriptiones, quarum una, quam *pag. 421.* delineari curavit, cum nostro monstro accuratam habet similitudinem." Schmid, "Observatio CLXIII: De Monstro Vitulino," 208.

165. "Epistola Invitatoria," 6.

166. Garliep, "Observatio IV: De Osse Monstroso," 4.

167. See Bausch, "Epistola invitatoria and draft statutes of 1652," 344. Cf. Daston, "Die Akademien," 22–23; and Toellner, "Im Hain des Akademos," 34.

168. Kenny, *The Uses of Curiosity*, 183. Unlike Hans Blumenberg's historical analysis of "theoretical curiosity," Kenny's study has not yet received the recognition it deserves. It is to Kenny's credit that he has pointed out and taken seriously the diversity of early modern usages of *curiositas* in various contexts and by different actors and institutions, rather than merging them into one "grand narrative" of the rise of curiosity as a scholarly emotion in the early modern period. A criticism of Blumenberg from this perspective can be found in Boehm, "Akademie-Idee und *Curiositas*," 83n71.

169. Kenny, *The Uses of Curiosity*, 186.

170. See Kenny, *The Uses of Curiosity*, 188.

171. Cf. Kenny, *The Uses of Curiosity*, 189.

172. Kenny, *The Uses of Curiosity*, 428.

173. Cf. Kenny, *The Uses of Curiosity*, 191.

174. Garliep, "Observatio XX: De Conyza Monstrosa."

175. Garliep, "Observatio XX: De Conyza Monstrosa," 68. The fact that this list of similar earlier cases appears at the end of the *observatio* and not in a scholium is probably due to the author of the report having, in this case, the necessary library at his disposal to prepare it. The fact that the article was not accompanied by an additional scholium also suggests that the scholium is anticipated, so to speak, by this last section of the *observatio*.

176. See Garliep, "Observatio XX: De Conyza Monstrosa," 68.

177. Four such cases, and the reactions of the academy and its publisher, are summarized in the section of the "Protocollum" covering 1689. The individual cases, however, do not all date from this year. See "Protocollum," 41–42.

178. See "Protocollum," 71–72.

179. See "Protocollum," 47–49.

180. The paratext does not include any information on its author, suggesting that the Academia Naturae Curiosorum is, as it were, speaking here.

181. See "Benevolo Lectori Salutem & Officia," 341–342.

182. On this whole section, see *Tagebuch*, LAH P1–03, 42–43. A bilingual edition is now also available: Müller, Weber, and Berg, *Protocollum*.

183. *Tagebuch*, 43.

184. *Tagebuch*, 43.

185. See Hartmann, "Observatio LXXVI: Anatome Monstrosi Crediti Fœtûs," 176.

186. See Hartmann, "Observatio LXXVI: Anatome Monstrosi Crediti Fœtûs," 176. On the scholarly debate on the satyr in the early modern period, see Roling, *Drachen und Sirenen*, 289–391.

187. See Hartmann, "Observatio LXXVI: Anatome Monstrosi Crediti Fœtûs," 176–177.

188. See Hartmann, "Observatio LXXVI: Anatome Monstrosi Crediti Fœtûs," 177–178. The sex organs are not addressed; Hartmann explains that he wishes to save discussion of them for a "Tract. de generation. viviparorum ex ovo" (178).

189. See Hartmann, "Observatio LXXVI: Anatome Monstrosi Crediti Fœtûs"; original italics. On the whole section, see pp. 178–179.

190. See Céard, *La nature et les prodiges*, 438, where Céard also points out the prevalence of this way of thinking around 1600. For more on this, cf. Kraemer, *Ein Zentaur in London*, 216–217.

191. Telling evidence of this can be seen, for example, in the differences between the Latin and vernacular versions of the same text on monsters published around 1600. See Kraemer, *Ein Zentaur in London*, 185–191.

192. See Daston and Park, *Wonders*, 187–189; they draw on Barnes, *Prophecy and Gnosis*, 87–91 and 253.

193. See the *essai* entitled "D'un enfant monstrueux" in the second book of his *Essais*, first published in 1580; Villey, *Les essais de Michel Seignevr de Montagne*, 348–349. On this *essai*, cf. Céard, *La nature et les prodiges*, 432. On Montaigne's rejection of divination in general, see pp. 415–421.

194. See Daston and Park, *Wonders*, 189.

195. On this last point, see Daston and Park, *Wonders*, 176.

196. See, among others, U. Müller, "Die Leopoldina," 68–78.

197. On this last point, see Daston and Park, *Wonders*, particularly 208.

198. Daston and Park come to the same conclusion with a view to wonders in general; *Wonders*, 349.

CHAPTER 4: A Centaur in London

1. The date is indicated by a remark in the pamphlet itself referring to "this present Month of *March*." Bentley, *A True and Faithful Account*, 21. The dating of the brochure is confirmed by

a reference to it in the *Royal Magazine* and elsewhere as one of the new publications on the English book market in the months of January to March 1751. See "A *Catalogue of* Books," 466; on the *Gentleman's Magazine*, where the pamphlet is also mentioned as one of the new publications in March 1751, see below.

2. According to the front page, the pamphlet, purchasable for the small price of sixpence, had been printed "for M. Cooper, in Pater-Noster-Row." The M. Cooper in question was Mary Cooper, the widow of the London bookseller and printer Thomas Cooper, who was one of the most productive printers and dealers on the market for pamphlets. After Thomas's death in ca. 1740, Mary continued to run the company until 1761. See the entries on Cooper (Mary) and Cooper (Thomas) in Plomer, Bushnell, and Dix, *A Dictionary of Printers and Booksellers*, 60–61.

3. Bentley, *A True and Faithful Account*, front page.

4. Bentley, *A True and Faithful Account*, front page.

5. Bentley, *A True and Faithful Account*, front page.

6. The European origins of April Fools' Day are generally dated back to the sixteenth century. "April Fools' Day," *Encyclopædia Britannica*, accessed September 30, 2013, http://www.britannica.com/EBchecked/topic/30821/April-Fools-Day.

7. "Books Publish'd March 1751," 142. The pamphlet is listed tenth under the heading "Miscellaneous."

8. De Montluzin, *Attributions of Authorship*, does not contain any information about the author of this passage.

9. "An Authentick Account," 153. There is no information about the author of this entry, either, in de Montluzin, *Attributions of Authorship*.

10. The author of the satire uses "idle" in the sense of the second meaning given in the *Oxford English Dictionary*: "Of actions, feelings, thoughts, words, etc.: Void of any real worth, usefulness, or significance; leading to no solid result; hence, ineffective, worthless, of no value, vain, frivolous, trifling. Also said of persons in respect to their actions, etc." "Idle," *Oxford English Dictionary* (n.d.), accessed June 12, 2012, www.oed.com/view/Entry/91064?rskey=WvGUeB&result=1&isAdvanced=false#eid.

11. See the entry on Richard Bentley in Matthew and Harrison, in assoc. with the British Academy, *Oxford Dictionary of National Biography*, 298–299.

12. See introduction.

13. See Daston, "The Empire of Observation," particularly 81 and 101–102.

14. On this last point, cf. Daston and Park, *Wonders*, 330–331.

15. Both Bentley, *A True and Faithful Account*, 3.

16. Bentley, *A True and Faithful Account*, 4.

17. See Bentley, *A True and Faithful Account*, 4.

18. Bentley, *A True and Faithful Account*, 4.

19. See Bentley, *A True and Faithful Account*, 5; original italics.

20. On this whole section, see Bentley, *A True and Faithful Account*, 4–5.

21. See Bentley, *A True and Faithful Account*, 14.

22. Bentley, *A True and Faithful Account*, 5; original italics.

23. See Kloosterhuis, *Legendäre "Lange Kerls,"* xxix; and Schobeß, *Die Langen Kerls*, 11. On the canvassing of recruits for the royal infantry, see Kloosterhuis, *Legendäre "Lange Kerls,"* xxxv–xxxix; and Schobeß, *Die Langen Kerls*, 85–103.

24. Samuel Pepys, for instance, relates his impressions of two gigantic children from Ireland who were exhibited at Charing Cross. They made such a lasting impression on him that he went back to see them for a second time no more than a month later. Robert Hooke's journals also tell of a giant woman he claims to have seen in London. On this and other giants in seventeenth- and eighteenth-century London, see Altick, *The Shows of London*, 36 and 42; and on a giant

advertised as the "German giant" who toured England in 1660, Burnett, *Constructing "Monsters" in Shakespearean Drama*, 11. Moreover, the "tall lads" still belonged in a sense to the tradition of the monsters—such as people with hirsutism, or indeed giants—sometimes encountered at medieval and early modern courts. Their functions were, after all, representative as well as military. On these functions, see Kloosterhuis, *Legendäre "Lange Kerls,"* xxii–xxix.

25. Bentley, *A True and Faithful Account*, British Library 12330.g.26; cf. Lindsay, *Index of English Literary Manuscripts*, 343. Walpole first met Richard Bentley Junior in 1750, and they remained friends for many years; see the entry on Richard Bentley in Matthew and Harrison, in assoc. with the British Academy, *Oxford Dictionary of National Biography*, 298–299.

26. Bentley, *A True and Faithful Account*, 7.

27. Bentley, *A True and Faithful Account*, 7; original italics.

28. See chapter 3.

29. Bate, *The Rambler*, 316; cited in Benedict, *Curiosity*, 186.

30. Benedict, *Curiosity*, 186.

31. See Benedict, *Curiosity*, 186.

32. "Wonder-Monger."

33. Benedict, *Curiosity*, 185.

34. The website of the Jockey Club provides a brief history of the horseraces on Newmarket heath; Jockey Club, "Our Heritage" (n.d.), accessed August 21, 2022 https://www.thejockeyclub .co.uk/the-racing/our-heritage/.

35. Bentley, *A True and Faithful Account*, 7–8; original italics.

36. See Johns, "The Ambivalence of Authorship."

37. Peacham, *The Compleat Gentleman*, 12–13; cited in Smith, *The Body of the Artisan*, 7. On this whole section, see p. 7.

38. Bentley, *A True and Faithful Account*, 8; original italics.

39. It is supported by the publication of a prologue and an epilogue written by Christopher Smart for this performance. Note the use of the phrase "persons of distinction" by both Smart and Bentley, which is an important indication that this is the performance to which Bentley is alluding. Smart, *An occasional prologue and epilogue*.

40. Bentley, *A True and Faithful Account*, 8.

41. Bentley, *A True and Faithful Account*, front page; original italics.

42. See Altick, *The Shows of London*, 34–49 and, on the area around Charing Cross in particular, 36.

43. See "Charing Cross" (accessed September 30, 2013).

44. On this whole section, see Altick, *The Shows of London*, 34.

45. Altick, *The Shows of London*, 36.

46. Handbill in British Library, 551.d.18, a collection of seventy-nine handbills and promotional texts published in newspapers for so-called freak shows from the late seventeenth and early eighteenth century; see Altick, *The Shows of London*, 36. On the times at which the Angolan hermaphrodite was exhibited, see below.

47. On these newspaper advertisements and on the fact that the use of Latin was due to the sexual nature of the content of the passage in question, see da Costa, " 'Mediating Sexual Difference,' " 129.

48. Bentley, *A True and Faithful Account*, 9–10; original italics.

49. See Bentley, *A True and Faithful Account*, 10.

50. Bentley, *A True and Faithful Account*, 13.

51. The chapter "Des dispositions des Suisses pour les sciences & les arts" in Johann Georg Altmann's *L'état et les délices de la Suisse* is a particularly interesting testament to the widely held view that Swiss scholarship lagged behind that of other European countries. Here Altmann

argues against the notion that the Swiss lacked the necessary *ésprit* for exceptional achievement in the sciences and in art. See Altmann, *L'état et les délices*, 1, chapter 19, 379–397.

52. See a remark to this effect by the "Proprietors" in Bentley, *A True and Faithful Account*, 13.

53. Bentley, *A True and Faithful Account*, 13, 14–17.

54. On this whole section, see Altick, *The Shows of London*, 43–44.

55. See da Costa, "'Mediating Sexual Difference,'"129.

56. See Bentley, *A True and Faithful Account*, 19 and 21–22.

57. Bentley, *A True and Faithful Account*, 8–9. The text goes on to name a number of individuals, some of them real; see pp. 8–9, 18, 19, and 20–21.

58. See chapter 3.

59. With regard to wonders in general, cf. Daston and Park, *Wonders*, 349.

60. See Parsons, *A mechanical and critical inquiry*, liv. Cf. also da Costa, "'Mediating Sexual Difference,'" 133.

61. See da Costa, "'Mediating Sexual Difference,'" 129–130.

62. His Royal Society membership is mentioned on the front page of the polemic, so he had already been admitted by the time of its publication.

63. Cf. da Costa, "'Mediating Sexual Difference,'"133.

64. On the shifts in the medical discourse on hermaphrodites in the eighteenth century, see Klöppel, *XXoXY ungelöst*, chapter 2.

65. Parsons, *A Mechanical and Critical Inquiry*, 145.

66. See Parsons, *A Mechanical and Critical Inquiry*, vii–viii.

67. Parsons, *A Mechanical and Critical Inquiry*, 10–11. On the early modern discourse on tribadism, see Park, "The Rediscovery of the Clitoris,"

68. Such as, for example, in the introduction, where he writes, "by which Means, I hope, it will not be doubted, but that the Truth, which hitherto has been clouded and obscured on this Head, may be said at least to begin to dawn, and by abler Hands may hereafter be brought to a clearer Light." Parsons, *A Mechanical and Critical Inquiry*, ix.

69. Parsons, *A Mechanical and Critical Inquiry*, 9.

70. Parsons, *A Mechanical and Critical Inquiry*, xv–xvi.

71. Cf. Daston, "Afterword."

72. On the various contemporary positions on the sex of the "Angolan hermaphrodite," cf. da Costa, "'Mediating Sexual Difference.'"

73. Da Costa, "'Mediating Sexual Difference,'" 133.

74. On the relationship between "vulgarity" and enlightenment in the discourse on wonders since the early eighteenth century, see Daston and Park, *Wonders*, 343–360; and, with regard to the discourse on monsters, Kraemer, "Why There Was No Centaur."

75. Wolff, *Gedanken über das ungewöhnliche Phœnomenon*, 3.

76. Cf. Wolff, *Gedanken über das ungewöhnliche Phœnomenon*, esp. 4.

77. Cf. Wolff, *Gedanken über das ungewöhnliche Phœnomenon*, esp. 14.

78. On the entire paragraph, see Wolff, *Gedanken über das ungewöhnliche Phœnomenon*, 3–4. Cf. also Daston and Park, *Wonders*, 337. On Johann Jacob Scheuchzer's reception of the published version of Wolff's lecture, see Mauelshagen, *Wunderkammer auf Papier*, 249–256.

79. See Daston and Park, *Wonders*, particularly 413.

80. Cf. Gilbert, *Early Modern Hermaphrodites*, 146.

81. De Beer, *The Diary of John Evelyn*, vol. 2, 40. On Evelyn, cf. Daston and Park, *Wonders*, 217.

82. De Beer, *The Diary of John Evelyn*, vol. 3, 198.

83. De Beer, *The Diary of John Evelyn*, vol. 3, 255–256; original italics.

84. De Beer, *The Diary of John Evelyn*, vol. 5, 592. The portrait still exists. It was received by the Royal Society on April 25 of that year (592n1).

85. See Parsons, *A Mechanical and Critical Inquiry*, 20–21. Parsons is referring here to Allen, "An Exact Narrative of an Hermaphrodite." Evelyn noted in his diary that "divers curious persons" had looked at the hermaphrodite; he himself, however, had not joined them. De Beer, *The Diary of John Evelyn*, vol. 3, 492.

86. See Parsons, *A Mechanical and Critical Inquiry*, 13–21.

87. Parsons, *A Mechanical and Critical Inquiry*, 21.

88. Parsons, *A Mechanical and Critical Inquiry*, 2–3; original italics. Parsons's characterization of Bradley's observations as "curious" has nothing to do with the fleeting curiosity of the common folk. Parsons uses the adjective here in the older sense of the *diligentia*-like curiosity of the meticulous, tenacious scholar, which the members of the Academia Naturae Curiosorum had also invoked. Cf. chapter 3.

89. On the call for repeated observation in the early modern study of nature, see Daston, "The Empire of Observation," 93–95.

90. It was, in fact, not uncommon for scholars to ask their colleagues expressly to repeat the observations they had conducted. The Swiss naturalist and philosopher Charles Bonnet, for example, encouraged his younger Italian colleague Lazzaro Spallanzani to repeat the observations of others—explicitly including those of Bonnet himself. See Daston, "The Empire of Observation," 100.

91. See Nichols, *Literary Anecdotes*, 5, 487. Cf. Altick, *The Shows of London*, 49.

92. On this connection, see, for instance, Daston and Park, *Wonders*, 204; Moscoso, "Monsters as Evidence"; Hagner, "Enlightened Monsters"; and Hagner, "Vom Naturalienkabinett zur Embryologie." Contrary to a view held by earlier scholars, the Royal Society also continued to engage intensively with the subject of monsters in the eighteenth century. See da Costa, *The Singular*, particularly 10.

93. See Fontenelle, *De l'origine des fables*, esp. 11.

94. On this whole section, see Daston and Park, *Wonders*, 204.

95. Rheinberger and Müller-Wille, *Vererbung*, 48.

96. Rheinberger and Müller-Wille, *Vererbung*, 48.

97. Cf. Daston and Park, *Wonders*, 204.

98. Daston and Park, *Wonders*, 204–205.

99. See Daston and Park, *Wonders*, 205.

100. Haller, "Vorlesungen über die gerichtliche Arzneiwissenschaft," vol. 1, 191.

101. On these impressions, see Steinke, "Anatomie und Physiologie," 227.

102. Vorrede des Uebersezers [translator's preface] to Haller, *Vorlesungen über die gerichtliche Arzneiwissenschaft*, vol. 1, fol. *2r.

103. An enquiry with the Burgerbibliothek in Bern, which holds most of Albrecht von Haller's handwritten intellectual estate, revealed that the transcript is not mentioned in the finding aids for the estate of Albrecht or Gottlieb Emanuel von Haller, and has therefore very probably not survived. Neither is it mentioned in Haeberli, *Gottlieb Emanuel Von Haller*. I am grateful to Thomas Schmid of the Burgerbibliothek in Bern for drawing my attention to this. According to Thomas René Rohrbach, who undertakes an initial survey and assessment of the printed version of Haller's *Vorlesungen über die gerichtliche Arzneiwissenschaft* in his medicohistorical doctoral thesis, the transcript by Gottlieb Emanuel Haller is "lost." See Rohrbach, "Friedrich August Webers Edition," 127.

104. "Gerücht," col. 1206.

105. See the entries on *fama* and *rumor* in Georges, *Ausführliches lateinisch-deutsches und deutsch-lateinisches Handwörterbuch*, vol. 1, 1, cols. 2680 and 2425.

106. See "Gerücht," col. 3753.

107. "Vorbericht," fol. a4r.

108. "Gerücht," cols. 1206–1207.

109. "Gerücht," col. 3755.

110. See "Sage"; here: col. 1647.

111. See Haller, *Albrecht Hallers Tagebuch seiner Studienreise.*

112. Albrecht von Haller had a wide network of correspondents across much of Europe. Some 3 percent of the surviving letters written to him were from Great Britain; 298 letters have survived from London alone, making it ninth in terms of the provenance of Haller's correspondence, after Hanover, Bern, Lausanne, Zurich, Göttingen, Geneva, Basel, and Paris. See Stuber, Hächler, and Lienhard, *Hallers Netz,* 66 and 68.

113. See Stuber, Hächler, and Lienhard, *Hallers Netz,* 5.

114. See Stuber, Hächler, and Lienhard, *Hallers Netz,* 96, chart 6.16 on p. 97. Not all of these letters were sent from the British Isles, however. After London, the most common places of origin are Hanover and Basel. See chart 6.4 on p. 99.

115. Haller, *Vorlesungen über die gerichtliche Arzneiwissenschaft,* vol. 1, 187.

116. See Haller, *Vorlesungen über die gerichtliche Arzneiwissenschaft,* vol. 1, point (5) of translator's note (o) on p. 347.

117. The volumes of numerous English and German scholarly journals from the 1740s were examined in the preparation of this book, without finding any further trace of the rumor of the London centaur. I am grateful to Martin Stuber, who searched the Forschungsdatenbank zu Albrecht von Haller (research database on Albrecht von Haller, accessible at the Institute for the History of Medicine at the University of Bern, the Institute of History at the University of Bern, and the Burgerbibliothek in Bern) for any further traces of the rumor and for any evidence in Haller's letters, libraries, or writings that he had read Bentley's satire—with no success. I am also grateful to Hubert Steinke, who helped with this search and confirmed that Haller was a regular reader of the *Gentleman's Magazine,* though he rarely cited it.

118. "In der Arzneywissenschaft." On the 1748 and 1749 announcements, see Rohrbach, "Friedrich August Webers Edition," 2.

119. Teichmeyer, *Institutiones Medicinae Legalis vel Forensis* (1723).

120. Teichmeyer, *Institutiones Medicinae Legalis vel Forensis* (1740). See Rohrbach, "Friedrich August Webers Edition," 1–2.

121. See Rohrbach, "Friedrich August Webers Edition," 125.

122. On Weber's life and work, see Rohrbach, "Friedrich August Webers Edition," 15–73. It is unclear how the lecture notes taken by Albrecht von Haller's eldest son came into Weber's possession (127).

123. On the three levels of the work, see Rohrbach, "Friedrich August Webers Edition," 75.

124. See Rohrbach, "Friedrich August Webers Edition," 76. Weber does not always reproduce Teichmeyer's text in full, however. See p. 79.

125. See Rohrbach, "Friedrich August Webers Edition," 4 and 92.

126. Cf. Rohrbach, "Friedrich August Webers Edition," 4.

127. Rohrbach investigated the authenticity of the statements that Friedrich August Weber attributes to Haller on the basis of a number of samples chosen at random and concludes that Weber's edition is extremely reliable in this regard. See Rohrbach, "Friedrich August Webers Edition," 110–118 and, in summary, 125.

128. See Rohrbach, "Friedrich August Webers Edition," 76 and 77.

129. Haller, *Vorlesungen über die gerichtliche Arzneiwissenschaft,* vol. 1, 178.

130. Haller, *Vorlesungen über die gerichtliche Arzneiwissenschaft,* vol. 1, 178.

131. See Haller, *Vorlesungen über die gerichtliche Arzneiwissenschaft,* vol. 1, 179.

132. Haller, *Vorlesungen über die gerichtliche Arzneiwissenschaft*, vol. 1, 180.

133. On the "preternatural philosophy" of this period, which took the preternatural as its privileged subject, see Daston and Park, *Wonders*, 159–172.

134. Haller, *Vorlesungen über die gerichtliche Arzneiwissenschaft*, vol. 1, 180; original italics.

135. Friedrich August Weber omitted many of Teichmeyer's references in his edition of Haller's lectures, as is the case here. A glance at Teichmeyer's textbook shows which text he is referring to here; namely, Ludovici, *Usus Practicus Distinctionum Iuridicarum*. See Teichmeyer, *Institutiones Medicinae Legalis vel Forensis* (1740), 84.

136. Haller, *Vorlesungen über die gerichtliche Arzneiwissenschaft*, vol. 1, 180–181; original italics.

137. See "Mißgeburt"; here: cols. 489–490.

138. Haller, *Vorlesungen über die gerichtliche Arzneiwissenschaft*, vol. 1, 181.

139. Haller, *Vorlesungen über die gerichtliche Arzneiwissenschaft*, vol. 1, 181.

140. Haller, *Vorlesungen über die gerichtliche Arzneiwissenschaft*, vol. 1, 181.

141. See Haller, *Vorlesungen über die gerichtliche Arzneiwissenschaft*, vol. 1, 181.

142. Haller, *Vorlesungen über die gerichtliche Arzneiwissenschaft*, vol. 1, 182.

143. See Haller, *Vorlesungen über die gerichtliche Arzneiwissenschaft*, vol. 1, 182; cf. on this whole section pp. 181–182.

144. Haller, *Vorlesungen über die gerichtliche Arzneiwissenschaft*, vol. 1, 182–183.

145. Haller, *Vorlesungen über die gerichtliche Arzneiwissenschaft*, vol. 1, 183; cf. on this whole section pp. 182–183.

146. Haller, *Vorlesungen über die gerichtliche Arzneiwissenschaft*, vol. 1, 183.

147. Haller, *Vorlesungen über die gerichtliche Arzneiwissenschaft*, vol. 1, 183.

148. Haller, *Vorlesungen über die gerichtliche Arzneiwissenschaft*, vol. 1, 183.

149. Haller, *Vorlesungen über die gerichtliche Arzneiwissenschaft*, vol. 1, 183–184. On the changing function of the specimens of the Dutch anatomist Frederick Ruysch—renowned in his day for his skill as a preparator—from valuable collector's items and exhibits to objects of study by means of dissection, see Hagner, "Vom Naturalienkabinett zur Embryologie"; and Hagner, "Enlightened Monsters."

150. See chapter 3.

151. Haller, *Vorlesungen über die gerichtliche Arzneiwissenschaft*, vol. 1, 184.

152. Haller, *Vorlesungen über die gerichtliche Arzneiwissenschaft*, vol. 1, 184.

153. Haller, *Vorlesungen über die gerichtliche Arzneiwissenschaft*, vol. 1, 184.

154. On this whole section, see Haller, *Vorlesungen über die gerichtliche Arzneiwissenschaft*, vol. 1, 184–185.

155. Haller, *Vorlesungen über die gerichtliche Arzneiwissenschaft*, vol. 1, 185; on this whole section, see p. 185.

156. Haller, *Vorlesungen über die gerichtliche Arzneiwissenschaft*, vol. 1, 186; on this whole section, see pp. 185–186.

157. Haller, *Vorlesungen über die gerichtliche Arzneiwissenschaft*, vol. 1, 186; on this whole section, cf. p. 186.

158. As, for example, did Fortunio Liceti in his *De Monstrorum Natura*.

159. Haller, *Vorlesungen über die gerichtliche Arzneiwissenschaft*, vol. 1, 182.

160. Feyens compiled almost two hundred pages' worth of all the knowledge he could find about *phantasia* and *imaginatio* in the writings of both the ancients and more recent authors. His text, organized in *quaestiones*, compellingly documents the immense potential of *imaginatio* to explain monstrous births. See Fienus, *De Viribus Imaginationis Tractatus*.

161. The term "pathological imagination" was coined by Lorraine Daston and Katharine Park; see Daston and Park, *Wonders*, 339.

162. Cf. Daston and Park, *Wonders*, 341–342.

163. See Daston and Park, *Wonders*, 341; and Bundy, "The Theory of the Imagination," 260–274.

164. See Daston and Park, *Wonders*, 341.

165. Cf. Daston and Park, *Wonders*, 339.

166. Haller, *Vorlesungen über die gerichtliche Arzneiwissenschaft*, vol. 1, 7. Cf. also Rohrbach, "Friedrich August Webers Edition," 94.

167. Haller, *Vorlesungen über die gerichtliche Arzneiwissenschaft*, vol. 1, 86–87.

168. Haller, *Vorlesungen über die gerichtliche Arzneiwissenschaft*, vol. 1, 88.

169. Voltaire claims to have seen newborns himself who clearly showed the effect of the maternal imagination: "Les exemples en sont innombrables, & celui qui écrit cet article en a vû de si frappans, qu'il démentiroit ses yeux s'il en doutoit." Voltaire, "Imagination, Imaginer," 561. See also Daston and Park, *Wonders*, 330.

170. See Weber's reproduction of the passage from Teichmeyer's text; Haller, *Vorlesungen über die gerichtliche Arzneiwissenschaft*, vol. 1, 317nn.

171. Haller, *Vorlesungen über die gerichtliche Arzneiwissenschaft*, vol. 1, 317.

172. See Haller, *Vorlesungen über die gerichtliche Arzneiwissenschaft*, vol. 1, 88–89.

173. Haller, *Vorlesungen über die gerichtliche Arzneiwissenschaft*, vol. 1, 90.

174. "Imagination *des femmes*," 563.

175. "Imaginatio," cols. 571–572.

176. "Mißgeburt," col. 487.

177. "Einbildung," col. 533.

178. "Einbildung," col. 533.

179. "Einbildung," col. 533.

180. "Einbildungs-Krafft," col. 535.

181. Daston and Park, *Wonders*, 334.

182. Daston and Park, *Wonders*, 343; on this whole section, see pp. 334–343.

183. Daston, "The Empire of Observation," 106.

184. See Daston, "The Empire of Observation," 113n103.

185. See chapter 3.

186. Daston and Park, *Wonders*, 249; cf. Shapin, *A Social History of Truth*, 221 and 287.

187. Daston and Park, *Wonders*, 250–251; on this entire section, cf. pp. 248–251. On the changes that the concept of nature underwent in the course of the early modern period, see also Daston, "The Nature of Nature."

188. Daston and Park, *Wonders*, 331; on this whole section, cf. chapter 9; and Daston, "The Nature of Nature," 158–169.

189. Cf. also Daston and Park, *Wonders*, 252.

190. Daston and Park, *Wonders*, 347–348.

191. Serjeantson, "Proof and Persuasion," 162. See also the seminal study Hacking, *The Emergence of Probability*.

192. See Serjeantson, "Proof and Persuasion," 162.

193. Arnaud and Nicole, *La logique*, 340. See Daston and Park, *Wonders*, 251; and, on the transfer of statistical methods to historiography in *Logique*, Serjeantson, "Proof and Persuasion," 263.

194. Following a statement by Cicero, early modern rhetoric was intended to engender *fides*—"faith" or "trust"—in the listener. A distinction was made between the trust placed in the speaker and the trust placed in what was said. Both were regarded as essential to the persuasive power of the speech. See Serjeantson, "Proof and Persuasion," 147–149. For a fuller discussion of this, see chapter 1.

195. Locke, *An Essay concerning Human Understanding*, 4.15.4–6, 365–368. See Daston and Park, *Wonders*, 251.

196. Daston and Park, *Wonders*, 252–253; cf. Locke, *An Essay concerning Human Understanding*, 4.15.5, 366–367.

197. Daston and Park, *Wonders*, 252; they are referring here to Bayle, *Pensées diverses sur le comète*, 38.

198. Such an expression of skepticism regarding Lycosthenes's credibility, immediately followed by a reproduction of his account, can be found, for example, in Aldrovandi, *Monstrorum Historia*, 8, where Aldrovandi writes that Lycosthenes "multa vana scripsit."

199. Haller, *Vorlesungen über die gerichtliche Arzneiwissenschaft*, vol. 1, 87.

200. Haller, *Vorlesungen über die gerichtliche Arzneiwissenschaft*, vol. 1, 87–88.

201. Haller, *Vorlesungen über die gerichtliche Arzneiwissenschaft*, vol. 1, 88.

202. Haller's *Bibliotheca Botanica* (1771–1772), *Bibliotheca Chirurgica* (1774–1775), and *Bibliotheca Anatomica* (1774–1777) were completed and published during his lifetime, each comprising two volumes. A comprehensive *Bibliotheca Medicinae Practicae* was planned but not completed. On this, cf. Lehmann-Brauns, "Neukonturierung und methodologische Reflexion," 158n111; and, more recently, Kraemer, "Albrecht von Haller"; and Gantet and Kraemer, "Wie man mehr als 9000 Rezensionen schreiben kann."

203. On this position in time, and on the fact that the *historia literaria* constituted a primarily "German" phenomenon, see Zedelmaier, "'Conspectus Reipublicae Literariae,'" 85. Studies of the *historia literaria* from the perspective of the history of science and the history of ideas have only picked up pace in recent years.

Following early individual studies like Gierl, "Bestandsaufnahme im gelehrten Bereich"; Jaumann, "Jakob Friedrich Reimmanns Bayle-Kritik"; Zedelmaier, "'Historia literaria'"; Nelles, "'Historia Literaria' at Helmstedt"; and Grunert, Syndikus, and Vollhardt, "Ein Leitfaden," a first collected volume is now available: Grunert and Vollhardt, *Historia literaria*.

204. On the defining influence of Bacon's program for a *Historia Literarum*, which even the contemporary protagonists of the *historia literaria* viewed as such, see Syndikus, "Die Anfänge der Historia literaria," 6–14.

205. Schmidt-Biggemann, *Topica Universalis*, esp. 26, 28–29, and 221.

206. Zedelmaier, "'Conspectus Reipublicae Literariae,'" 80.

207. Zedelmaier, "'Conspectus Reipublicae Literariae,'" 82.

208. Cf. Lehmann-Brauns, "Neukonturierung und methodologische Reflexion," 131–132.

209. Zedelmaier, "'Conspectus Reipublicae Literariae,'" 71.

210. Lehmann-Brauns, "Neukonturierung und methodologische Reflexion," 154, making reference to Heumann.

211. See Lehmann-Brauns, "Neukonturierung und methodologische Reflexion," 158–159; on the terms *historia literaria universalis* and *specialis* as used by Heumann, see p. 138.

212. Steinke, "Anatomie und Physiologie," 246; on this whole paragraph, cf. pp. 246–248.

213. Haller, *Bibliotheca Anatomica*.

214. Steinke, "Anatomie und Physiologie," 249.

215. Zedelmaier, "Lesetechniken," 28.

216. On the size and content of the *Bibliotheca Anatomica*, cf. Steinke, "Anatomie und Physiologie," 249.

217. On Linné, see Müller-Wille and Scharf, "Indexing Nature"; Müller-Wille, "Vom Sexualsystem zur Karteikarte"; and Müller-Wille and Charmantier, "Natural History and Information Overload." On Albrecht von Haller, see Braun-Bucher, "Hallers Bibliothek und Nachlass," 521. Another worthwhile overview of the techniques that were utilized by scholars from the early modern period to the present day, and a discussion of the reasons for the use of freely arrange-

able slips of paper that was increasingly perceivable from the late seventeenth century onward, is provided by Zedelmaier, "Buch, Exzerpt, Zettelschrank, Zettelkasten."

218. See Stuber, Hächler, and Lienhard, *Hallers Netz*, 5. On the connection between review and excerpt in Haller's work, see the recent article Gantet and Kraemer, "Wie man mehr als 9000 Rezensionen schreiben kann."

219. Gantet and Kraemer, "Wie man mehr als 9000 Rezensionen schreiben kann."

220. Haller, *Vorlesungen über die gerichtliche Arzneiwissenschaft*, vol. 1, 187–190.

221. See Teichmeyer, *Institutiones Medicinae Legalis vel Forensis*, 87–89. Teichmeyer, too, can be a harsh judge of the early modern authors and the cases of barely credible monsters reported by them. Making reference to Gaspar Schott's *Physica Curiosa*, for example, he describes the Jesuit and naturalist as *superstitiosus*—superstitious. But this does not deter him from reproducing a considerable number of the cases of monstrosity that Schott reported. See p. 88.

222. Haller, *Vorlesungen über die gerichtliche Arzneiwissenschaft*, vol. 1, 187.

223. Haller, *Vorlesungen über die gerichtliche Arzneiwissenschaft*, vol. 1, 187.

224. On this whole section, cf. Haller, *Vorlesungen über die gerichtliche Arzneiwissenschaft*, vol. 1, 187–190.

225. Haller, *Vorlesungen über die gerichtliche Arzneiwissenschaft*, vol. 1, 190.

226. Haller, *Vorlesungen über die gerichtliche Arzneiwissenschaft*, vol. 1, 190. In itself, this class was indeed nothing new at that time. With regard to the monsters grouped within it, it corresponds to a large extent with the equivalent categories in the work of early modern scholars and their ancient authorities. Ambroise Paré, for example, describes a class of monsters distinguished by the absence of one or several body parts. He attributes them to a lack of seed at conception. See Paré, *Des monstres et prodiges*, chapter 8. The same is true up to a certain point of Haller's first class. Once again, Ambroise Paré serves as an example: the French surgeon writes of a class of monsters distinguished by a superfluity of specific body parts; he attributes them to an excess of seed at conception. See chapter 4.

227. Haller, *Vorlesungen über die gerichtliche Arzneiwissenschaft*, vol. 1, 190.

228. Haller, *Vorlesungen über die gerichtliche Arzneiwissenschaft*, vol. 1, 190.

229. "Rockenphilosophie," col. 1103; on this whole section, cf. col. 1103.

230. Praetorius, *Philosophia Colus*. "Phylosevieh," if spoken, may sound like the German word for philosophy. But it contains "Vieh," a common German term for cattle or livestock.

231. Haller, *Vorlesungen über die gerichtliche Arzneiwissenschaft*, vol. 1, 190–191.

232. See chapter 1.

233. Haller, *Vorlesungen über die gerichtliche Arzneiwissenschaft*, vol. 1, 191.

234. See the introduction.

235. Haller, *Vorlesungen über die gerichtliche Arzneiwissenschaft*, vol. 1, 191. On the fact that the boundary between humans and animals was presented in the Renaissance as comparatively fluid, see chapter 1.

236. See Haller, *Vorlesungen über die gerichtliche Arzneiwissenschaft*, vol. 1, 192.

237. See Haller, *Vorlesungen über die gerichtliche Arzneiwissenschaft*, vol. 1, 354nr.

238. Shaw, *Travels, or Observations*.

239. Haller, *Vorlesungen über die gerichtliche Arzneiwissenschaft*, vol. 1, 192–193.

240. Mauelshagen, *Wunderkammer auf Papier*, 253.

241. Mauelshagen, *Wunderkammer auf Papier*, 241; on Scheuchzer's critical reading of the *Wickiana*, 241–271.

242. On this whole section, see Daston and Park, *Wonders*, 343–344.

243. Browne, *Pseudodoxia Epidemica*, 15; cited in Daston and Park, *Wonders*, 344.

244. See Daston and Park, *Wonders*, 344.

245. See Daston and Park, *Wonders*, 348–349.

246. Haller, *Vorlesungen über die gerichtliche Arzneiwissenschaft*, vol. 1, 182.

247. Haller, *Operum Anatomici Argumenti Minorum Tomus Tertius*. The two books on monsters begin the volume.

248. See Haller, *Operum Anatomici Argumenti Minorum Tomus Tertius*, 1. Haller goes on to list these earlier publications.

249. See Haller, *Operum Anatomici Argumenti Minorum Tomus Tertius*, 1.

250. Haller, *Operum Anatomici Argumenti Minorum Tomus Tertius*, 3 and 131. The first book covers pp. 3–130, and the second pp. 131–173.

251. Steinke, "Anatomie und Physiologie," 226.

252. Steinke, "Anatomie und Physiologie," 226.

253. See Haller, *Elementa Physiologiæ Corporis Humani*, ix; cf. Steinke, "Anatomie und Physiologie," 229.

254. See Steinke, "Anatomie und Physiologie," 229.

255. See Haller, *Operum Anatomici Argumenti Minorum Tomus Tertius*, 3–4. On Liceti's etymologically inspired definition of *monstrum*, see Kraemer, *Ein Zentaur in London*, 207–210.

256. See Haller, *Operum Anatomici Argumenti Minorum Tomus Tertius*, 4. It would not be unreasonable to suspect Haller himself of being one of these anatomists: after all, he made in-depth comparative studies of the many variations exhibited by the vessels with regard to their origin, their course, and the frequencies of their occurrence. On this, see Steinke, "Anatomie und Physiologie," 232.

257. See the thirty-sixth chapter of the first *liber*, entitled "Monstra Plantarum"; Haller, *Operum Anatomici Argumenti Minorum Tomus Tertius*, 123–126.

258. See Haller, *Operum Anatomici Argumenti Minorum Tomus Tertius*, 4.

259. See Haller, *Operum Anatomici Argumenti Minorum Tomus Tertius*, 4nnb–g. On Hartmann's article on the alleged satyr, see chapter 3.

260. See chapter 2.

261. "Ce n'est pas une science de s'être rempli la tête de toutes les extravagances des Phéniciens et des Grecs; mais c'en est une de savoir ce qui a conduit les Phéniciens et les Grecs à ces extravagances." Fontenelle, *De l'origine des fables*, 40. According to current knowledge, Fontenelle began working on this text early. See Carré's introduction in this volume.

262. See, for example, chapters 29 ("Fetus Bicipites Artubus Duplicati") and 30 ("Exempla Aliarum Auctorum"); Haller, *Operum Anatomici Argumenti Minorum Tomus Tertius*, 98–108 and 109–115.

263. See Haller, *Operum Anatomici Argumenti Minorum Tomus Tertius*, 131.

264. See Haller, *Operum Anatomici Argumenti Minorum Tomus Tertius*, 131–133.

265. See particularly the concluding summary of his theses with regard to the individual kinds of monsters and monstrosities. Haller, *Operum Anatomici Argumenti Minorum Tomus Tertius*, 172.

266. See Haller, *Operum Anatomici Argumenti Minorum Tomus Tertius*, 173.

267. Daston, "The Empire of Observation," 104.

Conclusion

1. Wedel, "Observatio CCXXVIII: De Basilico."

2. Johannes Jähnisch was himself in possession of a curiosity cabinet. He was therefore presumably aware of the intense debate about forged specimens among collectors of *naturalia* in the European early modern period. On Jähnisch's natural history cabinet, see his outline biography in Barnett, "Medical Authority and Princely Patronage," 349.

3. See Wedel, "Observatio CCXXVIII: De Basilico," 204–205.
4. See Wedel, "Observatio CCXXVIII: De Basilico," 205.
5. See Daston and Park, *Wonders*, 248.
6. See particularly Johns, "Reading and Experiment."
7. Cf. Joas, Krämer, and Nickelsen, "History of Science."
8. In this vein: Heilbron, "History of Science"; Joas, Krämer, and Nickelsen, "Introduction."

Bibliography

Archival Sources

ARCHIVIO DI STATO DI BOLOGNA (ASB)

Assunteria di Studio, 100, fasc. 6, Carte relative allo Studio Aldrovandi.

BIBLIOTECA UNIVERSITARIA DI BOLOGNA (BUB)

Aldrovandi, Ulisse. "Discorso naturale di Ulisse Aldrovandi. Nel quale si tratta in generale del suo Museo, e delle fatiche da lui usate per raunare de varie parti del mondo, quasi in un Theatro di Natura tutte le cose sublunari, come piante, animali et altre cose minerali. Et parimente vi s' insegna come si de' venir nella certa et necessaria cognitione d'alcuni medicamenti incerti et dubbij, ad utilità grandissima non solo de' medici, ma d'ogni altro studioso [. . .]." 1572–1573. BUB Ms. Aldrovandi 91.
———. Guestbook. Ms. Aldrovandi 41 und 110.
———. "Pandechion Epistemonicon." Ms. Aldrovandi 105.
———. "Observationes." Ms. Aldrovandi 136.
———. "Tavole." Ms. Aldrovandi.

BIBLIOTHÈQUE ET ARCHIVES DU CHÂTEAU DE CHANTILLY

Dante. "Divine Comedy." Ms. 597.

BRITISH LIBRARY (BL)

Bentley, Richard. A True and Faithful Account of the Greatest Wonder Produced by Nature in these 3000 Years, in the Person of Mr. Jehan-Paul-Ernest Christian Lodovick Manpferdt; the Surprising Centaur, who Will Be Exhibited to the Publick, on the First of Next Month, at the Sign of the Golden Cross at Charing Cross. London: Printed for M. Cooper, in Pater-Noster-Row, 1751, 12330.g.26.
Cavalieri, Giovanni Battisti de'. Copperplate engraving of a centaur from the series Opera nel a quale vie molti Mostri de tute le parti del mondo antichi et moderni [. . .] von 1585. Prints & Drawings, 1841,0509.261.

LEOPOLDINA-ARCHIV, HALLE (LAH)

Aeltere Rechnungsstellung der Akademie 1677–1798 Sektio XIII No. 1. 26/21/1.

"Protocollum Academiae Caesareo-Leopoldinae Naturae Curiosorum, Inceptum ab Ejus Collega et Praeside, Celso I. A. O. R. M.DCXCIV." P1–01.

Sachs von Lewenhaimb, Philipp, to Johann Laurentius Bausch, 15. 5. 1660. MM 17, Matrikelmappe 17.

"Tagebuch der Kaiserlich-Leopoldinischen Akademie der Naturforscher, begonnen von ihrem Mitglied und Präsidenten Celsus I. Im Jahre des Heils 1694." Unpublished German translation of the "Protocollum," by Klaus Lämmel, Halle, P1–03.

LORD LEICESTER LIBRARY (LLL)

Dante. "Divine Comedy." MS Holkham 514.

STADTARCHIV SCHWEINFURT

Bausch, Johann Laurentius. *Collectaneorum chronologicorum Suinfurtensium.* Stadtarchiv Schweinfurt, Handschriften, Ha 102–104.

UNIVERSITÄTSBIBLIOTHEK ERLANGEN-NÜRNBERG (UBEN)

Fehr, Johann Michael, to Johann Georg Volckamer, June 18, 1674. In Briefsammlung Trew. Accessed through the database of Harald Fischer Verlag, www.haraldfischerverlag.de/hfv /sammlungen/trew_engl.php.

Fehr, Johann Michael, to Johann Georg Volckamer, July 30, 1679. In Briefsammlung Trew. Accessed through the database of Harald Fischer Verlag. www.haraldfischerverlag.de/hfv /sammlungen/trew_engl.php.

Fehr, Johann Michael, to Johann Georg Volckamer, January 25, 1681. In Briefsammlung Trew. Accessed through the database of Harald Fischer Verlag. www.haraldfischerverlag.de/hfv /sammlungen/trew_engl.php.

Fehr, Johann Michael, to Johann Georg Volckamer, October 10, 1682. In Briefsammlung Trew. Accessed through the database of Harald Fischer Verlag. www.haraldfischerverlag.de/hfv /sammlungen/trew_engl.php.

Mentzel, Christian, to Johann Georg Volckamer, May 28, 1681, in Briefsammlung Trew. Accessed through the database of Harald Fischer Verlag. www.haraldfischerverlag.de/hfv /sammlungen/trew_engl.php.

Primary Sources

Acta Eruditorum: Anno MDCLXXXII Publicata [. . .]. Leipzig: Grossius; Gletitschius; Güntherus, 1682.

Aldrovandi, Ulisse. *Discorso naturale di Ulisse Aldrovandi. Nel quale si tratta in generale del suo Museo, e delle fatiche da lui usate per raunare de varie parti del mondo, quasi in un Theatro di Natura tutte le cose sublunari, come piante, animali et altre cose minerali. Et parimente vi s' insegna come si de' venir nella certa et necessaria cognitione d' alcuni medicamenti incerti et dubbij, ad utilità grandissima non solo de' medici, ma d'ogni altro studioso* [. . .]. 1572–1573. www.filosofia.unibo.it/aldrovandi/pinakesweb/UlisseAldrovandi_discorsonaturale.asp.

————. *Quadrupedibus Solidipedibus Volumen Integrum Ioannes Cornelius Uterverius . . . Collegit, & Recensuit: Hieronymus Tamburinus in Lucem Edidit* [. . .]. Bologna: apud Victorium Benatium (sumpt. Tamburini), 1616.

————. *Quadrupedum Omnium Bisulcorum Historia. Ioannes Cornelius Uterverius Belga Colligere Incaepit. Thomas Dempsterus Baro a Muresk Scotus i.c. Perfecte Absoluit. Hieronymus Tamburinus in Lucem Edidit* [. . .]. Bologna: apud Sebastianum Bonhommium (impensis Hieronymi Tamburini), 1621.

————. *De Quadrupedibus Solidipedibus Volumen Integrum. Joannes Cornelius Uterverius . . . Collegit, & recensuit. Hieron. Tamburinus in Lucem Edidit* [. . .]. Frankfurt: Typis Ioan. Hoferi, impensis Ioannis Treudel, 1623.

————. *Serpentum, et Draconum Historiae Libri Duo Bartholomaeus Ambrosinus . . . Concinnavit* [. . .]. Bologna: sumptibus M. Antonij Berniae bibliopolae Bononiensis: apud Clementem Ferronium, 1640.

————. *De Quadrupedibus Solidipedibus Volumen Integrum. Ioannes Cornelius Uterverius . . . Collegit, & Recensuit. Marcus Antonius Bernia in Lucem Restituit* [. . .]. Bologna: apud Nicolaum Thebaldinum, 1649.

————. *Monstrorum Historia* [. . .].. Bologna: Typis Io(hannis) Baptistae Ferronij, 1658.

————. "Testamento di Ulisse Aldrovandi." In *Memorie della vita di Ulisse Aldrovandi Medico e Filosofo Bolognese, Con alcune Lettere scelte d'Uomini eruditi a lui scritte, e coll'Indice delle sue Opere Mss., che si conservano nella Bilblioteca dell'Istituto* [. . .]. Edited by Giovanni Fantuzzi, 67–85. Bologna: Lelio della Volpe, 1774 [1603].

————. *Monstrorum Historia: Cum Paralipomenis Historiæ Omnium Animalium* [. . .]. Theatrum Sapientiae. Textes 4. Paris: Société d'édition Les Belles Lettres; Turin: Société d'édition Les Belles Lettres, Nino Aragno Editore, 2002 [1642].

Allen, Thomas. "An Exact Narrative of an Hermaphrodite now in London." *Philosophical Transactions of the Royal Society of London* 32 (1667): 624.

Altmann, Johann Georg. *L'état et les délices de la Suisse, en forme de relation critique, par plusieurs auteurs célèbres* [. . .]. Le mouvement des idées au XVIIe siècle 3. 4 vols. Vol. 1. Amsterdam: Chez les Wetsteins et Smiths, 1730 [1714].

Arnaud, Antoine, and Pierre Nicole. *La logique ou l'art de penser: Contenant, outre les règles communes, plusieurs observations nouv., propres à former le jugement.* Critical ed. Edited by Pierre Clair et François Girba. Le mouvement des idées au XVIIe siècle 3. Paris: Presses Universitaires de France, 1965 [1662].

"An authentick Account of the surprising CENTAUR, the greatest Wonder produced by Nature these 3000 Years, lately proposed to be exhibited to public View, &c." *Gentleman's Magazine* 21 (1751): 153–54.

Bacon, Francis. *Sylva sylvarum. Or, a natural history in ten centuries. Whereupon is newly added the history natural and experimental of life and death, or of the prolongation of life. Published after the authors death, by William Rawley* [. . .]. London: Printed by J.R. for William Lee, 1670.

————. "Historia Ventorum." In *The Works of Francis Bacon, Lord Chancellor of England: With a Life of the Author*, edited by Basil Montagu, 13–97. London: Pickering, 1828.

————. "Historia Vitae et Mortis." In *The Works of Francis Bacon, Lord Chancellor of England: With a Life of the Author*, edited by Basil Montagu, 107–281. London: Pickering, 1828.

————. "Parasceve ad Historiam Naturalem, et Experimentalem." In *The Works of Francis Bacon, Lord Chancellor of England: With a Life of the Author*, edited by Basil Montagu, 405–25. London: Pickering, 1829.

————. *The Advancement of Learning.* Edited with an introduction by G. W. Kitchin. London: Dent, 1973.

———. *Novum Organum*. Translated by Graham Rees. In *The Instauratio Magna Part II: Novum Organum and Associated Texts*, edited by Graham Rees with Maria Wakely, 48–447. Oxford: Clarendon Press, 2004 [1620].

Bartholin, Thomas. *Acta Medica & Philosophica Hafniensia* [. . .]. Copenhagen: Sumptibus Petri Haubold Acad. Bibl. Typis Georgii Gödiani, Typogr. Reg., 1673.

Bate, Walter Jackson, ed. *The Rambler (The Third of Three Volumes)*. Vol. 5 of *The Yale Edition of the Works of Samuel Johnson*. 23 vols. New Haven, CT: Yale University Press, 1969.

Batman, Stephen. *The Doome Warning All Men to the Judgement* [. . .]. London: Ralphe Nubery, 1581.

Bauhin, Gaspard. *Institutiones Anatomicæ Corporis Virilis et Mulieribus Historiam Exhibentes* [. . .]. Basel: Apud Joann. Schröter, 1609.

———. *De Hermaphroditorum Monstrosorumque Partuum Natura ex Theologorum, Jureconsultorum, Medicorum, Philosophorum, & Rabbinorum Sententia Libri Duo* [. . .]. Oppenheim: Johan-Theodori de Bry, 1614.

———. *De Hermaphroditorum Monstrosorumque Partuum Natura: Ex Theologorum, Jureconsultorum, Medicorum, Philosophorum Et Rabbinorum Sententia: Libri Duo* [. . .]. Frankfurt: Apud Matthaeum Merianum, 1629.

Bausch, Johann Laurentius. "Epistola invitatoria and draft statutes of 1652." In Frances Mason Barnett, "Medical Authority and Princely Patronage: The Academia Naturae Curiosorum, 1652–1693," 343–44. PhD diss., University of North Carolina, 1995.

Bayle, Pierre. *Pensées diverses sur le comète*. Critical edition with an introduction and notes published by A. Prat. 2nd ed. 2 vols. Vol. 1. Paris: Droz, 1939 [1681].

"Benevolo Lectori Salutem & Officia." *Miscellanea Curiosa Medico-Physica Academiae Naturae Curiosorum sive Ephemeridum Medico-Physicarum Germanicarum Curiosarum* [. . .]. Annus Primus (1670): 341–44.

Bentley, Richard. *A True and Faithful Account of the Greatest Wonder Produced by Nature in these 3000 Years, in the Person of Mr. Jehan-Paul-Ernest Christian Lodovick Manpferdt; the Surprising Centaur, who Will Be Exhibited to the Publick, on the First of Next Month, at the Sign of the Golden Cross at Charing Cross*. London: Printed for M. Cooper, in Pater-Noster-Row, 1751.

Bernitz, Martin Bernhard von. "Observatio LII: Muscus Terrestris Repens Monstrosus." *Miscellanea Curiosa Medico-Physica Academiae Naturae Curiosorum sive Ephemeridum Medico-Physicarum Germanicarum Curiosarum* [. . .]. Annus Secundus (1671): 93.

Boaistuau, Pierre. *Histoires prodigieuses les plus memorables qui ayent esté observées, depuis la nativité de Jesus Christ, iusques a nostre siecle: Extraictes de plusieurs fameux autheurs, grecz, & latins, sacrez & prophanes* [. . .]. Paris: Pour Vincent Sertenas (par A. Brière), 1560.

"Books publish'd March 1751." *Gentleman's Magazine* 21 (1751): 142–43.

Browne, Thomas. *Pseudodoxia Epidemica; Or, Enquiries into Very Many Received Tenents and Commonly Presumed Truths*. Edited by Robin Robbins. 2 vols. Vol. 1, Oxford: Clarendon Press, 1981 [1646].

"A Catalogue of Books published in January, February, and March." *Royal Magazine; or, Quarterly Bee* 2 (January, February, and March 1751): 464–67.

"Classis IV. Von Eintzelnen Natur-Geschichten des Julii 1717. Artic. X. Historia eines Monstri." *Sammlung Von Natur-und Medicin- Wie auch hierzu gehörigen Kunst-und Literatur-Geschichten, So sich An. 1717. in den 3. Sommer-Monaten In Schlesien und andern Ländern begeben* [. . .]. (1718): 95–97.

Coberus, Tobias. *Observationum Castrensium et Ungaricum, Decas* [. . .]. Frankfurt: Palthenius, 1606.

Columbo, Realdo. *De Re Anatomica* [. . .]. Venice: Ex Typographia Nicolai Beulacquæ, 1559.

de Beer, Esmond Samuel, ed. *The Diary of John Evelyn: Now First Printed in Full from the Manuscripts Belonging to Mr. John Evelyn.* 6 vols. Vol. 2. Oxford: Clarendon Press, 1955.

———, ed. *The Diary of John Evelyn: Now First Printed in Full from the Manuscripts Belonging to Mr. John Evelyn.* 6 vols. Vol. 3. Oxford: Clarendon Press, 1955.

———, ed. *The Diary of John Evelyn: Now First Printed in Full from the Manuscripts Belonging to Mr. John Evelyn.* 6 vols. Vol. 5. Oxford: Clarendon Press, 1955.

"Dedicatory Epistle to Leopold I." *Miscellanea Curiosa Medico-Physica Academiae Naturae Curiosorum sive Ephemeridum Medico-Physicarum Germanicarum Curiosarum* [. . .]. Annus Primus (1670): fol. a4r–b2v.

de Graaf, Regnier. "Observatio CXXVIII: Monstrosus Uterus." *Miscellanea Curiosa Medico-Physica Academiae Naturae Curiosorum sive Ephemeridum Medico-Physicarum Germanicarum Curiosarum* [. . .]. Annus Primus (1670): 286–88.

de Heer, Henri. *Observationes Medicae* [. . .]. Liege: à Corswaremia, 1630.

———. *Observationes Medicae Oppido Rarae* [. . .]. Leipzig: Timotheus Hön; Andreas Kühne, 1645.

Duval, Jacques. *Traité des hermaphrodits, parties génitales, accouchements des femmes,* [. . .]. Rouen: Geuffroy, 1612.

———. *Responce au discours fait par le sieur Riolan, . . . contre l'histoire de l'hermaphrodit de Rouen.* Rouen: J. Courant, n.d.

"Einbildung." In *Grosses vollständiges Universallexicon Aller Wissenschaften und Künste* [. . .]. Edited by Johann Heinrich Zedler, col. 533. Halle: Johann Heinrich Zedler, 1734.

"Einbildungs-Krafft." In *Grosses vollständiges Universallexicon Aller Wissenschaften und Künste* [. . .]. Edited by Johann Heinrich Zedler, cols. 533–38. Halle: Johann Heinrich Zedler, 1734.

"Epistola Invitatoria ad Celeberrimos Europæ Medicos: Viri Magnifici, Amplißimi, Nobilißimi, Excellentißimi, Experientißimi Medicinæ Antistites Scrutatores Naturalium Arcanorum Solertißimi." *Miscellanea Curiosa Medico-Physica Academiae Naturae Curiosorum sive Ephemeridum Medico-Physicarum Germanicarum Curiosarum* [. . .]. Annus Primus (1670): 1–8.

Fabricius Hildanus, Wilhelm. *Observationum & Curationum Chirurgicarum Centuriae* [. . .]. Basel: Regis, 1606.

Fehr, Johann Michael, and Georg Hieronymus Welsch. *Epistolae Mutuae Argonautae ad Nestorem, et Nestoris ad Argonautam, de Thesauro Experientiae Medicae* [. . .]. Augsburg: Schönigkius, 1667.

Fienus, Thomas. *De Viribus Imaginationis Tractatus* [. . .]. Louvain: Rivius, 1608.

Fincel, Hiob. *Wunderzeichen: Warhafftige beschreibung und gründlich verzeichnus schrecklicher Wunderzeichen und Geschichten* [. . .]. 3 vols. Vol. 1. Jena: Christian Rödinger, 1556.

———. *Wunderzeichen: Warhafftige beschreibung und gründlich verzeichnus schrecklicher Wunderzeichen und Geschichten* [. . .]. 3 vols. Vol. 2. Leipzig: J. Berwald, 1559.

———. *Wunderzeichen: Warhafftige beschreibung und gründlich verzeichnus schrecklicher Wunderzeichen und Geschichten* [. . .]. 3 vols. Vol. 3. Jena: Richtzenhain & Rebart, 1562.

Fontenelle, Bernard de. *Histoire du renouvellement de l'Académie Royale des Sciences en M.DC.XCIX. et les éloges historiques.* Amsterdam: Pierre de Coup, 1709.

———. *De l'origine des fables.* Critical edition with with introduction, notes, and commentary by J.-R. Carré. Textes et traductions pour servir a l'histoire de la pensée moderne. Edited by Abel Rey. Paris: Librairie Felix Alcan, 1932 [1724].

Foxe, John. *Pandectae Locorum Communium Præcipua Rerum Capita & Titulos, Iuxta Ordinem Elementorum Complectentes.* London: John Day, 1572.

Garliep, Gustav Casimir. "Observatio XX: De Conyza Monstrosa, Laticauli, Cristata." *Miscellanea Curiosa sive Ephemeridum Medico-Physicarum Germanicarum Academiae Caesareo-Leopoldinae Naturae Curiosorum* [. . .]. Decuriae II. Annus Octavus (1690): 65–68.

———. "Observatio IV: De Osse Monstroso ex Cantho Sinistro Oculi Dextri Protuberante." *Miscellanea Curiosa sive Ephemeridum Medico-Physicarum Germanicarum Academiae Caesareo-Leopoldinae Naturae Curiosorum* [. . .]. Decuriae III. Anni Quintus et Sextus (1700): 4–8.

"Gerücht." In *Grosses vollständiges Universallexicon Aller Wissenschaften und Künste* [. . .]. Edited by Johann Heinrich Zedler, cols. 1206–7. Halle, Leipzig: Johann Heinrich Zedler, 1735.

Gessner, Konrad. *Pandectarum sive Partitionum Universalium Libri XXI.* Zurich: Froschouerus, 1548.

———. *De Quadrupedibus Oviparis. Adiectae Sunt Novae Aliquot Quadrupedum Figurae, in Primo Libro de Quadrupedibus Viviparis Desideratae, cum Descriptionibus Plerorunque Brevissimis: Item Oviparorum Quorundam Appendix.* Zurich: Froschouerus, 1554.

Greisel, Johann Georg. "Observatio LV: Anatome Monstri Gemellorum Humanorum." *Miscellanea Curiosa Medico-Physica Academiae Naturae Curiosorum sive Ephemeridum Medico-Physicarum Germanicarum Curiosarum* [. . .]. Annus Primus (1670): 152.

———. "Observatio LVI: Ren Succenturiatus Monstrosus, cum Ulcere." *Miscellanea Curiosa Medico-Physica Academiae Naturae Curiosorum sive Ephemeridum Medico-Physicarum Germanicarum Curiosarum* [. . .]. Annus Primus (1670): 152–53.

Hall, A. Rupert, and Marie Boas Hall, eds. *The Correspondence of Henry Oldenburg.* 13 vols. Vol. 6. Madison: University of Wisconsin Press, 1967.

———, eds. *The Correspondence of Henry Oldenburg.* 13 vols. Vol. 7. Madison: University of Wisconsin Press, 1970.

———, eds. *The Correspondence of Henry Oldenburg.* 13 vols. Vol. 8. Madison: University of Wisconsin Press, 1971.

———, eds. *The Correspondence of Henry Oldenburg.* 13 vols. Vol. 9. Madison: University of Wisconsin Press, 1973.

Haller, Albrecht von. *Elementa Physiologiæ Corporis Humani.* 8 vols. Vol. 1. Lausanne: Sumptibus Marci-Michael Bousquet & Sociorum, 1757.

———. *Operum Anatomici Argumenti Minorum Tomus Tertius Accedunt Opuscula Pathologica Aucta et Recensa.* Lausanne: Sumptibus Francisci Grasset & Socior., 1768.

———. *Bibliotheca Botanica* [. . .]. 2 vols. Zurich: Apud Orell, Gessner, Fuessli, et Socc., 1771–1772.

———. *Bibliotheca Chirurgica* [. . .]. 2 vols. Bern: Em. Haller und Joh. Schweighauser, 1774–1775.

———. *Bibliotheca Anatomica* [. . .]. 2 vols. Zurich: Apud Orell, Gessner, Fuessli, et Socc., 1774–1777.

———. *Vorlesungen über die gerichtliche Arzneiwissenschaft: Aus einer nachgelassenen lateinischen Handschrift übersetzt.* 2 vols. Vol. 1. Bern: bey der neuen typographischen Gesellschaft, 1782.

———. *Albrecht Hallers Tagebuch seiner Studienreise nach London, Paris, Straßburg und Basel, 1727–1728: Herausgegeben von Erich Hitzsche.* Berner Beiträge zur Geschichte der Medizin und der Naturwissenschaften, Neue Folge 2. 2nd ed. Bern: Huber, 1968.

Hartmann, Philipp Jakob. "Observatio LXXVI: Anatome Monstrosi Crediti Fœtûs." *Miscellanea Curiosa sive Ephemeridum Medico-Physicarum Germanicarum Academiae Caesareo-Leopoldinae Naturae Curiosorum* [. . .]. Decuriae II. Annus Quintus (1687): 176–79.

Heseler, Baldasar. *Andreas Vesalius' First Public Anatomy at Bologna 1540: An Eyewitness Report. Together with his Notes on Matthaeus Curtius' Lectures on anatomia mundini.* Lychnos-Bibliotek 18. Uppsala: Almquist & Wiksells, 1959.

Heunisch, Caspar. *Panacea Apostolica, Das ist/ Eine Christliche Leich-Predigt/ über den schönen Spruch deß heiligen Apostels Pauli I. Timoth.1/15 . . . Bey sehr volckreichen . . . Leichbegängniß Deß . . . Herrn Johan. Laurentii Bauschen/ Der Artzney vornehmen . . . Doctoris der . . . Stadt*

Schweinfurt . . . Welcher Sonnabends den 18. Novembris . . . 1665 . . . von dieser Welt selig abgeschieden [. . .]. Nuremberg: Endter, 1665.

Histoire de l'Académie Royale des Sciences: 1666 à 1698. Paris: Chez Panckoucke, Hôtel de Thou, rue de Poitevins, 1777.

"Imaginatio." In *Grosses vollständiges Universallexicon Aller Wissenschaften und Künste* [. . .]. Edited by Johann Heinrich Zedler, cols. 571–72. Halle: Johann Heinrich Zedler, 1739.

"Imagination des femmes enceintes sur le fœtus, pouvoir de l'." In *Encyclopédie, ou Dictionnaire raisonné des sciences, des arts et des métiers,* edited by Denis Diderot and Jean le Rond d'Alembert, 563–64. Paris: Samuel Faulche, 1765.

"In der Arzneywissenschaft." *Göttingische Zeitungen von gelehrten Sachen* 27 (March 18, 1751): 212.

Jänisch, Johannes. "Observatio CII: Buglossum Silvestre Monstrosum." *Miscellanea Curiosa Medico-Physica Academiae Naturae Curiosorum sive Ephemeridum Medico-Physicarum Germanicarum Curiosarum* [. . .]. Annus Primus (1670): 233–35.

Le journal des sçavans du lundy 10. fevrier M. DC. LXX. Paris: Chez Jean Cvsson, ruë S. Jacques, à l'Image de S. Jean-Baptiste, 1670.

Jung, Georg Sebastian. "Observatio CXV: Lapis Bezoar Monstrosus." *Miscellanea Curiosa Medico-Physica Academiae Naturae Curiosorum sive Ephemeridum Medico-Physicarum Germanicarum Curiosarum* [. . .]. Annus Primus (1670): 267–68.

Kircher, Athanasius. *Mundi Subterranei Tomi II.* [. . .]. Amsterdam: Janssonius & Wyerstrat, 1664.

Leibniz, Gottfried Wilhelm Freiherr von. "Bedenken von Aufrichtung einer Akademie oder Societät in Teutschland, zu Aufnehmen der Künste und Wissenschafften." In *Sämtliche Schriften und Briefe.* Part 4, vol. 1: *1667–1676,* edited by Zentralinstitut für Philosophie an der Akademie der Wissenschaften der DDR, 543–52. Berlin: Akademie-Verlag, 1983.

Liceti, Fortunio. *De Monstrorum Natura, Caussis, et Differentiis* [. . .]. Padua: Crivellarius, 1616.

———. *De Monstrorum Natura, Caussis, et Differentiis Libri Duo* [. . .]. Padua: Frambottus, 1634.

Locke, John. *An Essay concerning Human Understanding.* Collated and annotated, with prolegomena, biographical, critical, and historical by Alexander Campbell Fraser. 2nd ed. 2 vols. Vol. 2, New York: Dover, 1959 [1689].

Ludovici, Jacob Friedrich. *Usus Practicus Distinctionum Iuridicarum Iuxta Seriem Digestorum Adornatus* [. . .]. 3 vols. Halle: Schütz, 1603 [i.e., 1703].

Lusitanus, Amatus. *Curationum Medicinalium Centuriae Septem.* Bordeaux: Ex Typographia Gilberti Vernoy, 1620.

Lycosthenes, Conrad. *Prodigiorum ac Ostentorum Chronicon* [. . .]. Basel: per Henricum Petri, 1557.

———. *Wunderwerck oder Gottes unergründtliches vorbilden* [. . .]. Translated by Johannes Herold. Basel: Heinrich Petri, 1557.

Mandeville, Jean de. *Das buch des ritters herr hannsen von monte villa.* Translated by Michel Velser. Augsburg: Sorg, 1481.

Marlowe, Christopher. *Tamburlaine the Great: Who, from a Scythian shephearde, by his rare and woonderfull conquests, became a most puissant and mightye monarque. And (for his tyranny, and terrour in warre) was tearmed, the scourge of God* [. . .]. London: Printed by Richard Ihones: at the signe of the Rose and Crowne neere Holborne Bridge, 1590.

Micraelius, Johannes. *Lexicon Philosophicum Terminorum Philosophis Usitatorum* [. . .]. Stetini: Jeremiæ Mamphrasii, Michaelis Höpfneri, 1662 [1653].

Miscellanea Curiosa Medico-Physica Academiae Naturae Curiosorum sive Ephemeridum Medico-Physicarum Germanicarum Curiosarum Annus Primus [. . .]. Leipzig: Trescherus / Bauerus, 1670.

"Mißgeburt." In *Grosses vollständiges Universallexicon Aller Wissenschaften und Künste* [. . .]. Edited by Johann Heinrich Zedler, 486–92. Halle: Johann Heinrich Zedler, 1739.

Mollerus, Jacobus. *Discursus de Cornutis et Hermaphroditis Eorumque Jure* [. . .]. 3rd ed. Berlin: Johann Wilhelm Meyer, 1708 [1699].

Montuus, Hieronymus. *De Medica Theoresi Liber Primus* [. . .]. Lyon: Tornaesius Et Gazeius, 1556.

———. *Opuscula Juvenilia, D. Hieronymi Montui* [. . .]. Lyon: apud Joan. Tornaesium, et Guliel. Gazeium, 1556.

Müller, Uwe, Danny Weber, and Wieland Berg, eds. *Protocollum Academiae Caesareo-Leopoldinae Naturae Curiosorum. Edition der Chronik der Kaiserlich-Leopoldinischen Akademie.* Acta historica Leopoldina 60. Stuttgart: Wiss. Verlagsgesellschaft, 2013.

Mussatus, Albertinus. *Historia Augusta Henrici VII. Caesaris, et Alia, Quae Extant Opera. Laurentii Pignorii Spicilegio, necnon Felicis Osij et Nicolai Villani Castigationibus, Collationibus et Notis Illustrata* [. . .]. Venice: Typ. Pinelliana 1636.

Nichols, John. *Literary Anecdotes of the Eighteenth Century Comprizing Biographical Memoirs of William Bowyer Printer and Many of His Learned Friends, . . . and Biographical Anecdotes.* 9 vols. Vol. 5. London: Nichols et al., 1812.

Paré, Ambroise. *Des monstres et prodiges.* Critical ed. Edited with commentary by Jean Céard. Travaux d'humanisme et renaissance 115. Geneva: Droz, 1971 [1573].

Parsons, James. *A mechanical and critical inquiry into the nature of hermaphrodites.* London: Printed for J. Walthoe, over-against the Royal-Exchange in Cornhill, 1741.

Peacham, Henry. *The compleat gentleman: Fashioning him absolute in the most necessary & commendable qualities concerning minde or bodie that may be required in a noble gentlema[n].* London: Imprinted at London [by John Legat] for Francis Constable, and are to bee sold at his shop at the White Lio[n] in Paules churchyard, 1622.

Perondino, Pietro. *Magni Tamerlanis Scytharum Imperatoris Vita* [. . .]. Florence: s.n., 1553.

Perrault, Claude. *Memoires pour servir à l'histoire des animaux.* Paris: Imprimerie Royale, 1671.

Philosophical Transactions 5 (1670). Repr. New York: Johnson Reprint / Kraus Reprint, 1963.

Pictorius, Georg. *Sermonum Convivalium Libri X.* [. . .]. Basel: Petri, 1559.

———. *Sermonum Convivalium Apprimè Utilium, Libri X.* [. . .]. Basel: Henricpetrus, 1571.

Plinius Secundus, Gaius. *Natural History in Ten Volumes.* Books III–VII. Translated by H. Rackham. Loeb Classical Library 352. Edited by G. P. Goold. 4th ed. 10 vols. Vol. 2. Cambridge, MA: Harvard University Press, 1969.

Plutarch. *Plutarch's Moralia.* Translated by Frank Cole Babbitt. Loeb Classical Library 222. 16 vols. Vol. 2. Cambridge, MA: Harvard University Press / Putnam, Heinemann, 1971.

"Postloquium." *Miscellanea Curiosa, sive Ephemeridum Medico-Physicarum Germanicarum Academiae Naturae Curiosorum* [. . .]. Anni Nonus et Decimus (1680): 327–37.

Praetorius, Johannes. *Philosophia Colus oder Pfy/ lose vieh der Weiber: Darinnen gleich hundert allerhand gewöhnliche Aberglauben des gemeinen Mannes lächerig wahr gemachet werden . . ./ auffgesetzet durch MIciPSaM: Regem Numidiae.* Leipzig: Oehler, Freyschmied, 1662.

Rayger, Karl. "Observatio VII: Anatomia Monstri Bicipitis." *Miscellanea Curiosa Medico-Physica Academiae Naturae Curiosorum sive Ephemeridum Medico-Physicarum Germanicarum Curiosarum* [. . .]. Annus Primus (1670): 25–7.

Riolan, Jean. *Discours sur les hermaphrodites: Où il est demonstré contre l'opinion commune, qu'il n'y a point de vrays Hermaphrodites.* Paris: Pierre Ramier, 1614.

Rueff, Jacob. *De Conceptu et Generatione Hominis, et Iis Quae Circa Haec Potissimum Consyderantur Libri Sex* [. . .]. Zurich: C. Froschouerus, 1554.

———. *Ein schön lustig Trostbüchle von den empfengnussen vnd geburten der menschen* [. . .]. Zürych: Froschouer, 1554.

Sachs von Lewenhaimb, Philipp Jacob. *Ampelographia sive Vitis Viniferae Eiusque Partium Consideratio Physico-Philologico-Historico-Medico-Chymica: In Qua Tam de Vite in Genere, Quam in Specie de Eius Pampinis, Flore, Lachryma, Sarmentis, Fructu, Vini Multivario Usu, de Spiritu Vini, Aceto, Vini Fæce, & Tartaro, Curiosa Notata Plurima ad Normam Collegii Naturae Curiosorum Instituta Plurimis Jucundis Secretiis Naturae, Artisque Locupletata* [. . .]. Leipzig: Trescher 1661.

———. "Observatio CXVI: Radix Miranda Crucifixum cum Duabus Icunculis Exhibens." *Miscellanea Curiosa Medico-Physica Academiae Naturae Curiosorum sive Ephemeridum Medico-Physicarum Germanicarum Curiosarum* [. . .]. Annus Primus (1670): 268–72.

———. "Observatio XLVIII: Rapa Monstrosa Anthropomorpha." *Miscellanea Curiosa Medico-Physica Academiae Naturae Curiosorum sive Ephemeridum Medico-Physicarum Germanicarum Curiosarum* [. . .]. Annus Primus (1670): 139–44.

Schedel, Hartmann. *Register des Buchs der Croniken und Geschichten mit Figuren und Pildnussen von Anbeginn der Welt bis auf diese unnsere Zeit.* Munich: Kölbl, 1975 [1493].

Schenck von Grafenberg, Johann Georg. *Monstrorum Historia Memorabilis* [. . .]. Frankfurt: de Bry, Beckerus, 1609.

———. *Wunder-Buch von Menschlichen unerhörten Wunder- und Mißgebuhrten so wider den gemeinen Lauff der Natur erschröcklich frembd unnd seltzam gebildet* [. . .]. Frankfurt: bey Matthis Beckern: In Verlegung Dietrichs von Bry seeligen Wittib, sampt zweyer ihrer Söhnen, 1610.

Schenck von Grafenberg, Johannes. *Observationes Medicæ de Capite Humano* [. . .]. Basel: ex officina Frobeniana, 1584.

———. *Observationum Medicarum, Rararum, Novarum, Admirabilium et Monstrosarum, Liber Quartus, de Partibus Genitalibus Utriusque Sexûs* [. . .]. Freiburg im Breisgau: Martinus Böcklerus 1596.

Schmid, Johannes. "Observatio CLXIII: De Monstro Vitulino." *Miscellanea Curiosa Medico-Physica Academiae Naturae Curiosorum, sive Ephemeridum Medico-Physicarum Germanicarum* [. . .]. Anni Quartus et Quintus (1676): 207–8.

Schmidel, Ulrich. *Vera Historia, Admirandæ Cuiusdam Navigationis, Quam Huldericus Schmidel, Straubingensis, ab Anno 1534. usque ad Annum 1554. in Americam vel Novum Mundum, Iuxta Brasiliam & Rio della Plata, Confecit: Quid per Hosce Annos 19. Sustinuerit, Quam Varias & Quam Mirandas Regiones ac Homines Viderit* [. . .]. Nuremberg: Hulsius, 1599.

Schott, Gaspar. *Physica Curiosa, sive Mirabilia Naturae et Artis Libris XII. Comprehensa* [. . .]. 2 vols. Würzburg: Hertz, Endter, 1662.

———. *Physica Curiosa, sive Mirabilia Naturae et Artis Libris 12 Comprehensa* [. . .]. 2 vols. Nuremberg: Endter, Hertz, 1667.

———. *Physica Curiosa, sive Mirabilia Naturae et Artis Libris XII. Comprehensa* [. . .]. Würzburg: Hertz, Endter, 1697.

———. *Physica Curiosa, sive Mirabilia Naturae et Artis Libri XII.* [. . .]. 3. ed. Würzburg: Jobus Hertz, 1697 [1662].

Shaw, Thomas. *Travels, or Observations Relating to Several Parts of Barbary and the Levant.* Oxford: printed at the theatre, 1738.

Sinibaldi, Giovanni Benedetto. *Geneanthropeiae sive de Hominis Generatione Decateuchon, . . . Adjecta Est Historia Foetus Musipontani* [. . .]. Rome: Es Typographia Francisci Caballi, 1642.

Smart, Christopher. *An occasional prologue and epilogue to Othello, as it was acted at the Theatre-Royal in Drury-Lane . . . 7th of March, 1751, by persons of distinction for their diversion.* London: Printed for the author, 1751.

Teichmeyer, Hermann Friedrich. *Institutiones Medicinae Legalis vel Forensis* [. . .]. Jena: Bielckius, 1723.

———. *Institutiones Medicinae Legalis vel Forensis* [. . .]. Jena: sumptibus Ioh. Felicis Bielckii, 1740.

Valentini, Michael Bernhard. "Observatio XC: De Monstris Hassiacis Recens Natis." *Miscellanea Curiosa sive Ephemeridum Medico-Physicarum Germanicarum Academiae Naturae Curiosorum* [. . .]. Decuriae II. Annus Tertius (1685): 190–92.

Varchi, Benedetto. *La prima parte delle lezzioni di M. Benedetto Varchi nella quale si tratta della natura, della generazione del corpo humano, e de' mostri: Lette da lui publicamente nella accademia fiorentina.* Florence: appresso i Giunti, 1560.

Villey, Pierre, ed. *Les essais de Michel Seigneur de Montagne.* 3 vols. Vol. 2. Lyon: Pour François le Febure, 1595.

Vollgnad, Heinrich. "Observatio CCXIX: De Monstroso Foetu." *Miscellanea Curiosa Medico-Physica Academiae Naturae Curiosorum, sive Ephemeridum Medico-Physicarum Germanicarum* [. . .]. Annus Tertius (1673): 518–19.

Voltaire. "Imagination, Imaginer." In *Encyclopédie, ou Dictionnaire raisonné des sciences, des arts et des métiers,* edited by Denis Diderot and Jean le Rond d'Alembert, 560–63. Paris: Samuel Faulche, 1765.

"Vorbericht." *Sammlung Von Natur- und Medicin- Wie auch hierzu gehörigen Kunst- und Literatur-Geschichten, So sich An. 1717. in den 3. Sommer-Monaten In Schlesien und andern Ländern begeben* [. . .]., fol. a3r–b4v. 1718.

Waller, Robert, ed. *The Posthumous Works of Robert Hooke.* London: S. Smith and B. Walford, 1705.

Wedel, Georg Wolfgang. "Observatio CCXXVIII: De Basilico." *Miscellanea Curiosa Medico-Physica Academiae Naturae Curiosorum, sive Ephemeridum Medico-Physicarum Germanicarum* [. . .]. Annus Tertius (1673): 204–5.

———. "Progressus Academiae Naturae Curiosorum [. . .]." *Miscellanea Curiosa Medico-Physica Academiae Naturae Curiosorum, sive Ephemeridum Medico-Physicarum Germanicarum* [. . .]. Decuria II. Annus Primus (1683): cr–er.

Weinrich, Martin. *De Ortu Monstrorum Commentarius, in Quo Essentia, Differentiae, Causae et Affectiones Mirabilium Animalum Explicantur.* [Leipzig]: Heinricus Osthusius, 1595.

Welsch, Georg Hieronymus. *Sylloge curationum et observationum medicinalium centurias VI. complectens* [. . .]. Augsburg: Goebelius; Kuhnius, 1668.

Wolff, Christian. *Gedanken über das ungewöhnliche Phœnomenon, Welches den 17. Martii 1716, des Abends nach 7. Uhren zu Halle und an vielen andern Orten in und ausserhalb Deutschland gesehen worden / Wie er sie den 24. Martii in einer LECTIONE PUBLICA Auf der Universität zu Halle eröffnet.* Halle: n.p., 1716.

Wurfbain, Johannes Paul. "Observatio CCVI: De Folio Lactucæ Monstroso." *Miscellanea Curiosa sive Ephemeridum Medico-Physicarum Germanicarum Academiae Imperialis Leopoldinae Naturae Curiosorum* [. . .]. Decuriae II. Annus Decimus (1692): 411.

Secondary Sources

Adam, Wolfgang. *Poetische und kritische Wälder: Untersuchungen zu Geschichte und Formen des Schreibens "bei Gelegenheit."* Euphorion: Beihefte 5. Heidelberg: Winter, 1988.

Altick, Richard Daniel. *The Shows of London.* London: Belknap Press of Harvard University Press, 1978.

Antonelli, Ezio. "Ulisse Aldrovandi e la metamorfosi del mostruoso." *Studi e memorie per la storia dell'Università di Bologna* 3 (1983): 196–242.

Antonino, Biancastella, ed. *Animali e creature monstruose di Ulisse Aldrovandi.* Milano: Federico Motta, 2004.

Ashworth, William B., Jr. "Emblematic Natural History of the Renaissance." In *Cultures of Natural History*, edited by Nicholas Jardine, James A. Secord, and Emma C. Spary, 17–37. Cambridge: Cambridge University Press, 1996.

Austin, Greta. "Marvelous Peoples or Marvelous Races? Race and the Anglo-Saxon *Wonders of the East*." In *Marvels, Monsters, and Miracles: Studies in the Medieval and Early Modern Imaginations*, edited by Timothy S. Jones and David A. Sprunger, 25–51. Studies in Medieval Culture 42. Kalamazoo: Medieval Institute Publications Western Michigan University, 2002.

Bacchi, Maria Cristina. "Ulisse Aldrovandi e i suoi libri." *L'Archiginnasio* 100 (2005): 255–366.

Barnes, Robin B. *Prophecy and Gnosis: Apocalypticism in the Wake of the Lutheran Reformation.* Stanford, CA: Stanford University Press, 1988.

Barnett, Frances Mason. "Medical Authority and Princely Patronage: The Academia Naturae Curiosorum, 1652–1693." PhD diss., University of North Carolina, 1995.

Baroncini, Gabriele. *Forme di esperienza e rivoluzione scientifica.* Bibliotheca di Nuncius, Studi e testi 9. Florence: Leo S. Olschki, 1992.

Bates, Alan W. *Emblematic Monsters: Unnatural Conceptions and Deformed Births in Early Modern Europe.* Amsterdam: Rodopi, 2005.

Bäumer-Schleinkofer, Änne. "Ulisse Aldrovandi: Vollendung des Aristoteles in plinianischer Manier." *Berichte zur Wissenschaftsgeschichte* 17 (1994): 183–99.

Behrens, Rudolf, and Carsten Zelle, eds. *Der ärztliche Fallbericht: Epistemische Grundlagen und textuelle Strukturen dargestellter Beobachtung.* Culturae 6. Wiesbaden: Harassowitz Verlag, 2012.

Benedict, Barbara M. *Curiosity: A Cultural History of Early Modern Inquiry.* Chicago: University of Chicago Press, 2001.

Berg, Wieland. "Die frühen Schriften der Leopoldina—Spiegel zeitgenössischer 'Medizin und ihrer Anverwandten.'" *NTM* 22 (1985): 67–76.

Berg, Wieland, and Benno Parthier. "Die 'kaiserliche' Leopoldina im Heiligen Römischen Reich Deutscher Nation." In *Gelehrte Gesellschaften im mitteldeutschen Raum (1650–1820)*, edited by Detlef Döring and Kurt Nowak, 39–52. Stuttgart: Hirzel, 2000.

Berg, Wieland, and Jochen Thamm. "Die systematische Erfassung der Naturgegenstände. Zum Programm der *Academia Naturae Curiosorum* von 1652 und seiner Vorgeschichte." In *Die Gründung der Leopoldina—Academia Naturae Curiosorum—im historischen Kontext*, edited by Richard Toellner, Uwe Müller, Benno Parthier, and Wieland Berg, 285–304. Stuttgart: Wissenschaftliche Verlagsgesellschaft, 2008.

Beyer, Jürgen. "Tagungsbericht 'The Book Triumphant: The Book in the Second Century of Print, 1540–1640,' 09.09.2009–11.09.2009." Conference report, *H-Soz-u-Kult*, November 7, 2009. http://hsozkult.geschichte.hu-berlin.de/tagungsberichte/id=2839.

Biagioli, Mario, and Peter Galison, eds. *Scientific Authorship: Credit and Intellectual Property in Science.* New York: Routledge, 2003.

Blair, Ann. "Restaging Jean Bodin: The *Universae Naturae Theatrum* (1596) in Its Cultural Context." PhD diss., Princeton University, 1990.

———. "Humanist Methods in Natural Philosophy: The Commonplace Book." *Journal of the History of Ideas* 53 (1992): 541–51.

———. *The Theater of Nature: Jean Bodin and Renaissance Science.* Princeton, NJ: Princeton University Press, 1997.

———. "Reading Strategies for Coping with Information Overload ca. 1550–1700." *Journal of the History of Ideas* 64 (2003): 11–28.

———. "Note Taking as an Art of Transmission." *Critical Inquiry* 31, no. 1 (2004): 85–107.

———. "Scientific Readers: An Early Modernist's Perspective." *Isis* 95, no. 3 (2004): 420–30.

———. "*Historia* in Zwinger's *Theatrum humanae vitae*." In *Historia: Empiricism and Erudition in Early Modern Europe*, edited by Gianna Pomata and Nancy Siraisi, 269–96. Cambridge, MA: MIT Press, 2005.

———. "The Rise of Note-Taking in Early Modern Europe." *Intellectual History Review* 20, no. 3 (2010): 303–16.

———. *Too Much to Know: Managing Scholarly Information before the Modern Age.* New Haven, CT: Yale University Press, 2010.

Blumenberg, Hans. *Der Prozeß der theoretischen Neugierde.* Frankfurt am Main: Suhrkamp, 1973.

Boehm, Laetitia. "Studium, Büchersammlung, Bildungsreise: Elemente gelehrter Allgemeinbildung und individueller Ausprägung historisch-politischer Weltanschauung im konfessionellen Zeitalter." In *Die Bausch-Bibliothek in Schweinfurt: Wissenschaft und Buch in der Frühen Neuzeit*, edited by Menso Folkerts, Ilse Jahn, and Uwe Müller, 117–51. Acta Historica Leopoldina 31. Heidelberg: Deutsche Akademie der Naturforscher Leopoldina, 2000.

———. "Akademie-Idee und *Curiositas* als Leitmotiv der frühmodernen Leopoldina." In *Die Gründung der Leopoldina—Academia Naturae Curiosorum—im historischen Kontext*, edited by Richard Toellner, Uwe Müller, Benno Parthier, and Wieland Berg, 63–114. Stuttgart: Wissenschaftliche Verlagsgesellschaft, 2008.

Braun-Bucher, Barbara. "Hallers Bibliothek und Nachlass." In *Albrecht von Haller: Leben—Werk—Epoche*, edited by Hubert Steinke, Urs Buschung, and Wolfgang Proß, 515–26. Archiv des Historischen Vereins des Kantons Bern 85. Göttingen: Wallstein Verlag, 2008.

Brendecke, Arndt. "Papierfluten: Anwachsende Schriftlichkeit als Pluralisierungsfaktor in der Frühen Neuzeit." *Mitteilungen des Sonderforschungsbereichs* 573, no. 1 (2006): 21–30.

Bundy, Murray Wright. "The Theory of the Imagination in Classical and Mediaeval Thought." *University of Illinois Studies in Language and Literature* 12 (1927): 1–289.

Bürger, Thomas. "Die *Acerra Philologica* des Peter Lauremberg: Zur Geschichte, Verbreitung und Überlieferung eines deutschen Schulbuches des 17. Jahrhunderts." *Wolfenbütteler Notizen zur Buchgeschichte* 12 (1987): 1–24.

Burnett, Mark Thornton. *Constructing "Monsters" in Shakespearean Drama and Early Modern Culture.* Early Modern Literature in History. New York: Palgrave Macmillan, 2002.

Calabritto, Monica. "Examples, References and Quotations in Sixteenth-Century Medical Texts." In *Citation, Intertextuality, and Memory in the Fourteenth and Fifteenth Centuries*, edited by Yolanda Plumley, Giuliano Di Bacco, and Stefano Jossa, 58–73. Exeter: University of Exeter Press, 2011.

Canguilhem, Georges. "Monstrosity and the Monstrous." *Diogène* 40 (1962): 27–42.

———. "La monstruosité et le monstrueux." In *La connaissance de la vie*, edited by Georges Canguilhem, 219–36. Paris: Vrin, 2006.

Carlino, Andrea. "L'exception et la règle: A propos du XVe livre du *De Re Anatomica* de Realdo Colombo." In *Maladies, médecines et sociétés: Approches historiques pour le présent; Actes du VIe colloque d'Histoire au présent*, edited by François-Olivier Touati, 170–76. Paris: L'Harmattan et Histoire au présent, 1993.

———. *Books of the Body: Anatomical Ritual and Renaissance Learning.* Translated by John Tedeschi and Anne C. Tedeschi. Chicago: University of Chicago Press, 1999.

Carnevali, Rebecca. "The Publication of the *Monstrorum historia* (1642) by Ulisse Aldrovandi: From Popular to Academic Printing in Seventeenth-Century Bologna." MA thesis, University of London, 2014.

Cave, Terence C. "Copia and cornucopia." In *French Renaissance Studies 1540–70: Humanism and the Encyclopedia*, edited by Peter Sharratt, 52–69. Edinburgh: Edinburgh University Press, 1976.

———. *The Cornucopian Text: Problems of Writing in the French Renaissance*. Oxford: Clarendon Press, 1979.

Céard, Jean. Introduction to *Des monstres et prodiges*. Critical ed. Edited with commentary by Jean Céard, 9–46. Travaux d'humanisme et renaissance 115. Geneva: Droz, 1971.

———. *La nature et les prodiges: L'insolite au XVIe siècle*. 2nd ed. Genève: Droz, 1996.

———. "La *Monstrorum Historia* d'Ulisse Aldrovandi ou La nature, l'art et le monstre." In *Ulisse Aldrovandi: Monstrorum Historia*, vii–xliv. Theatrum sapientiae: Textes 4. Paris: Société d'édition Les Belles Lettres, Nino Aragno Editore, 2002.

Cevolini, Alberto, ed. *Thomas Harrison: The Ark of Sudies*. Turnhout: Brepols, 2017.

"Charing Cross." *Wikipedia* (n.d.). Accessed September 30, 2013, http://en.wikipedia.org/wiki/Charing_Cross.

Chartier, Roger. "Foucault's Chiasmus: Authorship between Science and Literature in the Seventeenth and Eighteenth Centuries." In *Scientific Authorship: Credit and Intellectual Property in Science*, edited by Mario Biagioli and Peter Galison, 13–31. New York: Routledge, 2003.

Ciseri, Lorenzo Montemagno. "A lezione con i mostri: Benedetto Varchi e la 'Lezzione sulla generazione dei mostri.'" *Rinascimento*, 2nd ser., 47 (2007): 301–45.

Clark, William. *Academic Charisma and the Origins of the Research University*. Chicago: University of Chicago Press, 2006.

Colclough, David. "'The Materials for the Building': Reuniting Francis Bacon's Sylva Sylvarum and New Atlantis." *Intellectual History Review* 20, no. 2 (2010): 181–200.

Cole, F. J. "The History of Albrecht Dürer's Rhinoceros in Zoological Literature." In *Science, Medicine, and History: Essays on the Evolution of Scientific Thought and Medical Practice, Written in Honour of Charles Singer*, edited by Edgar Ashworth Underwood, 337–56. London: Oxford University Press, 1953.

Costa, Rosa. "Fremde Wunder oder vertraute Fehler? Die Wundergeburtenberichte von Jakob Ruf im Spannungsfeld von Prodigiendeutung und naturkundlichen Erklärungen." Diploma thesis, Universität Wien, 2009.

———. "Fremde Wunder oder vertraute Fehler? Jakob Rufs Flugblatt zur Schaffhauser Wundergeburt im Spannungsfeld von Prodigiendeutung und naturkundlicher Erklärung." In *Von Monstern und Menschen: Begegnungen der anderen Art in kulturwissenschaftlicher Perspektive*, edited by Gunther Gebhard, 69–88. Bielefeld: Transcript, 2009.

Couzinet, Marie-Dominique. "La variété dans la philosophie de la nature: Cardan, Bodin." In *La varietas à la Renaissance: Actes de la journée d'étude organisé par l'École nationale des chartes (Paris, 27 avril 2000)*, edited by Dominique de Courcelles, 105–18. Études et rencontres de l'École des chartes 9. Paris: École des chartes, 2001.

Cox, Virginia. *The Renaissance Dialogue: Literary Dialogue in its Social and Political Contexts, Castiglione to Galileo*. Cambridge: Cambridge University Press, 1992.

Crawford, Julie. *Marvellous Protestantism: Monstrous Births in Post-Reformation England*. Baltimore: Johns Hopkins University Press, 2005.

Cunningham, Andrew. "Fabricius and the 'Aristotle Project' in Anatomical Teaching and Research at Padua." In *The Medical Renaissance of the Sixteenth Century*, edited by Andrew Wear, Roger K. French, and Ian M. Lonie, 195–222. Cambridge: Cambridge University Press, 1985.

Cunningham, Andrew, and Ole Peter Grell. *The Four Horsemen of the Apocalypse. Religion, War, Famine and Death in Reformation Europe*. Cambridge: Cambridge University Press, 2002.

da Costa, Palmira Fontes. "'Mediating Sexual Difference': The Medical Understanding of Human Hermaphrodites in Eighteenth-Century England." In *Cultural Approaches to the*

History of Medicine: Mediating Medicine in Early Modern and Modern Europe, edited by William de Blécourt and Cornelie Usborne, 127–47. Houndmills: Palgrave Macmillan, 2004.

———. *The Singular and the Making of Knowledge at the Royal Society of London in the Eighteenth Century*. Newcastle: Cambridge Scholars, 2009.

Daston, Lorraine. "The Nature of Nature in Early Modern Europe." *Configurations: A Journal of Literature, Science, and Technology* 6, no. 1 (Winter 1998): 149–72.

———. "Afterword: The Ethos of the Enlightenment." In *The Sciences in Enlightenment Europe*, edited by William Clark, Jan Golinski, and Simon Schaffer, 495–504. Chicago: University of Chicago Press, 1999.

———. "Warum sind Tatsachen kurz?" In "Cut and paste um 1900: Der Zeitungsausschnitt in den Wissenschaften." Special issue, *Kaleidoskopien: Zeitschrift für Mediengeschichte und Theorie* 4 (2002): 132–44.

———. "Die Akademien und die Neuerfindung der Erfahrung im 17. Jahrhundert." *Nova Acta Leopoldina* 87, no. 325 (2003): 15–33.

———. "The Empire of Observation, 1600–1800." In *Histories of Scientific Observation*, edited by Lorraine Daston and Elizabeth Lunbeck, 81–113. Chicago: University of Chicago Press, 2010.

Daston, Lorraine, and Katharine Park. "Unnatural Conceptions: The Study of Monsters in Sixteenth- and Seventeenth-Century France and England." *Past & Present* 92 (August 1981): 20–54.

———. "Hermaphrodites in Renaissance France." *Critical Matrix* 1, no. 5 (1985): 1–19.

———. "The Hermaphrodite and the Orders of Nature." *GLQ* 1 (1995): 419–38.

———. *Wonders and the Order of Nature 1150–1750*. New York: Zone Books, 1998.

Daston, Lorraine, and H. Otto Sibum. "Introduction: Scientific Personae and Their Histories." *Science in Context* 16, nos. 1/2 (2003): 1–8.

de Courcelles, Dominique, ed. *La varietas à la Renaissance: Actes de la journée d'étude organisé par l'École nationale des chartes (Paris, 27 avril 2000)*. Études et rencontres de l'École des chartes 9. Paris: École des chartes, 2001.

de Montluzin, Emily Lorraine. *Attributions of Authorship in the Gentleman's Magazine, 1731–1868: An Electronic Union List*. Charlottesville: University of Virginia, Bibliographical Society, University of Virginia Library Electronic Text Center, 2003 [distributed 2004].

Dear, Peter. "The Meanings of Experience." In *The Cambridge History of Science*, edited by Lorraine Daston and Katharine Park, 106–31. Cambridge: Cambridge University Press, 2006.

Doody, Aude, ed. *Pliny's Encyclopedia: The Reception of the "Natural History."* Cambridge: Cambridge University Press, 2010.

Duroselle-Melish, Caroline, and David Lines. "The Library of Ulisse Aldrovandi (d. 1605). Acquiring and Organizing Books in Sixteenth-Century Bologna." *Library* 16, no. 2 (2015): 133–61.

Egmond, Florike. "The *ad Vivum* Conundrum: Eyewitnessing and the Artful Representation of Naturalia in Sixteenth-Century Science." In *Zeigen—Überzeugen—Beweisen: Methoden der Wissensproduktion in Kunstliteratur, Kennerschaft und Sammlungspraxis der Frühen Neuzeit*, edited by Elisabeth Oy-Marra and Irina Schmiedel, 33–62. Heidelberg: arthistoricum.net, 2000. https://doi.org/10.11588/arthistoricum.697.c9495.

Egmond, Florike, and Sachiko Kusukawa. "Circulation of Images and Graphic Practices in Renaissance Natural History: The Example of Conrad Gessner." *Gesnerus* 73, no. 1 (2016): 29–72.

Fekadu, Sarah. "Variation: A. Rhetorisch-literarische Variation." In *Historisches Wörterbuch der Rhetorik*, edited by Gerd Ueding, cols. 1006–12. Darmstadt: Wissenschaftliche Buchgesellschaft, 2009.

Findlen, Paula. "Jokes of Nature and Jokes of Knowledge: The Playfulness of Scientific Discourse in Early Modern Europe." *Renaissance Quarterly* 43, no. 2 (Summer 1990): 292–331.

———. *Possessing Nature: Museums, Collecting, and Scientific Culture in Early Modern Italy.* Studies on the History of Society and Culture 20. Berkeley: University of California Press, 1994.

———. "Masculine Prerogatives: Gender, Space, and Knowledge in the Early Modern Museum." In *The Architecture of Science*, edited by Peter Galison and Emily Thompson, 29–57. Cambridge, MA: MIT Press, 1999.

———. "Inventing Nature: Commerce, Art, and Science in the Early Modern Cabinet of Curiosities." In *Merchants & Marvels: Commerce, Science, and Art in Early Modern Europe*, edited by Pamela H. Smith and Paula Findlen, 297–323. New York: Routledge, 2002.

———. "The Museum: Its Classical Etymology and Renaissance Genealogy." In *Museum Studies: An Anthology of Contexts*, edited by Bettina Messias Carbonell, 23–50. Malden, MA: Blackwell, 2004.

Fischel, Angela. "Zeichnung und Naturbeobachtung: Naturgeschichte um 1600 am Beispiel von Aldrovandis Bildern." In *Das Technische Bild: Kompendium zu einer Stilgeschichte wissenschaftlicher Bilder*, edited by Horst Bredekamp, Birgit Schneider, and Vera Dünkel, 212–23. Berlin: Akademie Verlag, 2008.

———. *Natur im Bild: Zeichnung und Naturerkenntnis bei Conrad Gessner und Ulisse Aldrovandi.* Humboldt-Schriften zur Kunst- und Bildgeschichte 9. Berlin: Gebr. Mann Verlag, 2009.

Föcking, Marc. "Mißverständnisse der Rezeption als Innovationsfaktoren in der Literatur der italienischen Renaissance." In *Innovation durch Wissenstransfer in der Frühen Neuzeit: Kultur- und geistesgeschichtliche Studien zu Austauschprozessen in Mitteleuropa*, edited by Johann Anselm Steiger, Sandra Richter, and Marc Föcking, 185–208. Chloe: Beihefte zum Daphnis 41. Amsterdam: Rodopi, 2010.

Föcking, Marc, and Bernhard Huss, eds. *Varietas und Ordo: Zur Dialektik von Vielfalt und Einheit in Renaissance und Barock.* Text und Kontext: Romanische Literaturen und Allgemeine Literaturwissenschaft 18. Stuttgart: Steiner, 2003.

Folkerts, Menso, Ilse Jahn, and Uwe Müller, eds. *Die Bausch-Bibliothek in Schweinfurt. Wissenschaft und Buch in der Frühen Neuzeit.* Acta Historica Leopoldina 31. Heidelberg: Deutsche Akademie der Naturforscher Leopoldina, 2000.

Folli, Irene Ventura. "La natura scritta, la libraria di Ulisse Aldrovandi (1522–1605)." In *Bibliothecae selectae*, edited by Eugenio Canone. Lessico intellettuale europeo 58, 495–506. Florence: Leo S. Olschki, 1993.

Foucault, Michel. "Was ist ein Autor?" Translated by Michael Bischoff, Hans-Dieter Gondek, and Hermann Kocyba. In *Michel Foucault: Schriften in vier Bänden*, edited by Daniel Defert and François Ewald, 1003–41. Frankfurt am Main: Suhrkamp, 2001.

———. *Order of Things.* Routledge Classics. London: Routledge, 2018 / Éditions Gallimard 1966.

French, Roger. *William Harvey's Natural Philosophy.* Cambridge: Cambridge University Press, 1994.

Friedman, John. *The Monstrous Races in Medieval Art and Thought.* Cambridge, MA: Harvard University Press, 1981.

Gantet, Claire, and Fabian Kraemer. "Wie man mehr als 9000 Rezensionen schreiben kann: Albrecht von Hallers wissenschaftliche Praxis, zwischen Lektüre und Rezensionen." *Historische Zeitschrift* 312, no. 2 (2021): 364–99.

Georges, Karl Ernst. *Ausführliches lateinisch-deutsches und deutsch-lateinisches Handwörterbuch: Aus den Quellen zusammengetragen und mit besonderer Bezugnahme auf Synonymik und Antiquitäten unter Berücksichtigung der besten Hülfsmittel.* Lateinisch-deutscher Teil. Repr. of the 8th ed. 2 vols. Vol. 1. Hannover: Hahn, 1972.

———. *Ausführliches lateinisch-deutsches und deutsch-lateinisches Handwörterbuch: Aus den Quellen zusammengetragen und mit besonderer Bezugnahme auf Synonymik und Antiquitäten unter Berücksichtigung der besten Hülfsmittel.* Lateinisch-deutscher Teil. Repr. of the 8th ed. 2 vols. Vol. 2. Hannover: Hahn, 1972.

Gerber, Georg. "Die Neu-Atlantis des Francis Bacon und die Entstehung der Academia Naturae Curiosorum (Leopoldina) und der Societät der Wissenschaften in Berlin." *Wissenschaftliche Annalen* 9 (1955): 552–60.

"Gerücht." In *Deutsches Wörterbuch*, edited by Jacob und Wilhelm Grimm, cols. 3751–58. Leipzig: Hirzel, 1897.

Gierl, Martin. "Bestandsaufnahme im gelehrten Bereich: Zur Entwicklung der 'Historia literaria' im 18. Jahrhundert." In *Denkhorizonte und Handlungsspielräume: Festschrift für Rudolf Vierhaus zum 70. Geburtstag*, edited by Martin Gierl, 53–80. Göttingen: Wallstein Verlag, 1992.

Giglioni, Guido. "From the Woods of Experience to the Open Fields of Metaphysics. Bacon's Notion of *Silva*." *Renaissance Studies* 28, no. 2 (2014): 242–61.

Gilbert, Ruth. *Early Modern Hermaphrodites: Sex and Other Stories.* Basingstoke, UK: Palgrave, 2002.

Giudicelli-Falguières, Patricia. "Invention et mémoire: Aux origines de l'institution muséographique, les collections encyclopédiques et les cabinets de merveilles dans l'Italie du XVIe siècle." PhD diss., Paris I, 1988.

Goesch, Andrea. *Diana Ephesia: Ikonographische Studien zur Allegorie der Natur in der Kunst des 16. bis 19. Jahrhunderts.* Europäische Hochschulschriften, Reihe 28: Kunstgeschichte 253. Frankfurt am Main: Peter Lang, 1996.

Grafton, Anthony. *The Footnote: A Curious History.* Cambridge, MA: Harvard University Press, 1997.

———. *Inky Fingers: The Making of Books in Early Modern Europe.* Cambridge, MA: Belknap Press of Harvard University Press, 2020.

Grond-Ginsbach, Caspar. "Georges Canguilhem als Medizinhistoriker." *Berichte zur Wissenschaftsgeschichte* 19 (1996): 235–44.

Grunert, Frank, Anette Syndikus, and Friedrich Vollhardt. "Ein Leitfaden durch das Labyrinth: Zur Funktion der Gelehrsamkeitsgeschichte in der Frühen Neuzeit." *Mitteilungen des Sonderforschungsbereichs* 573, no. 2 (2006): 35–42.

Grunert, Frank, and Friedrich Vollhardt, eds. *Historia literaria: Neuordnungen des Wissens im 17. und 18. Jahrhundert.* Berlin: Akademie Verlag, 2007.

———. "Einleitung." In *Historia literaria: Neuordnungen des Wissens im 17. und 18. Jahrhundert*, edited by Frank Grunert and Friedrich Vollhardt, 7–11. Berlin: Akademie Verlag, 2007.

Guthmüller, Bodo, and Wolfgang G. Müller, eds. *Dialog und Gesprächskultur in der Renaissance.* Wolfenbütteler Abhandlungen zur Renaissanceforschung 22. Wiesbaden: Harassowitz, 2004.

Hacking, Ian. *The Emergence of Probability: A Philosophical Study of Early Ideas about Probability, Induction, and Statistical Inference.* Cambridge: Cambridge University Press, 1975.

Haeberli, Hans. *Gottlieb Emanuel von Haller: Ein Berner Historiker und Staatsmann im Zeitalter der Aufklärung, 1735–1786.* Archiv d. Hist. Vereins d. Kantons Bern 41, 2. Bern: Feuz, 1952.

Hagner, Michael. "Vom Naturalienkabinett zur Embryologie." In *Der falsche Körper. Beiträge zu einer Geschichte der Monstrositäten*, edited by Michael Hagner, 73–107. Göttingen: Wallstein Verlag, 1995.

———. "Enlightened Monsters." In *The Sciences in Enlightened Europe*, edited by William Clark, Jan Golinski, and Simon Schaffer, 175–217. Chicago: University of Chicago Press, 1999.

Hammerl, Michaela. "Prodigienliteratur." *Lexikon zur Geschichte der Hexenverfolgung* (2007). www.historicum.net/no_cache/persistent/artikel/5523/.

Harms, Wolfgang. "Bedeutung als Teil der Sache in zoologischen Standardwerken der frühen Neuzeit (Konrad Gesner, Ulisse Aldrovandi)." In *Lebenslehren und Weltentwürfe im Übergang vom Mittelalter zur Neuzeit: Politik—Bildung—Naturkunde—Theologie; Bericht über Kolloquien der Kommission zur Erforschung der Kultur des Spätmittelalters 1983 bis 1987*, edited by Hartmut Boockmann, Bernd Moeller, and Karl Stackmann, 352–69. Abhandlungen der Akademie der Wissenschaften in Göttingen, Philologisch-Historische Klasse, Vol. 3, no. 179. Göttingen: Vandenhoeck & Ruprecht, 1989.

Harms, Wolfgang, and Michael Schilling, eds. *Die Sammlung der Zentralbibliothek Zürich: Kommentierte Ausgabe. Teil 1: Die Wickiana I (1500–1569)*. Deutsche Illustrierte Flugblätter des 16. und 17. Jahrhunderts 6. Tübingen: Niemeyer, 1997.

———, eds. *Die Sammlung der Zentralbibliothek Zürich: Kommentierte Ausgabe. Teil 2: Die Wickiana II (1570–1588)*. Deutsche Illustrierte Flugblätter des 16. und 17. Jahrhunderts 7. Tübingen: Niemeyer, 1997.

———. "Einleitung zu den Bänden VI und VII." In *Die Sammlung der Zentralbibliothek Zürich: Kommentierte Ausgabe. Teil 1: Die Wickiana I (1500–1569)*, edited by Wolfgang Harms and Michael Schilling, 7–12. Deutsche Illustrierte Flugblätter des 16. und 17. Jahrhunderts 6. Tübingen: Niemeyer, 1997.

Hartmann, Fritz. "Ärztliche Praxis und klinische Medizin in der Leopoldina." In *350 Jahre Leopoldina—Anspruch und Wirklichkeit: Festschrift der Deutschen Akademie der Naturforscher Leopoldina 1652–2002*, edited by Benno Parthier and Dietrich von Engelhardt, 381–418. Halle: Druck-Zuck GmbH, 2002.

Heilbron, John L. "History of Science or History of Learning." *Berichte zur Wissenschaftsgeschichte / History of Science and Humanities* 42, nos. 2/3 (2019): 200–219.

Hess, Volker, and J. Andrew Mendelsohn. "*Paper Technology* und Wissenschaftsgeschichte." *NTM* 1 (2013): 1–10.

Höfele, Andreas. *Stage, Stake, and Scaffold: Humans and Animals in Shakespeare's Theatre.* Oxford: Oxford University Press, 2011.

Höfele, Andreas, and Stephan Laqué. Introduction to *Humankinds: The Renaissance and its Anthropologies*, edited by Andreas Höfele and Stephan Laqué, 1–18. Pluralisierung und Autorität 25. Berlin, New York: de Gruyter, 2011.

Jahn, Ilse, ed. *Geschichte der Biologie: Theorien, Methoden, Institutionen, Kurzbiographien.* 3rd ed. Jena: Fischer, 1998.

Jaumann, Herbert. *Critica: Untersuchungen zur Geschichte der Literaturkritik zwischen Quintilian und Thomasius.* Brill's Studies in Intellectual History 62. Leiden: Brill, 1995.

———. "Jakob Friedrich Reimmanns Bayle-Kritik und das Konzept der Historia literaria: Mit einem Anhang über Reimmanns Periodisierung der deutschen Literaturgeschichte." In *Skepsis, Providenz, Polyhistorie: Jakob Friedrich Reimmann (1668–1743)*, edited by Martin Mulsow and Helmut Zedelmaier, 200–13. Tübingen: Niemeyer, 1998.

Joas, Christian, Fabian Kraemer, and Kärin Nickelsen. "Introduction: History of Science or History of Knowledge?" *Berichte zur Wissenschaftsgeschichte / History of Science and Humanities* 42, nos. 2/3 (2019): 117–25.

———, eds. "History of Science or History of Knowledge?" Special issue of *Berichte zur Wissenschaftsgeschichte / History of Science and Humanities* 42, nos. 2/3 (2019).

Johns, Adrian. "The Ambivalence of Authorship in Early Modern Natural Philosophy." In *Scientific Authorship: Credit and Intellectual Property in Science*, edited by Mario Biagioli and Peter Galison, 67–90. New York: Routledge, 2003.

———. "Reading and Experiment in the Early Royal Society." In *Reading, Society, and Politics in Early Modern England*, edited by Kevin Sharpe and Steven N. Zwicker, 244–71. Cambridge: Cambridge University Press, 2003.

Kefalas, Katherina. "Philipp Jacob Sachs von Lewenhaimb (1627–1672) und die Academia Naturae Curiosorum: Eine Edition seiner Briefe in chronologischer Reihenfolge sowie seines Reisetagebuchs." PhD diss., Universität Heidelberg, 2016.

Keller, Helmut. *Dr. Johann Laurentius Bausch (1605–1665): Gründer der Academia Naturae-Curiosorum*. Würzburg: Baer, 1955.

Keller, Vera. "Professionalizing Doubt: Johann Daniel Major's Observation 'On the Horn of the Bezoardic Goat,' Curiosity Collecting, and the Institutionalization of Natural History." In *Science in Early Modern Europe*, edited by Giulia Giannini and Mordechai Feingold, 199–235. Leiden: Brill, 2020.

Kenny, Neil. *The Uses of Curiosity in Early Modern France and Germany*. Oxford: Oxford University Press, 2004.

Kloosterhuis, Jürgen. *Legendäre "lange Kerls": Quellen zur Regimentskultur der Königsgrenadiere Friedrich Wilhelms I., 1713–1740*. Berlin: Geheimes Staatsarchiv Preußischer Kulturbesitz, 2003.

Klöppel, Ulrike. *XXoXY ungelöst: Hermaphroditismus, Sex und Gender in der deutschen Medizin; Eine historische Studie zur Intersexualität*. GenderCodes. Transkriptionen zwischen Wissen und Geschlecht 12. Bielefeld: Transcript, 2010.

Kraemer, Fabian. "'Under so viel wunderbarlichen und seltsamen Sachen ist mir nichts wunderbarlichers unnd seltsamers fürkommen': Vom 'Auftauchen' des Hermaphroditen in der Frühen Neuzeit." In *1–0–1 [one 'o one] intersex: Das Zwei-Geschlechter-System als Menschenrechtsverletzung; Katalog zur gleichnamigen Ausstellung vom 17. 6.–31. 7. 2005 in den Ausstellungsräumen der NGBK*, edited by Neue Gesellschaft für Bildende Kunst, 150–57. Berlin: NGBK, 2005.

———. "Die Individualisierung des Hermaphroditen in Medizin und Naturgeschichte des 17. Jahrhunderts." *Berichte zur Wissenschaftsgeschichte* 30, no. 1 (2007): 49–65.

———. "The Persistent Image of an Unusual Centaur: A Biography of Aldrovandi's Two-Legged Centaur Woodcut." *Nuncius* 24, no. 2 (2009): 313–40.

———. "Faktoid und Fallgeschichte: Medizinische Fallgeschichten im Lichte frühneuzeitlicher Lese- und Aufzeichnungstechniken." In *Die Sachen der Aufklärung—Matters of Enlightenment—La cause et les choses des Lumières*, edited by Frauke Berndt and Daniel Fulda, 525–36. Studien zum 18. Jahrhundert 34. Hamburg: Meiner, 2012.

———. "Hermaphrodites Closely Observed: The Individualisation of Hermaphrodites and the Rise of the *Observatio* Genre in Seventeenth-Century Medicine." In *L'hermaphrodite de la Renaissance aux lumières*, edited by Marianne Closson, 37–60. Paris: Classiques Garnier, 2013.

———. "Ein papiernes Archiv für alles jemals Geschriebene: Ulisse Aldrovandis *Pandechion epistemonicon* und die Naturgeschichte der Renaissance." *NTM* 1 (2013): 11–36.

———. "Why There Was No Centaur in Eighteenth-Century London: The Vulgar as a Cognitive Category in Enlightenment Europe." In *Wissenschaftsgeschichte und Geschichte des Wissens im Dialog—Connecting Science and Knowledge*, edited by Kaspar von Greyerz, Silvia Flubacher, and Philipp Senn, 317–45. Göttingen: V&R unipress, 2013.

———. "Ulisse Aldrovandi's *Pandechion Epistemonicon* and the Use of Paper Technology in Renaissance Natural History." *Early Science and Medicine* 19 (2014): 398–423.

———. *Ein Zentaur in London: Lektüre und Beobachtung in der frühneuzeitlichen Naturforschung*. Kulturgeschichten. Studien zu Frühen Neuzeit 1. Affalterbach: Didymos, 2014.

———. "*Richtig* beobachten: Zum zwiespältigen Verhältnis der *Academia Naturae Curiosorum* zu den Monstren." *Acta Historica Leopoldina* 65 (2015): 109–30.

———. "Albrecht von Haller as an 'Enlightened' Reader-Observer." In *Forgetting Machines: Knowledge Management Evolution in Early Modern Europe*, edited by Alberto Cevolini, 224–42. Leiden: Brill, 2016.

Kraemer, Fabian, and Helmut Zedelmaier. "Instruments of Invention in Renaissance Europe: The Cases of Conrad Gesner and Ulisse Aldrovandi." *Intellectual History Review* 24, no. 3 (2014): 321–41.

Kuhn, Heinrich C. *Venetischer Aristotelismus im Ende der aristotelischen Welt: Aspekte der Welt und des Denkens des Cesare Cremonini (1550–1631)*. Europäische Hochschulschriften. Reihe 20, Philosophie 490. Frankfurt am Main: Peter Lang, 1996.

Kusukawa, Sachiko. "The Sources of Gessner's Pictures for the *Historia Animalium*." *Annals of Science* 67, no. 3 (2010): 303–28.

Laeven, Hubertus. *The "Acta Eruditorum" under the Editorship of Otto Mencke: The History of an International Learned Journal between 1682 and 1707*. Translated by Lynne Richards. Amsterdam: Maarssen, 1990.

Lefèvre, Wolfgang. *Picturing the World of Mining in the Renaissance: The Schwarzer Bergbuch (1556)*. Preprints of the Max Planck Institute for the History of Science; 407. Berlin: Max Planck Institute for the History of Science, 2010.

Lehmann-Brauns, Sicco. "Neukonturierung und methodologische Reflexion der Wissenschaftsgeschichte: Heumanns *Conspectus Reipublicae Literariae* als Lehrbuch der aufgeklärten Historia literaria." In *Historia literaria: Neuordnungen des Wissens im 17. und 18. Jahrhundert*, edited by Frank Grunert and Friedrich Vollhardt, 129–60. Berlin: Akademie Verlag, 2007.

Leu, Urs. "Die Loci-Methode als enzyklopädisches Ordnungssystem." In *Allgemeinwissen und Gesellschaft: Akten des internationalen Kongresses über Wissenstransfer und enzyklopädische Ordnungssysteme, vom 18. bis 21. September 2003 in Prangins*, edited by Paul Michel, Madeleine Herren, and Martin Rüesch, 337–58. Berichte aus der Geschichtswissenschaft. Aachen: Shaker, 2007.

———. "Book and Reading Culture in Basle and Zurich, 1520 to 1600." Paper given at "The Book Triumphant: The Book in the Second Century of Print, 1540–1640," held at University of St Andrews, September 9–11, 2009.

———. "Reformation als Auftrag: Der Zürcher Drucker Christoph Froschauer d.Ä. (ca. 1490–1564)." In *Buchdruck und Reformation in der Schweiz*, edited by Urs Leu and Christian Scheidegger, 1–80. Zwingliana 45. Zurich: Theologischer Verlag Zürich, 2019.

Ley, Willy. *Dawn of Zoology*. Englewood Cliffs, NJ: Prentice Hall, 1968.

Lindsay, Alexander. *Index of English Literary Manuscripts, Sterne-Young*. Vol. 3, part 4. London: Continuum International, 1997.

Locher, Elmar. "Topos und Argument: Anmerkungen zur Verknüpfung des Monströsen—von Schedels *Weltchronik* zu Gaspard Schotts *Physica Curiosa*." In *Intertextualität in der frühen Neuzeit: Studien zu ihren theoretischen und praktischen Perspektiven*, edited by Wilhelm Kühlmann and Wolfgang Neuber, 195–223. Frühneuzeitstudien 2. Frankfurt am Main: Lang, 1994.

Long, Kathleen P. "The Cultural and Medical Construction of Gender: Caspar Bauhin." In *Hermaphrodites in Renaissance Europe*, edited by Kathleen P. Long, 49–75. Women and Gender in the Early Modern World. Aldershot: Ashgate, 2006.

Maclean, Ian. "Doctrines médicales à la Renaissance: Continuités et innovations." In *"Vera doctrina": Zur Begriffsgeschichte der "doctrina" von Augustinus bis Descartes / "Vera doctrina": L'idée de doctrine de saint Augustin à Descartes*, edited by Philipp Büttgen, Ruedi Imbach, Ulrich Johannes Schneider, and Herman J. Selderhuis. Herzog August Bibliothek Wolfenbüttel, 2006.

———. "La doctrine selon les médecins de la Renaissance." In *Vera doctrina: Zur Begriffsges-*

chichte der doctrina von Augustinus bis Descartes, edited by Philip Büttgen, 141–50. Wolfenbütteler Forschungen 123. Wiesbaden: Harassowitz, 2009.

Malcolm, Noel. "Thomas Harrison and His 'Ark of Studies': An Episode in the History of the Organisation of Knowledge." *Seventeenth Century* 19 (2004): 196–232.

Marcaida, José Ramón. "Portraying Technology in Gallery Paintings." *History and Technology* 25, no. 4 (2009): 391–97.

Matthew, Henry C. G., and Brian Harrison, in assoc. with the British Academy, eds. *Oxford Dictionary of National Biography*. 60 vols. Vol. 5. Oxford: Oxford University Press, 2004.

Mattirolo, Oreste. "Le lettere di Ulisse Aldrovandi a Francesco I e Ferdinando I Granduchi di Toscana e a Francesco Maria II Duca di Urbino." *Memorie della Reale Accademia delle Scienze di Torino* ser. 2, no. 54 (1903–1904): 355–401.

Mauelshagen, Franz. *Wunderkammer auf Papier: Die Wickiana zwischen Reformation und Volksglaube*. Frühneuzeit-Forschungen 15. Epfendorf am Neckar: biblioteca academica, 2011.

Meinel, Christoph. "Enzyklopädie der Welt und Verzettelung des Wissens: Aporien der Empirie bei Joachim Jungius." In *Enzyklopädien der frühen Neuzeit*, edited by Franz M. Eybl et al., 162–87. Tübingen: Niemeyer, 1995.

Minkowski. "Die Neu-Atlantis des Francis Bacon und die Leopoldina-Carolina." *Archiv für Kulturgeschichte* 26 (1936): 283–95.

Moscoso, Javier. "Vollkommene Monstren und unheilvolle Gestalten: Zur Naturalisierung der Monstrositäten im 18. Jahrhundert." Translated by Michael Hagner. In *Der falsche Körper: Beiträge zu einer Geschichte der Monstrositäten*, edited by Michael Hagner, 56–72. Göttingen: Wallstein Verlag, 1995.

———. "Monsters as Evidence: The Uses of the Abnormal Body during the Early 18th Century." *Journal of the History of Biology* 31 (1998): 355–82.

Moss, Ann. *Printed Commonplace-Books and the Structuring of Renaissance Thought*. 2nd ed. Oxford: Clarendon Press, 2002.

Mücke, Marion, and Thomas Schnalke. *Briefnetz Leopoldina: Die Korrespondenz der Deutschen Akademie der Naturforscher um 1750*. Berlin: de Gruyter, 2009.

Müller, Barbara. "Die Tradition der Tischgespräche von der Antike bis in die Renaissance." In *Martin Luthers Tischreden: Neuansätze der Forschung*, edited by Katharina Bärenfänger, Volker Leppin, and Stefan Michel, 63–78. Spätmittelalter, Humanismus, Reformation 71. Tübingen: Mohr Siebeck, 2013.

Müller, Uwe. "'Cometa Caudatus vel Barbatus': Anmerkungen zur Beurteilung des ersten Präsidenten der Academia Naturae Curiosorum Johann Laurentius Bausch als Naturforscher." In *Salve Academicum: Festschrift der Stadt Schweinfurt anläßlich des 300. Jahrestages der Privilegierung der Deutschen Akademie der Naturforscher Leopoldina durch Kaiser Leopold I. vom 7. August 1687*, edited by Uwe Müller, 40–68. Veröffentlichungen des Stadtarchivs Schweinfurt 1. Schweinfurt: Stadtarchiv Schweinfurt, 1987.

———. "Die Leopoldina unter den Präsidenten Bausch, Fehr und Volckamer (1652–1693)." In *350 Jahre Leopoldina—Anspruch und Wirklichkeit: Festschrift der Deutschen Akademie der Naturforscher Leopoldina 1652–2002*, edited by Benno Parthier and Dietrich von Engelhardt, 45–93. Halle: Druck-Zuck GmbH, 2002.

———. "Johann Laurentius Bausch und Philipp Jacob Sachs von Lewenhaimb: Von der Gründung der Academia Naturae Curiosorum zur Reichsakademie." In *Die Gründung der Leopoldina—Academia Naturae Curiosorum—im historischen Kontext: Laurentius Bausch zum 400. Geburtstag*, edited by Richard Toellner, Uwe Müller, Benno Parthier, and Wieland Berg, 13–41. Acta Historica Leopoldina 49. Stuttgart: Wissenschaftliche Verlagsgesellschaft, 2008.

———. "Die Leges der Academia Naturae Curiosorum 1652–1872." In *Die Gründung der*

Leopoldina—Academia Naturae Curiosorum—im historischen Kontext, edited by Richard Toellner, Uwe Müller, Benno Parthier, and Wieland Berg, 243–64. Acta Historica Leopoldina 49. Stuttgart: Wissenschaftliche Verlagsgesellschaft, 2008.

Müller, Uwe, Claudia Michael, Michael Bucher, and Ute Grad. *Die Bausch-Bibliothek in Schweinfurt. Katalog.* Acta Historica Leopoldina 32. Stuttgart: Wissenschaftliche Verlagsgesellschaft, 2004.

Müller-Wille, Staffan. "Vom Sexualsystem zur Karteikarte: Carl von Linnés Papiertechnologien." In *Nicht Fisch—nicht Fleisch: Ordnungssysteme und ihre Störfälle*, edited by Thomas Bäumler, Benjamin Bühler, and Stefan Rieger, 33–50. Sequenzia. Zürich: diaphanes, 2011.

Müller-Wille, Staffan, and Isabelle Charmantier. "Natural History and Information Overload: The Case of Linnaeus." *Studies in History and Philosophy of Science Part C: Studies in History and Philosophy of Biological and Biomedical Sciences* 43 (2012): 4–15.

Müller-Wille, Staffan, and Sara Scharf. "Indexing Nature: Carl Linnaeus (1707–1778) and His Fact-Gathering Strategies." *Working Papers on the Nature of Evidence: How Well Do "Facts" Travel?* 36, no. 8 (2009): 1–39.

Mulsow, Martin. *Prekäres Wissen: Eine andere Ideengeschichte der Frühen Neuzeit.* Berlin: Suhrkamp 2012.

Nauert, Charles G., Jr. "Humanists, Scientists, and Pliny: Changing Approaches to a Classical Author." *American Historical Review* 84, no. 1 (1979): 72–85.

Neigebaur, Johann Daniel Ferdinand. *Geschichte der Kaiserlichen Leopoldino-Carolinischen Deutschen Akademie der Naturforscher während des zweiten Jahrhunderts ihres Bestehens.* Jena: Friedrich Frommann, 1860.

Nelles, Paul. "'Historia Literaria' at Helmstedt: Books, Professors, and Students in the Early Enlightenment University." In *Die Praktiken der Gelehrsamkeit in der Frühen Neuzeit*, edited by Helmut Zedelmaier and Martin Mulsow, 147–75. Frühe Neuzeit 64. Tübingen: Niemeyer, 2001.

Neuber, Wolfgang. "Topik und Intertextualität: Begriffshierarchie und ramistische Wissenschaft in Theodor Zwingers *METHODUS APODEMICA*." In *Intertextualität in der frühen Neuzeit: Studien zu ihren theoretischen und praktischen Perspektiven*, edited by Wilhelm Kühlmann and Wolfgang Neuber, 253–78. Frühneuzeitstudien 2. Frankfurt am Main: Lang, 1994.

Niccoli, Ottavia. *Prophecy and People in Renaissance Italy.* Translated by Lydia G. Cochrane. Princeton, NJ: Princeton University Press 1990.

Ogilvie, Brian W. "Observation and Experience in Early Modern Natural History." PhD diss., University of Chicago, 1997.

———. "The Many Books of Nature: Renaissance Naturalists and Information Overload." *Journal of the History of Ideas* 64, no. 1 (2003): 29–40.

———. "Natural History, Ethics, and Physico-Theology." In *Historia: Empiricism and Erudition in Early Modern Europe*, edited by Gianna Pomata and Nancy G. Siraisi, 75–103. Cambridge, MA: MIT Press, 2005.

———. *The Science of Describing: Natural History in Renaissance Europe.* Chicago: University of Chicago Press, 2006.

Olmi, Giuseppe. "'Figurare e descrivere': Note sull'illustrazione naturalistica cinquecentesca." *Acta Medicae Historiae Patavina* 27 (1980–1981): 99–120.

———. *L'inventario del mondo: Catalogazione della natura e luoghi del sapere nella prima età moderna.* Annali dell'Istituto Storico Italo-Germanico 17. Bologna: Società editrice il Mulino, 1992.

———. "Die Sammlung: Nutzbarmachung und Funktion." In *Macrocosmos in Microcosmo. Die Welt in der Stube: Zur Geschichte des Sammelns 1450–1800*, edited by Andreas Grote, 169–89. Berliner Schriften zur Museumskunde 10. Opladen: Leske + Budrich, 1994.

———. "Ulisse Aldrovandi: Observation at First Hand." In *The Great Naturalists*, edited by Robert Huxley, 59–62. London: Thames & Hudson in association with the Natural History Museum, 2007.

Olmi, Giuseppe, and Fulvio Somoni, eds. *Ulisse Aldrovandi: Libri e immagini di Storia naturale nella prima Èta moderna*. Bologna: Bononia University Press, 2018.

Park, Katharine. "The Criminal and the Saintly Body: Autopsy and Dissection in Renaissance Italy." *Renaissance Quarterly* 47, no. 1 (1994): 1–33.

———. "The Rediscovery of the Clitoris: French Medicine and the Tribade, 1570–1620." In *The Body in Parts: Fantasies of Corporeality in Early Modern Europe*, edited by David Hillman and Carla Mazzio, 171–93. London: Routledge, 1997.

———. "Nature in Person: Medieval and Renaissance Allegories and Emblems." In *The Moral Authority of Nature*, edited by Lorraine Daston and Fernando Vidal, 50–73. Chicago: University of Chicago Press, 2004.

———. "Observation in the Margins, 500–1500." In *Histories of Scientific Observation*, edited by Lorraine Daston and Elizabeth Lunbeck, 15–44. Chicago: University of Chicago Press, 2010.

Park, Katharine, and Lorraine Daston. "Introduction: The Age of the New." In *The Cambridge History of Science*. Vol. 3: *Early Modern Science*, edited by Katharine Park and Lorraine Daston, 1–17. Cambridge: Cambridge University Press, 2006.

Parshall, Peter. "Imago Contrafacta: Images and Facts in the Northern Renaissance." *Art History* 16 (1993): 554–79.

Parthier, Benno. *Die Leopoldina: Bestand und Wandel der ältesten deutschen Akademie; Festschrift des Präsidiums der Deutschen Akademie der Naturforscher Leopoldina zum 300. Jahrestag der Gründung der heutigen Martin-Luther-Universität Halle-Wittenberg 1994*. Halle: Druck-Zuck GmbH, 1994.

Pinon, Laurent. "Entre compilation et observation: L'écriture de l'*Ornithologie* d'Ulisse Aldrovandi." *Genesis* 20 (2003): 53–70.

Plomer, Henry Robert, George Herbert Bushnell, and Ernest Reginald MacClintock Dix. *A Dictionary of Printers and Booksellers Who Were at Work in England, Scotland and Ireland from 1726 to 1775: Those in England by H. R. Plomer, Scotland by G. H. Bushnell, Ireland by E. R. McC. Dix*. Bibliographical Society Publication for the Year 27. Vienna: Oxford University Press, 1932.

Pomata, Gianna. "*Praxis Historialis*: The Uses of *Historia* in Early Modern Medicine." In *Historia: Empiricism and Erudition in Early Modern Europe*, edited by Gianna Pomata and Nancy G. Siraisi, 105–46. Transformations: Studies in the History of Science and Technology. Cambridge, MA: MIT Press, 2005.

———. "Observation Rising: Birth of an Epistemic Genre, ca. 1500–1650." In *Histories of Scientific Observation*, edited by Lorraine Daston and Elizabeth Lunbeck, 45–80. Chicago: University of Chicago Press, 2010.

———. "Sharing Cases: The *Observationes* in Early Modern Medicine." *Early Science and Medicine* 15 (2010): 193–236.

Pomata, Gianna, and Nancy G. Siraisi, eds. *Historia: Empiricism and Erudition in Early Modern Europe*. Transformations: Studies in the History of Science and Technology. Cambridge, MA: MIT Press, 2005.

———. Introduction to *Historia: Empiricism and Erudition in Early Modern Europe*, edited by Gianna Pomata and Nancy G. Siraisi, 1–38. Transformations: Studies in the History of Science and Technology. Cambridge, MA: MIT Press, 2005.

Popper, Nicholas. *Walter Ralegh's History of the World and the Historical Culture of the Late Renaissance*. Chicago: University of Chicago Press, 2012.

Poppi, Antonino. *Introduzione all'aristotelismo padovano*. Saggi e testi 10. 2nd ed. Padua: Editrice Antenore, 1991 [1970].

Reske, Christoph. *Die Buchdrucker des 16. und 17. Jahrhunderts im deutschen Sprachgebiet: Auf der Grundlage des gleichnamigen Werkes von Josef Benzing*. Wiesbaden: Harrassowitz, 2007.

Rey, Alain, ed. *Le grand Robert de la langue française*. 2nd ed. 9 vols. Vol. 1. Paris: Dictionnaires Le Robert, 1996.

Rheinberger, Hans-Jörg, and Staffan Müller-Wille. *Vererbung: Geschichte und Kultur eines biologischen Konzepts*. Frankfurt am Main: Fischer, 2009.

"Rockenphilosophie." In *Deutsches Wörterbuch*, edited by Jacob und Wilhelm Grimm. 33 vols. Vol. 14, cols. 1103–4. Leipzig: Hirzel, 1893.

Roger, Jacques. *The Life Sciences in Eighteenth-Century French Thought*. Edited by Keith R. Benson. Translated by Robert Ellrich. Stanford, CA: Stanford University Press, 1997.

Rohrbach, Thomas René. "Friedrich August Webers Edition von A. v. Hallers "Vorlesungen über die gerichtliche Arzneiwissenschaft (1782–1784)." Med. diss., Universität Bern, 2002.

Roling, Bernd. *Drachen und Sirenen: Die Rationalisierung und Abwicklung der Mythologie an den europäischen Universitäten*. Mittellateinische Studien und Texte 42. Leiden: Brill, 2010.

Rosenberg, Daniel. "Early Modern Information Overload." *Journal of the History of Ideas* 64, no. 1 (2003): 1–9.

Ruffini, Marco. "A Dragon for the Pope: Politics and Emblematics at the Court of Gregory XIII." *Memoirs of the American Academy in Rome* 54 (2009): 83–105.

"Sage." In *Deutsches Wörterbuch*, edited by Jacob und Wilhelm Grimm. 33 vols. Vol 16, cols. 1644–49. Leipzig: Hirzel, 1905.

Schenda, Rudolf. "Das Monstrum von Ravenna: Eine Studie zur Prodigienliteratur." *Zeitschrift für Volkskunde* 56 (1960): 209–25.

———. *Die französische Prodigienliteratur in der zweiten Hälfte des 16. Jahrhunderts*. Münchner Romanische Arbeiten 16. Munich: Hueber, 1961.

———. "Die deutschen Prodigiensammlungen des 16. und 17. Jahrhunderts." *Archiv für Geschichte des Buchwesens* 4 (1963): 637–710.

———. "Wunder-Zeichen: Die alten Prodigien in neuen Gewändern; Eine Studie zur Geschichte eines Denkmusters." *Fabula: Zeitschrift für Erzählforschung* 38 (1997): 14–32.

Schmidt-Biggemann, Wilhelm. *Topica Universalis: Eine Modellgeschichte humanistischer und barocker Wissenschaft*. Paradeigmata 1. Hamburg: Meiner 1983.

Schneider, Ulrich Johannes, ed. *Seine Welt wissen: Enzyklopädien in der Frühen Neuzeit; Katalog zur Ausstellung der Universitätsbibliothek Leipzig (Januar–April 2006) und der Herzog August Bibliothek Wolfenbüttel (Juni–November 2006)*. Darmstadt: Wissenschaftliche Buchgesellschaft, 2006.

Schobeß, Volker. *Die Langen Kerls von Potsdam: Die Geschichte des Leibregiments Friedrich Wilhelms I. 1713–1740*. Berlin: Trafo Verlag, 2007.

Schock, Flemming. *Die Text-Wunderkammer: Populäre Wissenssammlungen des Barock am Beispiel der Relationes Curiosae von E.W. Happel*. Beihefte zum Archiv für Kulturgeschichte 68. Cologne: Böhlau, 2011.

Scriba, Christoph J. "Auf der Suche nach neuen Wegen: Die Selbstdarstellung der Leopoldina und der Royal Society in London in ihrer Korrespondenz der ersten Jahre (1664–1669)." In *Salve Academicum: Festschrift der Stadt Schweinfurt anläßlich des 300. Jahrestages der Privilegierung der Deutschen Akademie der Naturforscher Leopoldina durch Kaiser Leopold I. vom 7. August 1687*, edited by Uwe Müller, 69–85. Veröffentlichungen des Stadtarchivs Schweinfurt 1. Schweinfurt: Stadtarchiv Schweinfurt, 1987.

Seeber, Hans Ulrich, ed. *Englische Literaturgeschichte: Unter Mitarbeit von Stephan Kohl,*

Eberhard Kreutzer, Annegret Maack, Manfred Pfister, Johann N. Schmidt und Hubert Zapf. 3rd ed. Stuttgart: Metzler, 1999.

Seifert, Arno. *Cognitio historica: Die Geschichte als Namengeberin der frühneuzeitlichen Empirie.* Historische Forschungen 11. Berlin: Duncker und Humblot, 1976.

———. *Der Rückzug der biblischen Prophetie von der Neueren Geschichte.* Beihefte zum Archiv für Kulturgeschichte 31. Cologne: Böhlau Verlag, 1990.

Serjeantson, Richard W. "Proof and Persuasion." In *The Cambridge History of Science,* edited by Lorraine Daston and Katharine Park, 132–75. Cambridge: Cambridge University Press, 2006.

Shapin, Steven. "The House of Experiment in Seventeenth-Century England." *Isis* 79 (1988): 373–404.

———. *A Social History of Truth: Civility and Science in Seventeenth-Century England.* Chicago: University of Chicago Press, 1994.

———. *The Scientific Revolution.* Chicago: University of Chicago Press, 1996.

Simpson, J. A., and E. S. C. Weiner, eds. *The Oxford English Dictionary.* Vol. 1. Oxford: Clarendon Press, 1991.

Siraisi, Nancy G. "Vesalius and Human Diversity in *De Humani Corporis Fabrica.*" *Journal of the Warburg and Courtauld Institutes* 57 (1994): 60–88.

———. "Anatomizing the Past: Physicians and History in Renaissance Culture." *Renaissance Quarterly* 53 (2000): 1–30.

———. *History, Medicine, and the Traditions of Renaissance Learning.* Cultures of Knowledge in the Early Modern World. Ann Arbor: University of Michigan Press, 2007.

Smith, Pamela H. *The Body of the Artisan: Art and Experience in the Scientific Revolution.* Chicago: University of Chicago Press, 2004.

Spinks, Jennifer. *Monstrous Births and Visual Culture in Sixteenth-Century Germany.* Religious Culture in the Early Modern World 5. London: Pickering & Chatto, 2009.

Steinke, Hubert. "Anatomie und Physiologie." In *Albrecht von Haller: Leben—Werk—Epoche,* edited by Hubert Steinke, Urs Boschung, and Wolfgang Pross. Archiv des Historischen Vereins des Kantons Bern 85, 226–54. Göttingen: Wallstein Verlag, 2008.

Stolberg, Michael. "Formen und Funktionen medizinischer Fallberichte in der Frühen Neuzeit (1500–1800)." In *Fallstudien: Theorie—Geschichte—Methode,* edited by Johannes Süßmann, Susanne Scholz, and Gisela Engel, 81–95. Frankfurter Kulturwissenschaftliche Beiträge 1. Berlin: Trafo Verlag, 2007.

Streudel, Johannes. "Leibniz und die Leopoldina." *Nova Acta Leopoldina* 16, no. 114 (1954): 465–74.

Stuber, Martin, Stefan Hächler, and Luc Lienhard, eds. *Hallers Netz: Ein europäischer Gelehrtenbriefwechsel zur Zeit der Aufklärung.* Basel: Schwabe Verlag, 2005.

Süßmann, Johannes, Susanne Scholz, and Gisela Engel, eds. *Fallstudien: Theorie—Geschichte—Methode.* Frankfurter Kulturwissenschaftliche Beiträge 1. Berlin: Trafo Verlag, 2007.

Swan, Claudia. "*Ad Vivum, naer het Leven,* from the Life: Defining a Mode of Representation." *Word & Image* 11, no. 4 (1995): 353–72.

Syndikus, Anette. "Die Anfänge der Historia literaria im 17. Jahrhundert: Programmatik und gelehrte Praxis." In *Historia literaria: Neuordnungen des Wissens im 17. und 18. Jahrhundert,* edited by Frank Grunert and Friedrich Vollhardt, 3–36. Berlin: Akademie Verlag, 2007.

Tagliaferri, Maria Christina, and Stefano Tommasini. "*Microcosmos naturae.*" In *Hortus pictus: Dalla raccolta di Ulisse Aldrovandi,* edited by Enzo Crea, 43–53. Rome: Edizioni dell'elevante, 1993.

te Heesen, Anke. "Cut and paste um 1900." In "Cut and paste um 1900: Der Zeitungsausschnitt in den Wissenschaften. " Special issue, *Kaleidoskopien: Zeitschrift für Mediengeschichte und Theorie* 4 (2002): 20–37.

———. "The Notebook: A Paper-Technology." In *Making Things Public: Atmospheres of Democracy*, edited by Bruno Latour and Peter Weibel, 582–89. Cambridge, MA: MIT Press, 2005.

Toellner, Richard. "Der Arzt als Gelehrter: Anmerkungen zu einem späthumanistischen Bildungsideal." In *Die Bausch-Bibliothek in Schweinfurt—Wissenschaft und Buch in der Frühen Neuzeit*, edited by Menso Folkerts, Ilse Jahn, and Uwe Müller, 39–59. Acta Historica Leopoldina 31. Heidelberg: Deutsche Akademie der Naturforscher Leopoldina, 2000.

———. "Im Hain des Akademos auf die Natur wissbegierig sein: Vier Ärzte der Freien Reichsstadt Schweinfurt gründen die Academia Naturae Curiosorum." In *350 Jahre Leopoldina—Anspruch und Wirklichkeit: Festschrift der Deutschen Akademie der Naturforscher Leopoldina 1652–2002*, edited by Benno Parthier and Dietrich von Engelhardt, 14–43. Halle: Druck-Zuck GmbH, 2002.

Tosi, Alessandro. *Portraits of Men and Ideas: Images of Science in Italy from the Renaissance to the Nineteenth Century*. Pisa: Edizioni Plus—University Press, 2007.

Tugnoli Pàttaro, Sandra. *Metodo e sistema delle scienze nel pensiero di Ulisse Aldrovandi*. Collana di studi epistemologici 3. Bologna: Clueb, 1981.

———, ed. *Lo studio Aldrovandi in Palazzo Pubblico (1617–1742)*. Collana di studi epistemologici 9. Bologna: CLUEB, 1993.

Ule, Willi. *Geschichte der Kaiserlichen Leopoldinisch-Carolinischen Deutschen Akademie der Naturforscher während der Jahre 1852–1887, mit einem Rückblick auf die frühere Zeit ihres Bestehens*. Halle: E. Blochmann und Sohn, Wilh. Engelmann in Kommission, 1889.

Vai, Gian Battista. "Aldrovandi's Will: Introducing the Term 'Geology' in 1603 / Il testamento di Ulisse Aldrovandi e l'introduzione della parola 'geologia' nel 1603." In *Four Centuries of the Word Geology: Ulisse Aldrovandi 1603 in Bologna*, edited by Gian Battista Vai and William Cavazza, 65–110. Bologna: Minerva Edizioni, 2003.

Vine, Angus. *Miscellaneous Order: Manuscript Culture and the Early Modern Organization of Knowledge*. Oxford: Oxford University Press, 2019.

Wahrig-Burfeind, Renate, ed. *Gerhard Wahrig, Deutsches Wörterbuch: Neu herausgegeben von Dr. Renate Wahrig-Burfeind, Mit einem "Lexikon der Deutschen Sprachlehre."* 7th ed. Gütersloh: Wissen Media Verlag GmbH, 2002.

Wallis, Faith. "The Experience of the Book: Manuscripts, Texts, and the Role of Epistemology in Early Medieval Medicine." In *Knowledge and the Scholarly Medical Traditions*, edited by Don G. Bates, 101–26. Cambridge: Cambridge University Press, 1995.

Wear, Andrew. "William Harvey and the 'Way of the Anatomists.'" *History of Science* 21, no. 3 (1983): 223–49.

Wellisch, Hans H. "How to Make an Index—16th Century Style: Conrad Gessner on Indexes and Catalogs." *International Classification* 8, no. 1 (1981): 10–15.

Werle, Dirk. "Die Bücherflut in der Frühen Neuzeit—realweltliches Problem oder stereotypes Vorstellungsmuster?" In *Frühneuzeitliche Stereotype: Zur Produktivität und Restriktivität sozialer Vorstellungsmuster*. Jahrbuch für Internationale Germanistik, A 99, edited by Miroslawa Czarnecka, 469–86. Bern: Lang, 2010.

Wilson, Adrian. *The Making of the Nuremberg Chronicle*. Amsterdam: N. Israel, 1976.

Wilson, Dudley. *Signs and Portents: Monstrous Births from the Middle Ages to the Enlightenment*. London: Routledge, 1993.

Wittkower, Rudolf. "Marvels of the East: A Study in the History of Monsters." *Journal of the Warburg and Courtauld Institutes* 5 (1942): 159–97.

———. "Marco Polo and the Pictorial Tradition of the Marvels of the East." In *Allegory and the Migration of Symbols*, edited by Rudolf Wittkower, 75–92. London: Thames and Hudson, 1977.

"Wonder-Monger." *Oxford English Dictionary*. Accessed April 27, 2011. www.oed.com/view dictionaryentry/Entry/229958.

Wooton, David. *The Invention of Science: A New History of the Scientific Revolution*. London: Allen Lane, 2015.

Wübben, Yvonne, and Carsten Zelle, eds. *Krankheit schreiben: Aufzeichnungsverfahren in Medizin und Literatur*. Göttingen: Wallstein Verlag, 2013.

Yachnin, Paul. "Shakespeare's Public Animals." In *Humankinds: The Renaissance and its Anthropologies*, edited by Andreas Höfele and Stephan Laqué, 185–98. Pluralisierung und Autorität 25. Berlin: de Gruyter, 2011.

Yates, Frances. *The Art of Memory*. Chicago: University of Chicago Press, 1966.

Yeo, Richard. *Notebooks, English Virtuosi, and Early Modern Science*. Chicago: Chicago University Press, 2014.

Zedelmaier, Helmut. *Bibliotheca universalis und Bibliotheca selecta: Das Problem der Ordnung des gelehrten Wissens in der frühen Neuzeit*. Beihefte zum Archiv für Kulturgeschichte 33. Cologne: Böhlau, 1992.

———. "'Historia literaria': Über den epistemologischen Ort des gelehrten Wissens in der ersten Hälfte des 18. Jahrhunderts." *Das Achtzehnte Jahrhundert* 22, no. 1 (1998): 11–21.

———. "De ratione excerpendi: Daniel Georg Morhof und das Exzerpieren." In *Mapping the World of Learning: The Polyhistor of Daniel Georg Morhof; Proceedings of a Conference in Cooperation with the Foundation for Intellectual History, London, Herzog August Bibliothek, Wolfenbüttel, September 10–11, 1998*, edited by Françoise Waquet, 75–92. Wiesbaden: Harassowitz, 2000.

———. "Lesetechniken: Die Praktiken der Lektüre in der Neuzeit." In *Die Praktiken der Gelehrsamkeit in der Frühen Neuzeit*, edited by Helmut Zedelmaier and Martin Mulsow, 11–30. Frühe Neuzeit 64. Tübingen: Niemeyer, 2001.

———. "Buch, Exzerpt, Zettelschrank, Zettelkasten." In *Archivprozesse: Die Kommunikation der Aufbewahrung*, edited by Hedwig Pompe and Leander Scholz, 38–53. Mediologie 5. Cologne: DuMont Literatur und Kunst Verlag, 2002.

———. "Johann Jakob Moser et l'organisation érudite du savoir à l'époque moderne." In *Lire, copier, écrire: Les bibliothèques manuscrites et leur usage au XVIIIe siècle*, edited by Elisabeth Décultot, 43–62. Paris: CNRS Ed., 2003.

———. "Wissensordnungen der Frühen Neuzeit." In *Handbuch Wissenssoziologie und Wissensforschung*, edited by Rainer Schützeichel, 835–45. Erfahrung—Wissen—Imagination. Schriften zur Wissenssoziologie 15. Konstanz: UVK Verlagsgesellschaft, 2007.

———. "Wissen erwerben: Lesen als Tätigkeit." In *Werkstätten des Wissens zwischen Renaissance und Aufklärung*, edited by Helmut Zedelmaier, 5–16. Historische Wissensforschung 3. Tübingen: Mohr Siebeck, 2015.

———. "Wissen repräsentieren: Die Bibliothek als Herrschaftsinstrument." In *Werkstätten des Wissens zwischen Renaissance und Aufklärung*, edited by Helmut Zedelmaier, 89–106. Historische Wissensforschung 3. Tübingen: Mohr Siebeck, 2015.

———. "'Conspectus Reipublicae Literariae': Besonderheit, Kontext, Grenzen." In *Christoph August Heumann (1681–1764): Gelehrte Praxis zwischen christlichem Humanismus und Aufklärung*, edited by Kasper Risbjerg Eskildsen, Martin Mulsow, and Helmut Zedelmaier, 71–89. Gothaer Forschungen zur Frühen Neuzeit 12. Stuttgart: Steiner, 2017.

Zedelmaier, Helmut, and Martin Mulsow, eds. *Die Praktiken der Gelehrsamkeit in der Frühen Neuzeit*. Frühe Neuzeit 64. Tübingen: Niemeyer, 2001.

———. "Einführung." In *Die Praktiken der Gelehrsamkeit in der Frühen Neuzeit*, edited by Helmut Zedelmaier and Martin Mulsow, 1–7. Tübingen: Niemeyer, 2001.

Index